DATA HANDLING IN SCIENCE AND TECHNOLOGY — VOLUME 8

Design and optimization in organic synthesis

DATA HANDLING IN SCIENCE AND TECHNOLOGY

Advisory Editors: B.G.M. Vandeginste and O.M. Kvalheim

DATA HANDLING IN SCIENCE AND TECHNOLOGY — VOLUME 8

Advisory Editors: B.G.M. Vandeginste and O.M. Kvalheim

Design and optimization in organic synthesis

ROLF CARLSON

Department of Organic Chemistry, Umeå University, S-901 87 Umeå, Sweden

ELSEVIER

Amsterdam — Oxford — New York — Tokyo 1992

ELSEVIER SCIENCE PUBLISHERS B.V.
Sara Burgerhartstraat 25
P.O. Box 211, 1000 AE Amsterdam, The Netherlands

Distributors for the United States and Canada:

ELSEVIER SCIENCE PUBLISHING C0MPANY INC.
655, Avenue of the Americas
New York, NY 10010, USA

ISBN 0-444-89201-X

© Elsevier Science Publishers B.V., 1992

This book is printed on acid-free paper.

Printed in The Netherlands

To my beloved

Contents

Foreword

1. Short background to why this book was written

Organic synthesis is an important area in chemical research. New synthetic methods are invented at an ever increasing rate which open up new ways to produce interesting chemical compounds. These factors are of tremendous importance for the practical use of synthetic methods both for academic research and for industrial applications.

All synthetic methods have emerged either as the result of an unforeseen observation or as the result of innovative thought. However, it is rare that the first experiments along a new train of thought give satisfactory results. Much tedious work is often required before a new idea can become established as a *synthetic method*. It is necessary to explore the reaction conditions to determine how they should be adjusted to obtain optimal results. It is also interesting to find out whether the reaction can be used as a general method for a number of similar substrates. To this end, it is necessary to determine the scope and limitations of the reaction. This in turn calls for more experimentation. Chemical phenomena are rarely the result of single causes. Instead, a number of factors are likely to be involved and, unfortunately, their influence will depend on still other factors. In order to be able to take such interactions into account, it is necessary to use multivariate methods which allow all pertinent factors to be considered *simultaneously*, both for designing experiments and for analyzing the result.

Knowledge of multivariate methods is not, however, widely spread in the community of synthesis chemists. Therefore, many new methods are still being investigated through poorly designed experiments and hence, new procedures are not properly optimized. Still, the most common method to carry out "systematic studies" is to consider "one factor at a time", although such an approach was shown by R.A. Fisher to be inappropriate over 60 years ago [1], when several factors are to be considered.

I believe that the reason why organic synthesis chemists do not apply statistical principles in their experiments is that they do not know how to use such methods. In general, they do not bother to read text-books on statistics because such books rarely describe how statistics may be relevant to their *chemical* problems. My personal experience may illustrate this: When I started my chemical career some 20 years ago

I was asked by my professor to study and optimize a chemical reaction. I asked how to proceed and he answered me "You must do it by trial and error". My immediate reaction to this was that there must be a more systematic way. A friend suggested that I should read a book on "Biological Analysis of Variance".[2] However, I was unable to translate *Latin squares*, *crop variation*, and *effects of fertilizers* into my world of three-necked flasks, dropping funnels and mechanical agitation. The pioneering work by George E.P. Box and coworkers published in the fifties [3] would have answered my questions at that time. Unfortunately, their work was obviously unknown to my professor and to the academic circles working with synthetic chemistry. The situation is probably the same today.

I am therefore convinced that there is a need for a book which spans a bridge between practical organic synthesis and statistics. Such a book must describe chemical problems *as they are seen by the synthesis chemist* and must introduce statistical tools so that the results obtained are chemically relevant to the chemist. The present book is a humble attempt to use these principles as guidelines.

Professional statisticians may complain that this is yet another introductory text on statistics, written by a non-statistician. This is true, my devotion and professional training are in the field of synthetic chemistry. However, in exploring this field I felt a need for statistical principles to guide my way. This book is an attempt to transmit my personal experiences to my fellow colleagues. If anyone among them should respond to the message and be stimulated to learn more about statistics, I would feel that my mission as an evangelist had been succesful.

The book is intended as an introductory text-book on multivariate methods in experimental organic synthetic chemistry. I have tried to describe how various statistical tools can be applied to common problems encountered when a chemical *reaction* is elaborated into a synthetic *method*. The methods are illustrated throughout by examples from organic synthesis. Many of the examples have been taken from my own experiments, not with a view to supporting our own results, but because the reasoning behind the experiment is known in detail. In the examples furnished by others, it is sometimes possible to trace the logic behind the experimental set-up, but quite often certain details remain obscured.

Statistical principles will be presented in the context of chemical examples to show how statistical inferences can be linked to chemical consequences. No previous knowledge of statistics is required.

Some of the reasoning in the book uses matrix formalism. This is for the sake of convenience, since some quantitative relations are more easily expressed in matrix language than otherwise. Readers without any previous experience of matrix calculus may skim those paragraphs in which matrix calculus is used without losing too much of the essential message. As matrix formalism is used in many contexts in physical and theoretical chemistry, it is my hope that this will not be a major obstacle to the reader. For those unfamiliar with matrix calculation, a short Appendix has been included at the end of this book.

2. Acknowledgements

First of all I would like to thank my wife *Inger* and my two sons, *Johan* and *Andreas* for their whole-hearted support during my preparation of the manuscript. They have all been very helpful, tolerant and understanding. I would also like to express my gratitude to the following people for their skillful assistance in helping me to complete the final version of the book: Mr. *Thomas Sigurdson* and Mr. *Alf Persson* who prepared the illustrations. *Thomas Sigurdson* also gave me invaluable advice on typographical layout as well as producing the front-cover illustration;
Mr. *Per Lundholm* who made the photographic reductions of the figures to fit the format of the camera-ready manuscript; Mr. *Lars Hübinette* who gave me excellent linguistic assistance; and my father *Carl-Eric Carlson*, who did the final proof-reading.

The following friends, colleagues, coworkers, and students also gave of their time to read preliminary versions of the manuscript, and I would like to express my gratitude to them for their criticism and helpful comments which I know have improved the final result. My sincere thanks to (in alphabetical order):
Anna-Karin Axelsson, Tanja Barth, Hans-René Bjørsvik, Lars Eklund, Erik Johansson, Ulf Larsson, Åsa Nilsson-Lindgren, Unni Marie Valle, and *Svante Wold.*

The examples given in the book have been taken from published works, and I would like to thank the authors and publishers for their kind permission to reproduce these examples. Several of the examples have been furnished by present and past coworkers, and their names appear in the references. Without their enthusiasm and achievements,it would have been impossible to produce this book.

Umeå, June 1991

Rolf Carlson

References

1. R.A. Fisher and W.A. MacKenzie
   J. Agr. Sci. 13 (1923) 311.

2. G. Bonnier and O. Tedin
   Biologisk Variationsanalys
   Svenska Bokförlaget (Bonniers), Stockholm 1940.

4

3. *(a)* G.E.P. Box and K.B. Wilson *J. Roy. Stat. Soc. Ser. B 13* (1951) 1;

(b) G.E.P. Box
Biometrics 10 (1954) 15;

(c) G.E.P. Box and P.V. Youle *ibid. 11* (1955) 287.

Chapter 1

Introduction: Strategies on different levels in organic synthesis

Organic synthesis is an exciting field of chemistry. The essence of synthetic chemistry is to find efficient ways to construct the desired, often complex, molecules using simple starting materials as building blocks. For long, organic synthesis was regarded as an "art in the midst of science". Despite the great achievements in the past, e.g. the syntheses of vitamin B_{12} [1], reserpine [2], quinine [3] it is astonishing that almost nothing can be found in the literature on how to establish multi-step synthetic procedures to complex molecules until 1967 when a first systematic analysis was presented in a paper by Corey [4].

An ideal synthesis would be as depicted in Fig. 1.1.

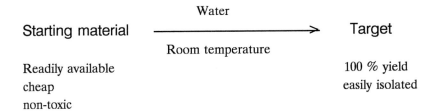

Fig.1.1: An ideal organic synthesis.

Unfortunately, such syntheses are rare. The most common situation is that the desired molecule, *target (molecule)*, must be constructed through joining different parts of the molecule in a sequence of reactions, *a synthetic path*. Criteria for selecting a suitable synthetic path are often formulated in terms of economy and linked with the overall chemical yield of the target. In academic research where the goal is to reach the desired target molecule, the overall yield is often the sole criterion. In industrial synthesis, the overall yield must be matched with other criteria, e.g. overall cost to produce a given amount of the target, toxicity of

6

chemicals and solvents involved, and environmental consequences of running the synthesis on production scale.

We shall see that it is possible to envisage strategies for solving problems at different levels of complexity in the area of organic synthesis. These strategic levels can be arranged hierarchically as given in Fig. 1.2. The levels are interrelated in such a way, that a higher level above imposes constraints on the possible solutions at lower levels.

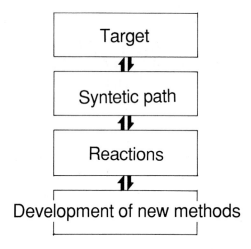

Fig.1.2: Strategies at different levels in organic synthesis.

1. The target

There may be many reasons for synthesizing a target molecule:

Natural products: Before the advent of spectroscopic methods (UV, IR, NMR, MS), total synthesis by independent methods was a way to achieve a proof of the proposed structure of a molecule. Today, spectroscopic methods solve most structural problems. Total synthesis by independent routes is still used occasionally as structural proofs, and recent examples are provided by Corey [5] in the synthesis of leukotriene B_4, and Kishi [6] in the synthesis of palytoxine. Today, most natural product synthesis is used to prepare sufficient amounts of interesting compounds, e.g.

to study their physiological effects. Often, interesting natural products are available only in tiny amounts from natural sources.

Theoretical or mechanistic hypotheses: To make experiments by which theoretical models or mechanistic hypotheses can be checked, the critical compounds must be available. If they cannot be found in nature, they must be synthesized.

Pharmaceuticals: In the pharmaceutical industry, often hundreds (sometimes thousands) of analogs to an interesting compound can be prepared in order to find a compound with desired pharmacological properties.

Useful chemicals: In our daily life we are surrounded by objects which are obtained through synthesis, e.g. polymers, dyes, textile fabrics, artificial flavour, and insecticides.

It is seen that there can be many different reasons for selecting a target. Often, interaction with other scientific disciplines, or the needs of the market furnish the specific target. In this respect, organic synthesis is an applied science.

2. The synthetic path

A fundamental principle is that a synthetic path should be short, i.e. contain as few reaction steps as possible. It is also desirable that the path should be convergent, which means that larger parts of the target structure are prepared through separate paths and then joined together in a later coupling step. This is a consequence of the economy of yield (and time).

As the number of known chemical reactions is very large and the number of known chemical compounds is even larger, the number of possible combinations of chemicals and reactions will be overwhelming. This points to a problem, viz. how to establish efficient paths to link available starting material to the desired target. Often more than one path can be conceived. The different paths will define the "tree" of available routes from the starting materials to the target shown in Fig. 1.3.

It is in principle possible to plan multistep synthesis in two directions: *Forwards*, from starting to the target, and *backwards*, form the target to the starting materials. Hitherto, it is the backward (retrosynthetic) search strategies which have been most rewarding. The target structure is analysed to discern suitable smaller building blocks, *synthons*, which can be joined together by known reactions. The term synthon is used to denote a fragment of the target structure such that the functionalities of the synthon fragment can be linked with features in the target structure by known reactions. The structure of a synthon can be rather vaguely described, e.g. a carbanion can be stabilized by an electron-withdrawing group, EWG, which can be nitro, cyano, keto, carboxylic ester ... The actual *precursors*, i.e.

8

the real compounds used in the synthetic reaction are then determined from the synthons so that the number of discrete chemical reaction steps, necessary to produce the target, can be minimized. The strategies along these principles have been suggested by Corey[7] and Hendrickson.[8] An instructive introductory text-book on retroanalysis of synthetic pathways has been written by Warren.[9]. Analysis by a totally different approach has been described by Ugi.[10] The method by Ugi uses a topological model for chemistry which makes it possible to search in both directions. An approach similar to the one by Ugi has been suggested by Zefirov.[11]

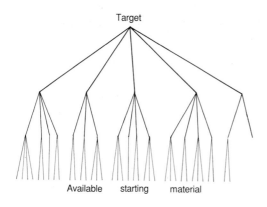

Fig.1.3: A tree of synthetic pathways from available starting materials to the target.

These strategies are thoroughly described in the references given, and we do not go into details on this here. These topics are beyond the scope of the present book which treats aspects on the study of synthetic *reactions*.

3. The synthetic reaction

3.1. Tactics

When a promising synthetic path has been found, the next phase will be to evaluate the individual synthetic reaction steps along this path. In this context, it is essential that the reactions can be controlled so that they produce the desired result. Several problems are encountered:

(1) Control of reactivity: Starting material and intermediary products must react in the right positions. Often there are several functionalities which may interfere, and various tricks are used to force the reaction to go in the desired direction.

(2) Control of stereochemistry: When the target has a defined stereostructure, it is necessary to be able to control the stereochemical outcome of the reaction.

It is impossible to give even a short account of these aspects in this introductory chapter. Only a few highlights on certain tactics will be given. We shall see, that the problems associated with this area will be a suitable starting point for the subjects developed in the remaining chapters of this book.

3.2   Control of reactivity

There are several tactics available which can be used to control reactivity, e.g. modification of the substrate, modification of the reagent(s), and/or modification of the medium in which the reaction is conducted.

Substrate modification is usually obtained by three principles:

(1) By using protective groups to block interfering functionalities.[12] The striving is to develop reagents which selectively block the desired function under mild conditions and which can be removed under mild conditions as well. However, the use of protective groups is often regarded as a necessary evil, since it lengthens the synthetic path. If possible, chemoselective reagents are preferred.

(2) Certain functional groups are used to activate a reaction site to afford a chemoselective reaction at this site. When the job is done, the activating group is removed, e.g. the classical β-ketoester alkylation followed by decarboxylation.

(3) The electrophilic or nucleophilic properties of a functional group is altered by Umpolung.[13]

Reagent modification as a means to control reactivity is an area where enormous efforts are spent on research. Two main directions can be discerned:

(1) Development of new reagents where the current trend is to probe deeper and deeper into the periodic table to explore new elements. The use of transition elements is such a field. [14] Among the non-metal elements, silicon and selenium have been extensively explored. [15] *(2) Modification of known reagents*: A rather extreme example is furnished by the complex hydrides which have been and are still being explored by H. C. Brown and coworkers.[16] By modifying the central atom (B, Al, Sn...), the counterion (Na, K, Li...) and the number and types of non-hydrogen ligands, a very large number of reducing agents has been developed. The structural modifications afford a high degree of selectivity. The principles used

by Brown in modifying the complex hydrides have been used by others to other types of reagents, although the examples are less abundant.

3.3. Control of stereochemistry

When the target has a defined stereostructure it is necessary to conduct the reaction steps so that a correct configuration can be obtained at double bonds, at rings and at chiral centers.

In the past, it was rather common that racemic mixtures were obtained in complex synthesis. The desired enantiomer was then obtained by resolving the racemic mixture. However, this reduces the overall yield by 50 %.

Today, the trend is to synthesize complex molecules in optically pure form. This is possible due to an ever increasing number of available optically active starting materials (the chiral pool) and the fact that new stereoselective procedures are being developed at a steadily increasing rate. The armoury to achieve stereocontrol is impressive and it is beyond the scope of this chapter to go into any details; only a few brief comments will be given.

Double bonds can be created by a number of available stereospecific reactions (E2 elimination, pyrolytic elimination, e.g. Cope, Chugaev reactions), stereoselective Wittig and related reactions, reduction of triple bonds, by *cis/trans* isomerisation of existing double bonds either photochemically or by wet chemistry, e.g. the Corey-Winter procedure.[17]

Ring stereochemistry can be efficiently created through multicenter reactions, and in this respect the Diels-Alder reaction is extremely useful. Other tricks are to use intramolecular reactions, e.g. iodolactonisation.[18]

Chiral centra are more difficult to create. Chirality must come from somewhere. It is either introduced into the target by a chiral precursor of natural origin, or it is introduced at a prochiral site by a chiral reagent. The Sharpless epoxidation of allylic alcohols is a good example of the latter.[19] The use of enzymes for stereoselective transformations is increasing, although much is still done by trial and error strategy. The use of enzymes to catalyze reactions in organic solvents is increasing and will probably open up new ways to selective transformations.[20]

When a complex target with several chiral centers is to be synthesized certain heuristics have been developed to ensure a good result:

(a) If the target is to be obtained in chiral form, at least one asymmetric center with the correct configuration must be present in the starting material. Of course, the efficiency will be increased if the starting material contains as much as possible of the asymmetry of the target. Sometimes there may be only a vague resemblance

between the final target and a chiral starting material, which nevertheless was used to efficiently create the desired final stereostructure. The ability to discern such patterns is the basis of the *chiron* concept, developed by Hanessian.[21] A computer programs for analysis by the chiron approach is available.

(b) If chiral centers are far apart in the target, optically active building blocks should be used, e.g. the synthesis of peptides where optically active aminoacids must be used.

(c) If stereocenters in the target are close to each other, it is often possible to use one chiral center in a precursor and to create new chiral centers by stereoselective reactions. In this context, the tactics based on double asymmetric induction developed by Masamune have proved to be especially useful.[22]

4. Strategies for elaborating synthetic reactions. An outline of the scope of the problem

For a proposed route to the target molecule to be successful, it is necessary that all intermediate reactions involved in the separate steps can be conducted with an acceptable result. To achieve an optimum overall result, all intermediary steps must also be optimized. In the paragraphs above it was briefly mentioned how different tactics can be used to control the synthetic steps. For this, a functioning reaction must be available. To find out whether or not a reaction can be used for the desired purpose, it is necessary to run experiments. The outcome of these experiments is sometimes a source of great satisfaction, sometimes, however, it can be very frustrating. Some common situations encountered at this very early stage of a synthetic endeavour and some questions which then arise are:

* A known synthetic procedure is to be used. However, the experimenter realizes immediately that an improved result is to be expected if a more polar solvent is used instead of toluene which was used in the original method. How to find the best polar solvent?

* A publised method is reported to give an excellent result. Attempts to use the method fail to give reproducible results. The first attempt afforded 80 % yield, but the second gave less than 20 %. Such situations occur when there are one or more important experimental factors which are out of control. How to identify the critical factors so that the reaction can be brought under control.

* A reaction gives a promising result. It is, however, strongly believed that the result can be considerably improved by adjusting the experimental conditions. How to do this in an efficient way?

* By a sudden flash of inspiration, the chemist can imagine a totally new principle to carry out a synthetic transformation. However, nothing is previously known on this. To find out whether or not the idea is fruitful it must be evaluated by experimental testing. It is not very probable that one single experiment can give a definite answer. It is more likely that it will be necessary to test different solvents, different modifications of the reagents, and to test the reactions on different substrates before the idea can be evaluated to any degree of certainty. It would be very unfortunate if an excellent idea was prematurely abandonned if the first experiments failed to produce the desired result due to a poor selection of test systems. The questions which arise will therefore be: Which solvents, reagents, and substrates should be chosen to give a reliable answer?

New synthetic reactions or new reagents will open up new possibilities. However, all new reactions have emerged, either as a result of a brilliant idea, or as a result of an unforeseen observation, and tedious work is often required before the idea of a new synthetic reaction has been developed into a synthetic method.

Chemistry is an experimental science. All knowledge in chemistry has ultimately been obtained by experimental observations. Hence, it is of the utmost importance that chemical experiments are designed in a proper way to furnish the required information. It is also desirable that this information is obtained without too much effort. This requires the experiments to be designed with a view to obtaining a maximum of information in a limited number of individual experimental runs. This is particularly important when new methods are being developed. Therefore, general strategies for the study of synthetic reactions must include various aspects of experimental design.

5. Theme and variations

Any synthetic reaction can be described as a theme:

Starting material $\xrightarrow[\text{Solvent}]{\text{Reagent}}$ Product

This theme can be varied in an infinite number of ways:

(a)   We can vary the *reaction system* by changing e.g. the substrate, the reagent(s) and/or the solvent. The union of all potential substrates, reagents, and solvents is called the *reaction space*.

(b)   For any given system there are several *experimental variables* (pH, temperature, concentration of reactants, flow rates etc.) which can be adjusted to influence the result. The union of all these possible variations is called the *experimental space*. In the experimental space there are always two main types of problems to solve: *(1) Which* experimental factors will have a real influence? *(2) How* should the significant factors be adjusted to achieve an optimum result?

(c)   Often several results are of interest: Yield, selectivity, cost etc. How should the result be evaluated in such cases?

A geometric illustration of the possible variations of a synthetic "theme" is shown in Fig.1.4. The different "axes" of the reaction space and the experimental space can for the moment be regarded just as "variations". In the chapters to follow, we shall see that these "axes" can be quantified to describe multidimensional variations.

14

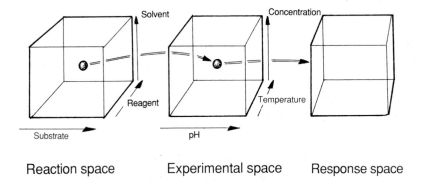

Fig.1.4: Geometric illustration of possible variations of a synthetic reaction.

The trouble is that these possible variations of the reaction theme are not independent.

*   A sluggish substrate can be brought to reaction by using a more aggressive reagent or by using forced reaction conditions.

*   Substrate and reagents will be differently solvated if the solvent is changed, and this will alter the reactivity pattern. These effects may sometimes be counteracted by adjusting the reaction temperature.

To be able to discern all the important factors involved in determining the result of any reaction, it is necessary to use proper experimental designs.

When a reaction mechanism is known, it is sometimes possible to deduce, by theoretical reasoning, how a good result could be achieved. However, with *new* reactions, mechanistic details are still obscured and it is not likely that such details will be revealed before the utility of the reaction has been demonstrated. This implies that the scope and limitations of the reaction and of the optimun experimental conditions must be determined before mechanistic details are known with certainty. Thus, the desired information must be obtained by inferences from experimental observations. It is therefore desirable that the required information on which factors are important can be adequately obtained, if possible with a minimum of effort. This implies requirements on the experimental design, both for exploring the reaction space and the experimental space. Examples of questions often encountered in practice are:

* Can the reaction be run with several similar substrates? What is the scope of the reaction in this respect?

* What is the general scope of the reaction, i.e. which variations are tolerated with respect to the substrate, the reagents and the solvent? A wide scope with regard to solvent variation means a high possibility to develop one-pot procedures.

* How can the optimum reagent and the optimum solvent for the desired transformation be found?

* How should test systems to obtain the desired information on the questions above be selected?

* How should a screening experiment be conducted in the view to determining *which* experimental variables are important to control?

* How should one proceed to determine the optimum experimental conditions?

* Sometimes the result of a synthetic reaction can be characterized by more than one single criterion. Often several criteria are of interest (yield, selectivity, ease of work-up, cost ...). How can a simultaneous optimization of several criteria be achieved?

* How will the optimum experimental conditions change if the reaction system is varied (other substrates, reagents or solvents)?

The remaining chapters of this book will describe how multivariate statistical methods can provide convenient and efficient methods for solving the problems outlined above. We shall see how these methods can be integrated into an overall strategy for developing synthetic reactions into useful synthetic methods.

16

References

1:     *(a)* R.B. Woodward,
Pure and Appl. Chem. 17 (1968) 519;

(b) R.B. Woodward,
Pure and Appl. Chem. 33 (1973) 145;

(c) A. Eschenmoser
Angew. Chem. 81 (1969) 301.

2.   R.B. Woodward, F.E. Bader, H. Bickel, A.J. Frey and R.W. Kierstaed
Tetrahedron 2 (1958) 1.

3.   R.B. Woodward and W.E. Doering
J. Am. Chem. Soc. 66 (1944) 849.

4.   E.J. Corey
Pure and Appl. Chem. 14 (1967) 19.

5.   E.J. Corey, A. Marfat, J.E. Munroe, K.S. Kim, P.B. Hopkins and F. Brion
Tetrahedron Lett. 22 (1981) 1077.

6.   R.W. Armstrong, J.-M. Beau, S.H. Cheon, W.J. Christ, H. Fujioka, W.-H. Ham,
L.D. Hawkins, H. Jin, S. H. Kang, Y. Kishi, M.J. Martinelli, M.W. MvWorther Jr,
M. Mizuno, M. Nakata, A.E. Stutz, F.X. Talmas, M. Taniguchi, J.A. Tino, K. Ueda,
J. Uenishi, J.B. White and M. Yonaga.
J. Am. Chem. Soc. 111 (1989) 7525, 7530.

7.   E.J. Corey and X.-M. Cheng
The Logic of Chemical Synthesis
Wiley, New York 1989.

8.   J.B. Hendrickson
Topics Curr. Chem. 62 (1976) 49.

9.   S. Warren
Organic Synthesis. The Disconnection Approach
Wiley, Chichester 1982.

10.   *(a)* I. Ugi, J. Bauer, J. Brandt, J. Friedrich, J. Gasteiger, C. Jochum and
W. Schubert
Angew. Chem. 91 (1979) 94;

(b) J. Bauer and I. Ugi,
J. Chem. Res. (M) (1982) 3101;

(c) J. Bauer, R. Herges, E. Fontain, and I. Ugi
Chimia 39 (1985) 43.

11.   N.S. Zefirov
Acc. Chem. Res. 20 (1987) 237.

12.   F.W. McOmnie
Protective Groups in Organic Chemistry
Plenum Press, London 1973.

13.   T.A. Hase (Ed.)
Umpoled Synthons
Wiley, New York 1987.

14. See, for instance,
    S.G. Davies
    Organotransition Metal Chemistry. Application to Organic Synthesis
    Pergamon Press, Oxford 1982.

15. See, for instance,
    (a) E. Colvin
    Silicon in Organic Synthesis
    Butterworths, London 1981;

    (b) C. Paumier
    Selenium Reagents and Intermediates in Organic Synthesis
    Pergamon Press, Oxford 1986.

16. H.C. Brown and S. Krishnamurthy
    Tetrahedron 35 (1979) 567.

17. E.J. Corey and R.A.E. Winter
    J. Am. Chem. Soc. 87 (1965) 934.

18. M.D. Dowle and D.I. Davies
    Chem. Soc. Rev 8 (1979) 171.

19. *(a)* K.B. Sharpless and R.C. Michaelson
    J. Am. Chem. Soc. 95 (1973) 6136;

    (b) A. Pfenninger
    Synthesis (1986) 89.

20. A.M. Klibanov
    Trends Biochem. Sci. 14 (1989) 141.

21. S. Hanessian
    Total Synthesis of Natural Products: The Chiron Approach
    Pergamon Press, Oxford 1983.

22. S. Masamune
    Heterocycles 21 (1984) 107.

Chapter 2

Experimental study of reaction conditions. Initial remarks

Organic synthesis and experimental design

When a synthetic procedure is run for the first time at the bench, a satisfactory result is obtained only rarely. If the purpose of the experiment was to prepare a limited quantity, e.g. for spectroscopic characterization or for toxicological screening, it is probably of minor importance if the yield was low. If, on the other hand, the purpose was to check a new method, or to run a pilot experiment for larger-scale preparation, a poor yield is often a starting point for tedious work to develop the procedure.

It is evident that experiments run at random also will give results at random. It is necessary to use planned experiments. It is, however, of tremendous importance *how* the experiments are planned and executed. There are no computational methods, no statistical tricks available which can extract the desired chemical information from experimental data, if the experiments have been run in such a way that there is no such information in the data.

Planning of experiments and analysis by statistical methods will furnish precise and detailed answers *only* if precise and detailed questions are posed to the experimental system. For chemical problems, such questions are of chemical origin. Experimental design can therefore never substitute chemical reason or knowledge. With a good experimental design, statistics will provide the chemist with efficient tools:

* To determine if a change in the reaction conditions will have a significant influence on yield, selectivity, purity of product etc.

* To predict suitable conditions for improved results in future experiments.

* To run a few experiments in addition to experiments already performed with a view to obtaining *complementary information*.

* To rapidly determine optimum experimental conditions.

An advantage of statistically designed experiments is that, as a rule, rather *few* experiments are needed to furnish the desired information. Hence, the chemist must decide beforehand, *what* the desired information is.

1. How to approach the problem?

1.1. State the objective clearly

An experiment is run to furnish data which can be used to solve a problem. The problem arises since it is an obstacle to achieving the objective. To clearly identify the problem, it is therefore necessary to express the objective as concrete as possible:

* *   What do I want to achieve?

Example: If the goal is to determine experimental conditions which give a yield > 85 %, the search is terminated when this yield is attained. If the goal is to achieve the highest possible yield, the study must include checks to make sure that there are not any other conditions which give an even higher yield. The experimental strategies are likely to be different in these cases.

The author's experience as an experimental design consultant is that a common cause of failures is that the objectives of the project have not been clearly stated.

1.2. Analyse the problem

* *   What is already known?

* *   What is *not* known?

* *   What *do I need to know* to solve the problem?

Then, run the experiment to fill in the gap and to obtain the necessary information. Hence, we can see that the role of an explorative experiment is to furnish *new* information. Thus, two essential questions that should be posed prior to any experiment are:

What do I want to know?

Is the experiment properly designed to provide the answer?

1.3 Step-wise strategy

It is a common experience, that the best time to plan an experiment is when the results are known. "If this had been known to us when the project started, we would have reached the final solution much faster."

In experimental research it is beneficial if new experiments can be planned in the light of the experience gained in previous experiments. Often, the questions will be more precise when the knowledge and experience increase.

Therefore: Decompose the overall, often complex, problem into a set of smaller sub-problems, which can be analyzed through experimental studies. Then plan the next step in the investigation in the view of the results already obtained.

Example: Step-wise approach to the study of a new reducing reagent for ketones

1. Will the reaction go at all?

2. Can the reaction be run in different ways: (different solvents, different reaction vessels, different ways to mix the reactants etc.) Will there be many reaction systems where the reaction can be used?

3. Which system is the most promising for future development?

4. Which are the best experimental conditions?

5. Will these conditions apply to all types of ketones? What are the scope and limitations with regard to possible substrates?

6. What about the reaction mechanisms?

7. etc. etc.

To find answers to all these questions, different types of experiments are needed, and we shall see that there are useful strategies for the design of such experiments.

A step-wise, iterative, approach to problem solving can be illustrated as in Fig. 2.1. This figure has been borrowed from George E. P. Box [1] and shows the essential features of the philosophy of empirical science.

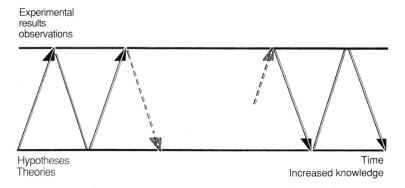

Fig.2.1: Interactive approach to problem solving in experimental sciences.

In this framework, an experiment is a source for *new* information as opposed to the idea of *"the crucial experiment"*, which is a consequence of a Newtonian reductionistic view of science. In this context, scientific hypotheses should be logical deductions from given postulates. Accordingly, in the scientific paradigm supported by this spirit, Popper [2] says that the sole role of an experiment is to falsify a hypothesis. However, my experience of organic synthesis is that crucial experiments are not the only means to progress. For any synthetic method, there are so many unknown factors, so many drastic assumptions to be made, so many simplifications of known "laws of nature" to make them fit non-ideal systems, so many unforeseenable causes of failure, that a pure deduction from known facts is not likely to result in a new synthetic *method*, which could be evaluated by a single "crucial experiment".

2.   Concretization of the problem: Independent variables, experimental domain, responses

A synthetic problem can often be stated as follows: One or several measurable results are influenced by a number of external factors controlled by the experimenter and we wish to determine; *(a) how* these factors influence the result, or *(b) which combination of the factor settings* will give the best result.

A self-evident statement is that the result of an experiment *(yield, selectivity, etc)* will depend on the detailed experimental conditions. Hence, we can assume that there is a functional relationship

Result = f(experimental conditions)

The experimental conditions will be defined by all the factors which can be altered or varied to determine the detailed experimental procedure. However, even in simple cases it will be impossible or, at least, very difficult to derive an analytical expression for this functional relation by a pure deduction from chemical theory. How such functions can be *experimentally* established is described in the next chapter.

2.1. Independent variables

The term *independent variables* will be used to denote those experimental factors which, *(a)* we believe will have an influence on the experimental result, and *(b)* can be controlled by the experimenter and independently set to predetermined values or levels in the experiment. Independent variables will be denoted by x_i (x is the value of the variable "i" in the experiment).

Examples of independent variables are: *pH, reaction temperature, concentrations of reactants, stirring rate, type of catalyst, rates of adding reagents.*

There are two types of independent variables:

(1) *Continuous* (quantitative) variables, which can be adjusted to any value over their range of variation, e.g. *pH, temperature, concentrations.*

(2) *Discrete* (qualitative) variables, which describe non-continuous variation, e.g. *type of catalyst* (Pd on Carbon or Pt on Alumina), *type of solvent* (carbon tetrachloride or hexane), *type of equipment* (reactor A, reactor B).

2.2. Experimental domain

To explore an experimental procedure, the experimenter chooses a range of variation for all the experimental variables considered: *(a)* for all *continuous* (quantitative) variables, the upper and lower bounds for their variation are specified; *(b)* for the *discrete* (qualitative) variables, types of equipment, types of catalysts, nature of solvents etc. are specified. Assume that each experimental variable defines a coordinate axis along which the settings of the variables can be marked. Assume also that these axes of variation are orthogonal (perpendicular) to each other. When there are three experimental variables to consider, these axes will span a three-dimensional space which is easily imagined. When there are many variables to consider, the axes will define a multi-dimensional space, the experimental space, see Fig.1.4. Multidimensional spaces are more difficult to imagine, but they can be regarded as analogs to the three-dimensional space. The part of the experimental space which is confined by the lower an upper bounds of the variation of the experimental variables and hence, include all possible combinations of their settings is called *the experimental domain*.

2.3. Responses

The measured values of the result is called the *response*.

A response can be, e.g. the yield (expressed in percent, grams, tons etc.), purity of crude product, cost of producing a ton of purified product, or reaction time for obtaining a 90 % conversion of starting material etc. etc.

The letter y will be used to denote a measured response. If there are several measured responses, indices will be used to distinguish between them, e.g. y_1, y_2 ...

However, the measured response, y, will never be the "true" response, usually denoted by the Greek letter η. At best, y will be a good *estimator* of η. The following relation applies

$$y = \eta + e$$

where e is an error (noise) component. The term "error" is used as a technical term in this context. It should not be confused with the normal meaning of the word, i.e. mistake, although a mistake made in the execution of an experiment will, of course, give an erroneous response. The concept of error is further discussed in the next chapter.

If the functional relation between the true response, η, and the experimental variables was perfectly known we could write

$$\eta = f(x_1, x_2, x_3..., x_k)$$

The function f describes the variation in response induced by each of the experimental variables, x_i. However, in the presence of error, this relation must be replaced by

$$y = f(x_1, x_2, x_3..., x_k) + e$$

To be able to discern whether or not a change in the experimental conditions will produce a systematic change in the measured response, the systematic variation, which is described by f, must be above the noise level described by e. The presence of an error component in all experimentally determined data necessitates the use of statistical methods in the evaluation of experiments. In the next chapter, we shall see that useful approximations of the function f can be established which will make it possible to determine the influence of each experimental variable, x_i, on the response y.

3. Two common problems: Screening of important variables, optimization

3.1. Screening of important variables

In any synthetical procedure there is a number of variables which, in principle, might have an influence on the result, e.g. pH, time of addition of the various reagents, reaction temperature, and many others. The problem is that it will not be possible to predict, by mere theoretical considerations, *which* factors or combination of factors will be necessary to control to achieve an acceptable result . The answer must be found through experimental studies. It is likely that at least some of these factors will have a significant influence on the result, while others will exert only a minor influence or no influence at all. The problem is to determine *which ones* are important. We can see this as a screening procedure, where variables having a minor or neglible influence will pass through, whereas the important variables will be caught on the screen.

It is evident that random and haphazardous variations of the experimental conditions are not a proper way to improve the result. It is also evident that any synthetic procedure can be influenced by a large number of variables. It would, however, lead to a prohibitively large number of experiments to check *all* combinations of possible experimental conditions. In such a situation it is rather natural to use an approach by which each factor is "systematically" investigated in "controlled experiments" by examining each variable one at a time while maintaining all the remaining variables constant at fixed levels. This approach is a very poor strategy, since it will most often lead to the wrong conclusions. The reasons for this are discussed below.

To determine which variables are important to control, it is absolutely necessary to use strategies which allow *all* variables to be varied *simultaneously* over the set of experiments. In such experiments it is equally important that the variation of each variable should be *uncorrelated* to the variation of other variables, i.e. that two or more variables should not be changed in the same direction over the set of experiments.

This can be avoided if *factorial designs* or *fractional factorial designs* are used for designing screening experiments. Such experimental designs, and some other types of designs which can be used for screening experiments, will be discussed in Chapters 5 - 8.

3.2. Optimization

A common problem when a synthetic method is being developed is how to improve the result with a view to obtaining, for instance, an increased yield, a higher selectivity, a lower cost, a shorter reaction time etc. We will use the term *optimization* in this context to signify "a systematic search for improvement".

With *new* synthetic methods, mechanistic details are still obscured. It is not likely that such details will be revealed until the preparative utility of the procedure has been demonstrated. This means that an optimization of the experimental conditions must generally *precede* a mechanistic understanding. Hence, the optimum conditions must be inferred from experimental observations. The common method of adjusting one-variable-at-a-time, is a poor strategy, especially in optimization studies (see below). It is necessary to use multivariate strategies also for determining the optimum experimental conditions. There are many useful, and very simple strategies for this: sequential simplex search, the method of steepest ascent, response surface methods. These will be discussed in Chapters 9 - 12.

3.3. Do not use the "one-variable-at-a-time" (OVAT) method!

This approach is seducingly simple but it is a poor strategy. It is seemingly successful, but will often lead to totally erroneous conclusions. The method can be used only if the variables are independent of each other. Unfortunately, in most cases the variables are independent only in that sense that they can be *adjusted* independently. When they exert their influence on the chemical system, the level of one variable may well modify the influence of other variables. This can lead both to compensatory and to amplifying effects.

Example: The influence of variation of *pH* and the *reaction temperature* on the yield of a chemical reactions was studied. In a first series of experiments, the *reaction temperature* was fixed at $T = 40\ °C$, and the *pH* was varied. A plot of the observed variation in yield *vs. pH* is shown in Fig. 2.2a. A maximum is observed for *pH* = 5.0

In a second series of experiments, the system is buffered to *pH* = 5.0, and the *reaction temperature* is varied. The plot of yield variation *vs. temperature* is shown in Fig. 2.2b. A maximum yield of 70 % is observed at $T =\ 55\ °C$. By this it is often concluded that the optimum operation conditions would be: *pH* = 5.0, and $T = 55\ °C.$

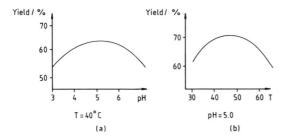

Fig.2.2: Investigation of one variable at a time.

This conclusion may be totally wrong if there is an interaction between *pH* and *T*. The situation may well be as depicted in Fig. 2.3 where the variations in both variables are given and the function describing the yield variation will be described by a response surface over the plane spanned by the two variables. The topography of the surface is given by the isoresponse contours which show the levels of the yields for different settings of the experimental variables. It is seen that the optimum yield will be 97 %, and that the conditions for this will be at *pH* = 3.6 and $T = 55\ °C$, which does not at all correspond to the results obtained by the one-variable-at-a-time study.

By varying one variable at a time, a false maximum was found. The search terminated at a point on a rising ridge of the response surface. If the initial values of the fixed variable had been different, the search would have found another point on the ridge. This is probably the reason for the common belief that experimental procedures have several local optima. The one-variable-at-a-time strategy is therefore often said to be pseudo-convergent, it will hit a false optimum. Ridges of the kind shown in Fig. 2.5 occur when the response surface is curved *and when there*

28

are interaction effects between the experimental variables. Unfortunately such interaction effects are common. A proper multivariate experimental design (i.e. a design which considers more than one variable simultaneously) allows interaction to be detected and avoids the pitfalls.

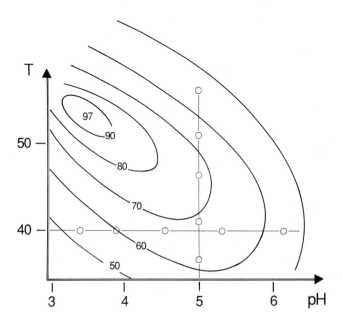

Fig.2.3: Studying one variable at a time will not establish the optimum conditions.

4. When is it possible to use a multvariate design to explore experimental conditions?

In the paragraphs above it is claimed that it is necessary to use multvariate designs in the study of experimental procedures when there is more than one experimental variable involved. Which requirements should be fulfilled to allow such designs?

* It is assumed that the problem is clearly stated and that the experimenter has defined what he/she wants to know, e.g. how the yield is influenced by the variables which are considered important to the outcome of the experiment.

* It is assumed that the experimenter is familiar with the experimental system under study and that it is possible to reproduce the result upon repetition of an experiment, i.e. that a *similar* result is obtained if a given experiment is repeated. There will always be a random variation but there should not be any dramatic variations in the result in repeated runs. This will enable the experimenter to check that he/she has the whole experimental system under control. A failure to reproduce a given experiment is likely to be due to some experimental factor out of control. Under such conditions it is impossible to obtain any reliable information whatsoever.

* It is assumed that the experimenter can approximately assign which range of variation of the experimental variables should be considered, i.e. the experimental domain of interest is limited.

Temperature can vary from -273 °C to infinity. In organic synthesis the *possible domain* is often limited to the interval from -196 °C (liquid nitrogen) to ca. 500 °C (pyrolytic decomposition). In practice, the range of variation which is of interest is often considerable narrower, e.g. from 0 °C (ice-bath) to the reflux temperature of the solvent. Often the limits are set by the experimenter to achieve a convenient experimental procedure. This means that the experimental domain of immediate interest is often considerably smaller than the possible experimental domain. In a limited experimental domain it is possible to describe the variation in response by means of very simple mathematical models.

* It is assumed that the experimental variables can be controlled by the experimenter and can be accurately set to predetermined values.

* It is assumed that the chemical phenomenon under study does not exhibit a discontinuous behaviour in the explored domain, e.g. by the emergence of a different reaction mechanism. This is not a strict requirement.

If there are discontinuities they will appear, sooner or later. However, this is often a cause of confusion and frustration and it will generally impose a lot of extra work before things can be clarified. In general, a response can be assumed to be continuous if all experimental variables are continuous. If one or several experimental variables are discrete (qualitative), e.g. different types of catalysts, one cannot safely assume a continuous response. It is often assumed that the different systems defined by the discrete variations show some kind of similarity, and in that case it is possible to make comparisons. If a similarity cannot be assumed it would be rather pointless to make any comparisons whatsoever. More detailed strategies for exploring discrete variations are discussed in the context of *principal properties* in Chapter 15.

It is seen from the above points that the requirements on the experimental system for multivariate design are almost always fulfilled.

References

1. G.E.P. Box and P.V. Youle
   Biometrics 11 (1955) 319.

2. K.R. Popper
   The Logic of Scientific Discovery
   Basic Books, London 1959.

Chapter 3

Models as tools

What is a model?

The word "model" has different meanings. In a scientific context, it usually means a picture or an image of certain aspects of the real world. The model will thus in some way reproduce these aspects in a form which is useful to us. This means that all models are artefacts fabricated by us, to facilitate our comprehension of the complex real world. Our knowledge of chemistry is, in essence, expressed by models for chemistry. Certain models have been found to be rather general and such models are therefore often referred to as "laws of nature" or "theories". It should, however, be borne in mind that they are just models for our comprehension. All knowledge of chemistry is based on experimental observations in the past. Models can be established, when patterns in the observed data allow certain generalizations to be made. Some of these generalizations will be related to previously known details or models, while others will be new. With increasing experience and when new observations become available, old models will be modified, revised or completely rejected. This is a continuous process, usually referred to as the progress of science. An interesting review of the concepts of models is given in the book *Chemistry through Models* by Suckling et al. [1]. As all models have been derived under a given set of premises, assumptions or constraint, and as it is impossible to *a priori* cover *all* possible interfering factors which may cause perturbations of the models, the general position must be: "All models are wrong, but some are useful" (George E. P. Box, 1954).

There are many types of models used in chemistry, and they can be divided into different categories:

Iconic models, which are picture-like images of the original. An example of such a model is:

Molecular models (ball and stick models, space-filling models) which reproduce typical bond lengths and bond angles and hence stereochemical and congestional effects. However, certain aspects are not covered by the model; they are, for instance, not soluble in cyclohexane in the same way as the original.

Heuristic models, "Rules of thumb", which are usually expressed in verbal form, e.g. "Like dissolves like"

"If a not too large molecule contains three or more polar functional groups, it cannot efficiently be extracted into an organic solvent from an aqueous layer by mere shaking in a separatory funnel"

"An increase in the reaction temperature by 10 °C will approximately double the reaction rate."

Quantitative models is a heterogeneous group of models expressed in mathematical language. This includes what can be called *hard models* of general applicability, e.g. thermodynamic models, quantum mechanical models, absolute rate theory, as well as *soft models* or *local models*, usually expressed in terms of analogy and similarity, e.g. linear free energy relationships (LFERs), correlations for spectroscopic structural determination, empirically determined kinetic models, and as we shall see, models obtained by statistical treatment of experimental data from properly designed experiments.

1.   Synthetic chemistry and quantitative models

1.1. Hard models and synthesis

Two types of responses are of general interest in exploring a synthetic reaction: the *yield* of the reaction and the *selectivity*, if more than one product can be formed.

The yield can be expressed as an integral of the rate over time

Yield = ∫ (Rate) dt

Selectivity is usually expressed as some function of the distribution of the products. If the products are interconvertible, selectivity can often be modelled by means of an equilibrium relation. If the products are formed by parallel irreversible reactions, the selectivity can often be described as a ratio of the different rates of formation.

This will invoke both thermodynamics and theories of the kinetics of chemical reactions. Equilibria are related to thermodynamics and will depend on the Gibbs energy of the system. Rates are related to activation energies of the various reaction steps. By absolute rate theory [2], and in particular by Linear Free Energy Relationships (LFERs), kinetic models can be related to thermodynamic parameters.[3] The energetics of a system will define a potential energy surface. Reaction mechanisms will then be different paths over this potential energy surface. Temperature dependence of rate and equilibrium constants can be derived from thermodynamic and kinetic models. The free energy of any entity "i" is related to its chemical potential, μ_i, which can be defined as the partial derivative of the free energy of the species with respect to the number of this species in the system. The chemical potential is then related to the concentration of the species by the corresponding activity coefficient, γ_i, to take deviations from ideality into account.

An overall rate expression for the formation of a product will take the form

$$Rate = k^{(Observed)} [A]^a [B]^b [C]^c....[Z]^z$$

The observed (phenomenological) rate constant, $k^{(Observed)}$, will be a function of the specific rates and equilibria involved in elementary steps of the reaction mechanism.

In principle, it would be possible to determine the outcome of any chemical reaction if: *(a)* The reaction mechanisms were known in detail, i.e. if all equilibrium constants and all rate constants of intermediary steps were known; and *(b)* the initial concentrations of the reactants and the activity coefficients of all species involved were perfectly known. However, this is never the case in practice. It would be impossible to derive such a model by deduction from physical chemical theory without introducing drastic assumptions and simplifications. A consequence of this is, that the precision of any detailed prediction from such hard models will be low. In addition to this, physical chemical models rarely take interaction effects between experimental variables into account, which means that, *in practice*, such models will not be very useful for analysing the influence of experimental variables on synthetic operations.

There is another way of overcoming this problem, and that is to use *soft, local models* which can be established through experimental observations. The remaining chapters of this book will deal with these aspects of the problem.

1.2. Soft models and synthesis

In Chapter 2 it was stated that it is reasonable to assume that there is some functional relation between the experimental variables, $x_1, x_2...x_k$ and the observed response, y.

$$y = f(x_1, x_2...x_k) + e$$

As was seen in the preceeding section on hard models it is evident that it will be very difficult to derive an analytical expression for the function f by theoretical means. For practical purposes, it is, however, possible to use experiments to establish sufficiently good *local* approximations of the function f.

2. Local models by Taylor expansions of the response function

2.1. Continuous experimental variables

Practical constraints on any experimental procedure imply that the experimental domain of interest will always be smaller than the theoretically possible domain. The unknown response function f describes a chemical phenomenon. As chemical events depend on the energetics of the system it is reasonable to assume that the function f is smooth and several times differentiable with respect to the experimental variables.

Under these conditions, it will be possible to approximate f in the *experimental domain of interest* by a Taylor expansion. A Taylor expansion will have a form

$$y = f(0) + \sum_{i=1}^{k} \frac{\partial f(0)}{\partial x_i} \cdot x_i + \sum_{i=1}^{k} \sum_{j=1}^{k} \frac{1}{1 \cdot 2} \cdot \frac{\partial f(0)}{\partial x_i \partial x_j} \cdot x_i x_j + \dots + R(x) + e$$

The experimental variables have been scaled so that $0 = (x_1 = x_2 = x_3 \dots = x_k = 0)$, will be the center of the experimental domain. The model also contains a rest term, $R(x)$, which becomes smaller and smaller as more terms are included in the model. The rest term describes the deviation between the observed response, y, and the variation described by the terms in the Taylor expansion. This means that $R(x)$ will contain a systematic error due to the truncation of the Taylor expansion. We will consider the approximation sufficiently good if the deviation $R(x)$ is not significantly larger than the experimental error. We will then include $R(x)$ in the overall error term, e.

The Taylor expansion is more conveniently written as

$$y = B_0 + B_1 x_1 + B_2 x_2 + \dots + B_k x_k + B_{ij} x_i x_j + \dots + B_{11} x^2 + \dots B_{ii} x_i^2 + \dots R(x) + e$$

For almost all chemical applications, it will be sufficient to include up to second degree terms in the Taylor expansion, ($R(x)$ is small), provided that the experimental domain is not too large.

The coefficients, B_0, B_1, B_2 ... B_{ij} ... etc., in the polynomial model will be called *parameters* of the model.[1]

---

[1] Some authors use the term *parameter* synonymously to denote an *experimental variable*. This terminology is not used here. In this book, the word *parameter* will denote something which has been estimated from experimental data.

These parameters can be estimated by *multiple linear regression*. This method is described below. By this procedure, the polynomial model is fitted to known experimental results so that the deviations between the observed responses and the corresponding responses calculated from the model are as small as possible. How these calculations are done and how the experiments should be laid out to obtain good estimates of the model parameters is treated in detail in the chapters that follow.

The polynomial model will be called a *response surface model* or shortly a *response surface*, since it will define a surface in the space spanned by $\{y, x_1, x_2...,x_k\}$. Response surface models can be used to assess the influence of all experimental variables on the measured response. This can now be translated into the concepts of experimental design:

The experiments which will furnish the best information on how the response, y, is influenced by the experimental variables, x_1, x_2, ... x_k, are those experiments from which a response surface model can be estimated with good precision, i.e. which allow good estimates of the corresponding model parameters.

* The constant term, β_0, will estimate the response when all variables are set to zero value. Usually the experiments are run so that this corresponds to the center of the experimental domain.

* The linear coefficients, β_1,...., β_k, will be direct measures of the linear dependence of the corresponding variables.

* A cross-product coefficient, β_{ij}, will be a measure of the interaction effect between the variables x_i and x_j and the model terms $\beta_{ij}x_ix_j$ together with the linear terms β_ix_j and β_jx_i will describe the joint influence of x_i and x_j on the response.

* Square terms, e.g. $\beta_{11}x_1^2$, will describe a non-linear influence of x_1 on the response.

2.2. Geometric interpretation of response surface models

Response surface models can be interpreted in geometrical terms. For the sake of simplicity, this will be illustrated by an example with only two experimental variables, x_1 and x_2. The principles can be extended to include any number of experimental variables. It is, however, difficult to visualize geometry in more than three dimensions. The interpretations of model in higher dimension will be completely analogous.

36

Assume that the "theoretical" response function

$$\eta = f(x_1, x_2)$$

can be depicted as in Fig. 3.1.

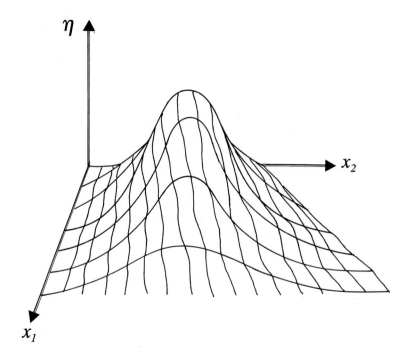

Fig.3.1: Theoretical, but unknown, response function.

We are not primarily interested in obtaining a description of the response function over all possible settings of the experimental variables. We need only to describe the variation in *the experimental domain of interest*, as shown in Fig. 3.2. This is a much easier problem to solve.

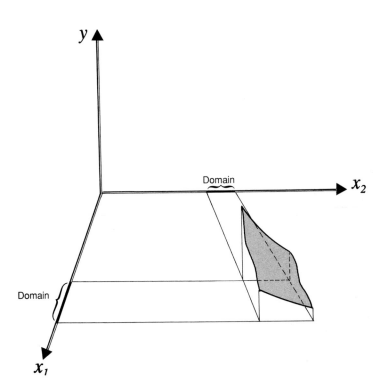

Fig. 3.2: Features of the response function in the experimental domain of interest.

2.3. Linear models

By a linear model

$$y = \beta_0 + \beta_1 x_1 + \beta_2 x_2 + e$$

the response function is approximated by a *plane*, Fig. 3.3.

The coefficient β_1 is the slope of the plane along the x_1 axis, and β_2 is the slope along the x_2 axis. These coefficients will therefore describe the sensitivity of the response, y, to the variations of the corresponding experimental variables. A variable which has a large influence on y will thus have a large coefficient. This means that a comparison of the numerical values of estimated model parameters offers a means of assessing the relative importance of the experimental variables with respect to their influence on the response. Significant variables will yield a variation above the noise level, e.

Linear response surface models will seldom give a perfect description of the variation in y, but they are very useful when rough estimates of the influence of the experimental variables are sufficient, e.g. for screening experiments.

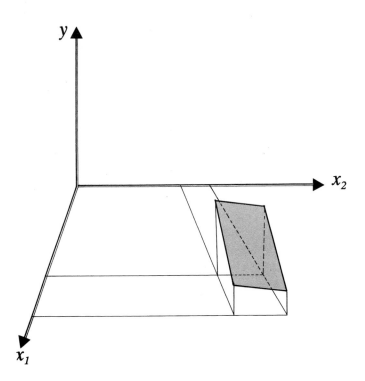

Fig.3.3: Linear response surface model.

2.4. Second-order interaction models

If the response surface looks like the one given in Fig. 3.4., a linear model will not yield a good description. The response surface is *a twisted plane*. The slope of the surface along the x_1 axis will depend on the value of variable x_2. This means that the influence of variable x_1 will depend on the settings of variable x_2, i.e. there is an interaction effect between x_1 and x_2. In such cases, the model will be improved if a *cross-product term*, $\beta_{12}x_1x_2$ is included:

$$y = \beta_0 + \beta_1 x_1 + \beta_2 x_2 + \beta_{12} x_1x_2 + e$$

The cross-product coefficient, β_{12}, will describe the twist of the plane.

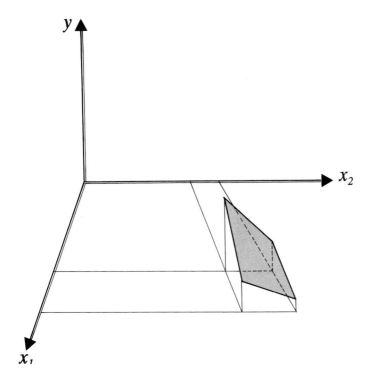

Fig. 3.4: A second-order interaction model can describe interactions between the experimental variables. The surface is a twisted plane. The twist is described by the cross-product term.

40

2.5. Quadratic models

Close to an optimum response, neither a linear model nor a model with interaction terms will yield a good description of the variation of y. The situation may be as shown in Fig. 3.5.

The response surface is curved, and there is a best setting of x_1 and x_2 which gives the highest response, i.e. *the optimum conditions*. In such cases it will be necessary to describe the curvature of the response surface. This can be accomplished by introducing also the *quadratic terms*:

$$y = \beta_0 + \beta_1 x_1 + \beta_2 x_2 + \beta_{12} x_1 x_2 + \beta_{11} x_1^2 + \beta_{22} x^2 + e$$

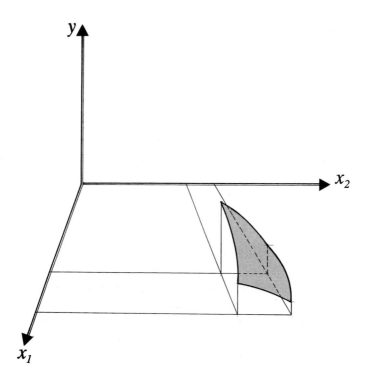

Fig.3.5: Close to an optimum, the response surface is curved. The curvature can be described by including also quadratic terms in the model.

The quadratic coefficient β_{11} describes the curvature along the x_1 axis, and β_{22} describes the curvature along the x_2 axis. The quadratic coefficients can be positive or negative and they can have different numerical values. Quadratic response surface models can therefore describe a variety of curved surfaces allowing analysis of surfaces which are convex and have a maximum or which are concave and have a minimum response as well as surfaces which are saddle-shaped.

2.6. Step-wise strategy

From the example above it is evident how the exploration of the experimental conditions can be approached in a step-wise manner:

Start with a model with only linear terms. If this gives sufficient information for the choice of suitable experimental conditions, there is no need to try a more complicated model.[2] If the linear model gives a poor description of the response, it is alway possible to run a set of complementary experiments to augment the model by adding also *interaction terms*. If this still gives a poor fit, then yet another set of complementary experiments can be run to include *quadratic terms*. This means that it will be possible to adjust the number of individual experiments to cope with the *slope*, the *twist*, and the *curvature* of the response surface.

2.7. A note on experimental design

The parameters in the response surface model will allow for an evaluation of each variable if the estimated value of each model parameter is *independent* of the estimated value of other model parameters. (For certain experimental designs, this is not possible to attain, but the estimates are as independent as possible.)

The parameters should measure the influence of the corresponding variables, i.e. the *slopes* of the surface, β_i, as measures of linear dependencies of the variables; the *twists* of the surface, β_{ij}, as measures of interactions between variables, and the *curvatures*, β_{ii}, as measures or non-linear influences of the variables, *and nothing else!* This calls for careful spacing of the variable settings in the experimental domain to determine the series of experiments used to estimate the parameters, i.e. the experimental design. These aspects will be treated in detail in the following chapters: Chapters 5 - 7, which deal with screening experiments based on linear and second order interaction models, and Chapter 12, which describes quadratic models for optimization.

---

[2] This is according to the principle of *Occham's razor*, which says that if there are several possible explanations, we should choose the most simple one.

2.8. Interpretation of response surface models

Provided that a proper experimental design has been used to establish the model parameters, and that the model adequately describes the variation of y in the experimental domain, we can use the model to evaluate the influence of each variable and assess the significance of each term in the model. This can be accomplished through statistical tests which compare the estimated parameters to estimates of the experimental error. This will answer the *Which?* and *How?* questions.[4]

Will it also be possible to answer the *Why?* questions, which implies that we give the estimated parameters a physical meaning? The answer is *yes*, although great care should be exercised in such interpretations.

The model parameters $(\beta_i, \beta_{ij}, \beta_{ii})$ are partial derivatives of the "theoretical" response function f and as such they will have physical meanings.

Examples

An increased yield, y, achieved through an increase in the reaction temperature, x_i, will show up as a positive value of β_i. This parameter is likely to be related to the temperature dependence of the kinetics of the reaction.

A significant interaction effect, β_{ij}, where x_i is the reaction temperature, and x_j is the initial concentration of a reactant, is probably related to the heat capacity of the reaction system.

A significant interaction effect, β_{ik}, where x_i is the reaction temperature, and x_k is the stirring rate, will indicate that the dissipation of heat from the system is important. This observation can be linked with the enthalpy of the reaction.

Although the response surface models are established from experimental data, they are not to be regarded as purely empirical models. Since it is possible to give a physical interpretation of the model parameters, response surface models are better described as semi-empirical models.

2.9. Response surfaces are local models

The polynomial approximation of f is a local model which is only valid in the explored experimental domain. It is not possible to extrapolate and draw any conclusions outside this domain. It is therefore important to determine a good experimental domain prior to establishing a response surface model with many parameters involved. By the methods of *Steepest ascent*, Chapter 10, and *Sequential simplex search*, Chapter 11, it is possible to locate a near-optimum experimental domain by means of a limited series of experiments.

2.10. Modelling with discrete (qualitative) variables

When one or several experimental variables are discrete, we cannot rely on a geometrical interpretation of smooth and continuous response functions. A change of catalyst, e.g. *Pd on carbon* to *Pt on alumina*, may well change the influence of other *continuous* variables.

2.11. Discrete variations on two levels

Often the discrete variations are an alternative choice between two varieties, *e.g.* two different catalysts, two different solvents, two different types of equipment, two different reagents etc. In such cases, the approach is straightforward and it will be possible to establish polynomial response functions to evaluate the effects also of discrete variations. The trick is to use a dummy variable and assign an arbitrary value of $+1$ or -1 to the alternative choices. It is then possible to fit a polynomial model to the experimental results obtained for different settings of the experimental variables.

Models which can be used to explore discrete variations are

the linear model

$$y = \beta_0 + \Sigma \, \beta_i \, x_i + e$$

and *the second order interaction model*

$$y = \beta_0 + \Sigma \, \beta_i \, x_i + \Sigma\Sigma \, \beta_{ij} \, x_i x_j + e$$

A systematic variation of y due to a discrete variation will be picked up as a significant linear coefficient. A significant cross-product coefficient for a discrete variable and a continuous variable will describe how the influence on y of the continuous variable is modified by the discrete change. It is also possible to have interaction effects between discrete variables. For instance, one catalyst may work well in one solvent, but may be totally useless in another.

It is not possible to use quadratic models which contain square terms of the discrete variables. It is impossible to detect any curvature from only two settings, and quadratic terms would therefore make no sense.

2.12. Discrete variations on more than two levels

In cases where there are more than two levels of a discrete change, it is not advisable to adopt polynomial models to account for the effects of variations, although questionable suggestions in this direction can sometimes be found in the literature.[3] It is, of course, possible to evaluate discrete variations, two at a time, and match the winning candidates against each other. For such purposes, polynomial models can be safely used. A better strategy will be to use a *block design* in which the discrete settings are studied in separate experimental blocks. Block designs are briefly discussed in Chapter 6.

In screening experiments where a large number of discrete selections can be made, e.g. to select a suitable solvent for a new procedure, it is advisable to use a design based on *principal properties*. This strategy is discussed in Chapters 15, 16.

3. Some initial aspects on modelling from experimental data

3.1. Errors

The "true" response η can never be observed. The observed response y will always be contaminated by an error term e.

$$y = \eta + e$$

There will always be two different kind of contributions to the error term: *systematic* and *random* errors.

A *systematic error* will occur if the method used to determine the response, y, underestimates or overestimates the true response. A well known example of this is that an isolated yield is always lower than the actual conversion due to losses during work-up. Other examples are: when chromatographic methods are used to determine the composition of a reaction mixture and it is assumed that isomeric compounds will have the same detector response, which may not be exactly true; different batches of chemicals may contain impurities which causes a systematic variation in yield.

Systematic errors are difficult to detect *within* the experimental system in which they occur, but they can be detected if the method used to determine the response is calibrated against other, independent methods. Certain types of systematic errors can be eliminated through a proper experimental design. In Chapter 6 is described how

---

[3] In order not to upset the responsible authors, the present author prefers not to give direct references to the papers.

block designs can be used to isolate systematic variation due to, for instance, different batches of chemicals, or different types of equipment. In the presence of a systematic error, the measured response, y, will not be a good estimator of the true response η, and will will say that there is a *bias* in the estimation of the response. A method which determines a response without bias is called an *accurate* method, and this property of the method is often referred to as *accuracy*.

A *random error* will be present in all experimental procedures. If an experiment is repeated several times, the results of the individual runs will not be identical. They may show a similar result but there will be a variation around an average value. Some of the runs will have a result above the average value, some will be lower. It is not possible to foresee if an experimental run will give a measured response above or below the average result. The variation observed is a random variation. The presence of a random variation makes it necessary to use statistical methods to draw correct conclusions from any experimentally determined data. The degree of the random variation is often referred to as *precision*.

3.2. Why will there always be a random variation?

There are many reasons: When the response is measured there is an analytical error of the method (errors of integration in gas chromatograms or in NMR spectra due to signal/noise ratio, temperature variation in detectors, errors in transferring samples). However, the most important sources of the experimental error are the unknown factors which are not controlled in the experiment. Although often assumed to be negligible, there is also a contribution from minor variations in the adjustment of the controlled experimental factors. It is not possible to exactly reproduce identical settings between experimental runs. There will always be small variation in, for instance, temperature control, weighing when starting materials are dosaged, adjusting the rates of addition of reagents to the reaction mixture (dropping funnels are notoriously difficult to calibrate), stirring in heterogeneous systems etc.

A common feature of these errors is that they are likely to occur independently of each other. As we shall see, this simplifies the analysis of experiments, since it is often possible to use known statistical distributions to assess the significance of the results obtained, e.g. the *normal distribution*, or other distributions related to the *normal distribution*, such as the *t distribution* or the *F distribution*. Significance tests based on these distributions are discussed later in this chapter.

46

3.3. Error distribution

To understand how random variations in the experimental conditions, as well as in the determination of the response, will influence the experimental results, the following discussion should be illuminating:

Assume that there is only one source of random variation when an experiment is run, e.g. two pipettes are used to apportion a liquid reagent. Each pipette is used at random and approximately equally frequent. Other experimental factors are meticulously controlled and the experimental conditions will be perfectly reproducible in this respect upon repeated runs. If the two pipettes are slighlty different, and this difference influences the result, the outcome of a series of experiments will show two different responses as illustrated in Fig. 3.6a.

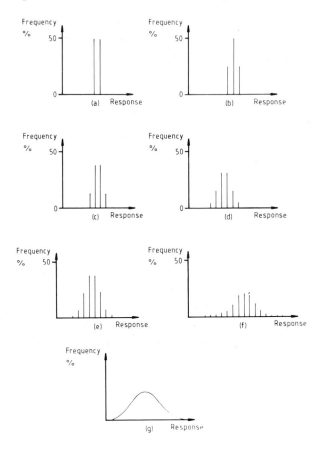

Fig.3.6: Error distributions.

If there are two sources of random variation which are approximately equally frequent and which produce a similar variation in the experimental response, the observed result will be as in Fig. 3.6b.

The observed frequencies with 3, 5, 7 and 14 different and equally probable random error sources are shown in Fig. 3.6c-f. With *many* different sources of experimental error, it is seen that the frequency of the experimental response data can be approximately described by the bell-shaped curve in Fig. 3.6g.

Such bell-shaped curves can be seen as approximations of the *normal distribution*. An analytical expression for the frequency, $f(y)$, of a response, y, which obeys the *normal distribution* is

$$f(y) = (2\pi\sigma^2)^{-\frac{1}{2}} \exp [(y - \mu)^2 / 2\sigma^2]$$

The shape of the curve, and hence the frequency, depends on two parameters only: the mean, μ, and the variance, σ^2. The mean, μ, will correspond to the "true" response η.

The dispersion or spread of observed y around the mean is described by the variance, σ^2. The variance is calculated as the average of the squared differences between observed y and the mean μ.

$$\sigma^2 = \Sigma (y_i - \mu)^2 / N$$

The positive square root of the variance is called the *standard deviation*.

$$\sigma = \sqrt{\sigma^2}$$

The frequency distribution is often called the *probability density function*, since the probability of finding a response y in the interval ($a < y < b$) is proportional to the surface under the curve in the interval ($a - b$), Fig. 3.7. This means that we can assess the probability of finding a response in a given interval. An important example is that the probability that a response is within an interval $\pm$ two standard deviations from the mean is greater than 95 %.

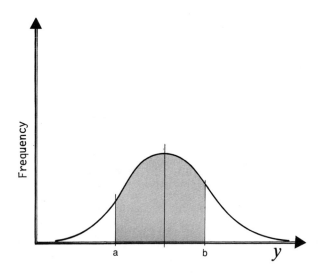

Fig 3.7: The probability is proportional to the surface under the frequency distribution curve.

The mean, μ, and the *variance, σ^2*, refer to an infinite *population* and are not known to the experimenter. However, it is possible to obtain estimates of these parameters through experiments. In the absence of systematic errors, the average response obtained by n repetitions *(replications)* of an experiment gives an unbiassed estimate of μ or η. In statistical terms, in can be said that the *expectation value* of the average response is η

$$\bar{y} = (\Sigma y_i) / n$$

$$E[\bar{y}] = \eta$$

The divisor n is the total number of replicated experiments.

An unbiassed estimate, s^2, of the error variance, σ^2, can also be obtained from the replicated experiment.

$$s^2 = \Sigma(y_i - \bar{y})^2 / (n - 1)$$

The divisor of the sum of squared deviations between each observation and the mean should be $(n-1)$ to give an unbiassed estimate of σ^2. Since all observations have already been used once to calculate the mean, there are only $(n-1)$ independent values left to describe the variation. This number is called the *degrees of freedom (d.f.)* for the variance estimation. The degrees of freedom will play an important role in statistical tests which can be used to evaluate experimental data.

There is another argument for justifying the assumption of normally distributed errors in synthetic chemical experiments. Generally, the responses of interest would be related to yields and selectivities of the reactions. These responses depend on the energetics of the system. The overall error will be a composite of all possible random perturbations of the system. These perturbations will affect the behaviour of the individual molecular entities in the system, but it is unlikely that any such perturbation would give any single molecule an infinite energy or some other infinite property, i.e. the energy distribution over the molecules will have a limited variance. The observed response would be an average of the contribution of the molecules in the system. The *Avogadro number*, $6.02 \cdot 10^{23}$, is tremendously large, which implies that even in a tiny chemical sample there will be a very large number of the individual molecules. This means that the error which can be observed would be the spread in observed responses, which in turn will be an average property of a sample drawn from population with a finite variance. A *central limit*[4] consequence will be that the observed error distribution would well be approximated by the *normal distribution*.

4. Modelling

4.1. Errors and models

Replication of a given experiment with fixed settings of the experimental variables will give an estimate of the average response, y, in this particular experiment, together with an estimate of the error variance, s^2. However, this is not enough for our purposes. We wish to determine how a *change* in the experimental conditions will influence the result, i.e. to determine, *(a) which variables* are responsible for the change, and *(b) how* they exert their influence.

Thus we wish to determine *systematic variations* in the presence of *noise*:

---

[4] *The central limit theorem:* If a population has a finite variance, σ^2, and a mean, μ, then the distribution of the sample mean approaches the normal distribution with the variance σ^2/n and the mean μ as the sample size n increases. [5]

$$y = f(x_1, \ldots x_k) + e$$

where f describes the systematic variation, and e will contain a random experimental error. As we have seen, the general features of the function f can be portrayed by a response surface model. The shape of the response surface is determined by the model parameters, β_i, β_{ik}, β_{ii}. The effects of changing experimental variables are transmitted via the corresponding model parameters into the systematic variation of y. To be significant, this variation must be above the noise level. Hence, to determine the influence of the experimental variables, we need to estimate the values of their model parameters and then compare these values to an estimate of the experimental error. For this purpose t statistics can be used (see below).

To be of any practical value, a response surface model should give a satisfactory description of the variation of y in the experimental domain. This means that the model error $R(x)$ should be negligible, compared to the experimental error. By multiple linear regression, least squares estimates of the model parameters would minimize the model error. Model fitting by least squares multiple linear regression is described in the next section.

From the response surface model it is possible to calculate (*predict*) the response for any settings of the experimental variables. This makes it possible to compute the residuals, e_i, as the difference between the observed response, y_i^{Obs}, in an experiment "i", and the response predicted from the model, y_i^{Pred}, for the corresponding experimental conditions.

$$e_i = y_i^{Obs} - y_i^{Pred}$$

If the model were absolutely perfect, the residuals would be a measure of the experimental error *only*. If the model is adequate, the variance of the residuals should not be significantly greater than the variance of experimental error. This can be checked by an F-test, provided that an independent estimate of the experimental error variance is available. Such an estimate can be obtained through replication of one or more experiment.

In the sections below a brief outline is given of how experimental data can be used to fit a response surface model and how statistical methods can be used to evaluate the results. The author's intensions in this introductory chapter is to give the reader a feeling for how statistical tools can be used in an experimental context. Detailed descriptions follow in subsequent chapters.

4.2. Scaling of variables

For response surface modelling, scaled variables, x_i, will be used intstead of the "natural" variable, u_i. The range of variation of each continuous variable, u_i, in the experimental domain will be linearly transformed into a variation of x_i centered around zero and usually spanning the interval $[-1 \leq x_i \leq +1]$. (For discrete variables on two levels, each level will be arbitrarily assigned the value -1 or $+1$). The scaling is done in the following way:

Calculate the average, u^0, of the high level, u^{High}, and the low level, u^{Low}. Then, determine the step of variation, δu, from the average to the high level.

$$u^0 = (u^{High} + u^{Low}) \: / \: 2 \delta u = u^{High} - u^0$$

The scaled value x of u is calculated as follows

$$x = (u - u^0) \: / \: \delta u$$

Example: The influence of T, temperature variation, in the interval $20 - 60 \, °C$ is studied.

The average temperature is $40 \, °C$. The step of variation from the average to the high level is $20 \, °C$. A scaled temperature variable, x_T, will thus be:

$$x_T = (T - 40) \: / \: 20$$

It is seen that the low level corresponds to $x_T = -1$, and the high level corresponds to $x_T = +1$.

An advantage of using scaled variables for response surface modelling is that the estimated model parameters can be directly compared to each other. Their numerical value will be a direct measure of how the ranges of variation in the experimental domain are prone to influence the response. *This is exactly what is desired.*

Scaling of the variables also has an historical background. In the past, (B.C. before computers), all calculations had to be carried out by hand or by mechanical devices. Least squares fitting of models was a cumbersome process with the natural variables, and more easily done with scaled variables. Natural variables would also have lead to large rounding-off errors.

4.3. Multiple linear regression. Least squares fitting of response surface models

Assume that we wish to test a model with linear terms and an interaction term to describe the variation in y in an experimental domain spanned by the scaled variables x_1 and x_2, $[-1 \leq x_i \leq 1]$.

$$y = \beta_0 + \beta_1 x_1 + \beta_2 x_2 + \beta_{12} x_1 x_2 + e$$

If only one experiment is performed, e.g. with all variables at their high level, $(x_1 = x_2 = 1)$, and the response y_1 is obtained in this experiment and if these values are put into the response surface model, we obtain

$$y_1 = \beta_0 + \beta_1 + \beta_2 + \beta_{12} + e_1$$

The observed response is a linear combination of the model parameters and the experimental error. Now, if an experiment is run with another setting of the variables (e.g $x_1 = -1$, $x_2 = 1$), which afforded the response y_2, we obtain

$$y_2 = \beta_0 - \beta_1 + \beta_2 - \beta_{12} + e_2$$

which is another linear combination of the model parameters. It is seen that other settings of the experimental variables would give yet other linear combinations.

To be able to estimate values of each model parameter, we need to run at least as many experiments as there are unknown model parameters. The trouble is the error terms, e_i, which contain an unknown random experimental error which is not constant in the different experimental runs. We need to take these error terms into account and try to assign values to the model parameters in such a way that the error terms can be kept as small as possible *over the whole set of experiments*.

This will be done by the method of least squares multiple linear regression. This method will estimate the model parameter so that the sum of the squared error terms (residuals), Σe_i^2, will be as small as possible. How this is done is illustrated by an example.

4.4. Example. An elimination reaction

The influence of *reaction temperature, T*, and the *concentration of base*, *[Base]*, on the yield of an elimination reaction was studied. The ranges of variation of the experimental variables were: 20 °C $\leq T \leq$ 60 °C, and $1.0 \leq [Base] \leq 2.0$.

It was assumed that a response surface model with linear terms augmented with a term for the interaction effect would be sufficient to describe the variation in yield in the experimental domain. Table 3.1 shows the experimental design[5] (in natural variables) which was used to determine the model. The yields obtained are also given.

Table 3.1: Experimental design in natural variables and yields obtained in the elimination experiment

Exp no	T (°C)	[Base] (M)	Yield (%)
1	20	1.0	78.1
2	60	1.0	92.8
3	20	2.0	77.8
4	60	2.0	96.8
5	40	1.5	83.7

The corresponding design expressed in scaled variables, x_1 (temperature), x_2 (base concentration) is given in *Table 3.2*. Such tables will be called *design matrices*.

Table 3.2: Design matrix, D, of the elimination experiment

$$
\mathbf{D} = \begin{array}{ccc}
x_1 & x_2 & \textit{Exp. no} \\
\left[\begin{array}{cc}
-1 & -1 \\
1 & -1 \\
-1 & 1 \\
1 & 1
\end{array}\right] & & \begin{array}{c}
1 \\
2 \\
3 \\
4
\end{array}
\end{array}
$$

The suggested model is

$$y = \beta_0 + \beta_1 x_1 + \beta_2 x_2 + \beta_{12} x_1 x_2 + e$$

The model can also be written in matrix notation as below. It is very convenient to use matrices to describe the calculations involved in the modelling process. A short summary of matrix calculus is given in an appendix at the end of this book.

---

[5] The experimental design shown in Table 3.1 is actually a *two-level factorial design* completed with one experiment at the center of the experimental domain. Properties of factorial designs are discussed in Chapter 5.

$$y = \begin{bmatrix} 1 & x_1 & x_2 & x_1x_2 \end{bmatrix} \begin{bmatrix} \beta_0 \\ \beta_1 \\ \beta_2 \\ \beta_{12} \end{bmatrix} + e$$

$$y = x\,\beta + e$$

where x is a row vector of variables in the model and β is a column vector of the "true" model parameters, i.e. the true coefficient of the Taylor expansion.

If we put the values of the scaled variables in the model for the five experimental runs we obtain

$$78.1 = \beta_0 - \beta_1 - \beta_2 + \beta_{12} + e_1$$
$$92.8 = \beta_0 + \beta_1 - \beta_2 - \beta_{12} + e_2$$
$$77.8 = \beta_0 - \beta_1 + \beta_2 - \beta_{12} + e_3$$
$$96.8 = \beta_0 + \beta_1 + \beta_2 + \beta_{12} + e_4$$
$$83.7 = \beta_0 \qquad\qquad\qquad\quad + e_5$$

In matrix notation this corresponds to

$$78.1 = \begin{bmatrix} 1 & -1 & -1 & 1 \end{bmatrix} \beta + e_1$$
$$92.8 = \begin{bmatrix} 1 & 1 & -1 & -1 \end{bmatrix} \beta + e_2$$
$$77.1 = \begin{bmatrix} 1 & -1 & 1 & -1 \end{bmatrix} \beta + e_3$$
$$96.5 = \begin{bmatrix} 1 & 1 & 1 & 1 \end{bmatrix} \beta + e_4$$
$$83.8 = \begin{bmatrix} 1 & 0 & 0 & 0 \end{bmatrix} \beta + e_5$$

These can be combined in an overall matrix equation

$$
\begin{bmatrix} 78.1 \\ 92.8 \\ 77.8 \\ 96.8 \\ 83.7 \end{bmatrix}
=
\begin{bmatrix}
1 & -1 & -1 & 1 \\
1 & 1 & -1 & -1 \\
1 & -1 & 1 & -1 \\
1 & 1 & 1 & 1 \\
1 & 0 & 0 & 0
\end{bmatrix}
\begin{bmatrix} \beta_0 \\ \beta_1 \\ \beta_2 \\ \beta_{12} \end{bmatrix}
+
\begin{bmatrix} e_1 \\ e_2 \\ e_3 \\ e_4 \\ e_5 \end{bmatrix}
$$

$$
\mathbf{y} \qquad = \qquad \mathbf{X} \qquad\qquad \mathbf{\beta} \quad + \quad \mathbf{e}
$$

i.e. $\mathbf{y} = \mathbf{X\beta} + \mathbf{e}$

The matrix $\mathbf{X}$ is called the model matrix. The columns in $\mathbf{X}$ correspond to the variables in the model. The model matrix is obtained from the design matrix, $\mathbf{D}$, by adding a column of ones (corresponds to β_0), columns of the cross product (and squared terms if there are such terms in the model).

In the following, the symbols β_0, β_i, β_{ij} and β_{ii} will be used to denote the "true" model parameters, i.e. the coefficients of the Taylor expansion. For the *estimated* values of these parameters, the symbols b_0, b_i, b_{ij} and b_{ii} will be used. If the attempted model is correct, the expectation values of the estimated model parameters would be the true model parameters, i.e.

$E[b_i] = \beta_i$

and the vector of estimated model parameters

$b = [b_0 \ b_1 \ ... \ b_i \ ... \ b_{ij} \ ...]'$ would be an unbiassed estimator of the vector of the true model parameters,

$\beta = [\beta_0 \ \beta_1 \ ... \ \beta_i \ ... \ \beta_{ij}]'$

(the prime, ', denotes the transposed matrix).

$E[b] = \beta$

A least squares fit of the model parameters is obtained by the following important relation. More details of why this gives the least squares estimate is given in Appendix 3A.

$$(\mathbf{X'X})^{-1}\mathbf{X'y} = \mathbf{b}$$

The calculations necessary to obtain the least squares estimates in the elimination experiment will thus be: First, compute the matrix $\mathbf{X'X}$

$$\mathbf{X'X} = \begin{bmatrix} 1 & 1 & 1 & 1 & 1 \\ -1 & 1 & -1 & 1 & 0 \\ -1 & -1 & 1 & 1 & 0 \\ 1 & -1 & -1 & 1 & 0 \end{bmatrix} \begin{bmatrix} 1 & -1 & -1 & 1 \\ 1 & 1 & 1 & 1 \\ 1 & 1 & -1 & -1 \\ 1 & -1 & 1 & -1 \\ 1 & 0 & 0 & 0 \end{bmatrix} = \begin{bmatrix} 5 & 0 & 0 & 0 \\ 0 & 4 & 0 & 0 \\ 0 & 0 & 4 & 0 \\ 0 & 0 & 0 & 4 \end{bmatrix}$$

Then, compute the inverse matrix $(\mathbf{X'X})^{-1}$. This will be possible only if the determinant $|\mathbf{X'X}|$ is different from zero.

In this case, the inverse matrix $(\mathbf{X'X})^{-1}$ is easily obtained. Since $\mathbf{X'X}$ is a diagonal matrix, the inverse matrix will also be a diagonal matrix in which the elements will be the inverted values of the elements in $\mathbf{X'X}$.

$$(\mathbf{X'X})^{-1} = \begin{bmatrix} 1/5 & 0 & 0 & 0 \\ 0 & 1/4 & 0 & 0 \\ 0 & 0 & 1/4 & 0 \\ 0 & 0 & 0 & 1/4 \end{bmatrix}$$

In this case, the calculations given by $(\mathbf{X'X})^{-1}\mathbf{X'y} = \mathbf{b}$ correspond to

$$
\begin{bmatrix} 1/5 & 0 & 0 & 0 \\ 0 & 1/4 & 0 & 0 \\ 0 & 0 & 1/4 & 0 \\ 0 & 0 & 0 & 1/4 \end{bmatrix}
\begin{bmatrix} 1 & 1 & 1 & 1 & 1 \\ -1 & 1 & -1 & 1 & 0 \\ -1 & -1 & 1 & 1 & 0 \\ 1 & -1 & -1 & 1 & 0 \end{bmatrix}
\begin{bmatrix} 78.1 \\ 92.8 \\ 77.8 \\ 96.8 \\ 83.7 \end{bmatrix}
=
\begin{bmatrix} b_0 \\ b_1 \\ b_2 \\ b_{12} \end{bmatrix}
$$

$$
\begin{bmatrix} 1/5 & 0 & 0 & 0 \\ 0 & 1/4 & 0 & 0 \\ 0 & 0 & 1/4 & 0 \\ 0 & 0 & 0 & 1/4 \end{bmatrix}
\begin{bmatrix} 78.1 + 92.8 + 77.8 + 6.8 + 83.7 \\ -78.1 + 92.8 - 77.8 + 96.8 \\ -78.1 - 92.8 + 77.8 + 96.8 \\ 78.1 - 92.8 - 77.8 + 96.8 \end{bmatrix}
=
\begin{bmatrix} b_0 \\ b_1 \\ b_2 \\ b_{12} \end{bmatrix}
$$

$$
\begin{bmatrix} 1/5 & 0 & 0 & 0 \\ 0 & 1/4 & 0 & 0 \\ 0 & 0 & 1/4 & 0 \\ 0 & 0 & 0 & 1/4 \end{bmatrix}
\begin{bmatrix} 429.2/2 \\ 33.7 \\ 3.7 \\ 4.3 \end{bmatrix}
=
\begin{bmatrix} b_0 \\ b_1 \\ b_2 \\ b_{12} \end{bmatrix}
$$

$$
\begin{bmatrix} 1/5 \cdot 429.0 \\ 1/4 \cdot 33.7 \\ 1/4 \cdot 3.7 \\ 1/4 \cdot 4.3 \end{bmatrix}
=
\begin{bmatrix} 85.84 \\ 8.42 \\ 0.925 \\ 1.075 \end{bmatrix}
=
\begin{bmatrix} b_0 \\ b_1 \\ b_2 \\ b_{12} \end{bmatrix}
$$

If the estimated coefficients are rounded to two decimals the model will be

$$y = 85.84 + 8.43\, x_1 + 0.92\, x_2 + 1.08\, x_1 x_2 + e$$

The model suggests that the variation in temperature, x_1, is important, but that the variation in base concentration, x_2, is not. However, it is not possible to draw any safe conclusions without a statistical analysis of the model. We will return to this

58

model after a discussion of how such statistical analysis can be done. A three-dimensional projection of the model is shown in Fig.3.8.

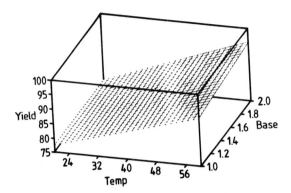

Fig.3.8: Three-dimensional plot of the estimated response surface in the elimination reaction.

Summary: Suggest a response surface model. Design a series of experiments. From the design matrix, **D**, in scaled variables, construct the model matrix, **X**, by augmenting **D** with a column of ones and columns for the cross-products (and the squares). Let **β** be the vector of model parameter to be estimated, and let **y** be the vector of measured responses obtained by running the experiments in **D**, and let **e** be the vector of errors such that $e' = [e_1 \, e_2.....e_n]$. The results can be summarized as

$$y = X\beta + e$$

A least squares estimate **b** of the true parameter vector **β** is obtained by computing

$$(X'X)^{-1}X'y = b$$

In Appendix 3A it is shown why this will give the least squares estimates of the model parameters.

A least squares estimate is no guarantee whatsoever, that the model parameters will have *good* properties, i.e. that they will measure what we want them to do, viz. the influence of the variables. The quality of the model parameters is governed by the properties of the *dispersion matrix* $(X'X)^{-1}$, and hence it depends ultimately on the experimental design used to determine the model. The requirements for a good design will be discussed in Chapter 5.

4.5. Lack-of-fit

Any experimental procedure will be afflicted by a random error. The error variance, σ^2, is not known to the experimenter, but it is possible to obtain an unbiassed estimate, s^2, of the variance by replication of an experiment. For synthetic chemical systems, we have seen that it is reasonable to assume that the experimental errors are normally and independently distributed. It is also reasonable to assume that the variance is constant in a limited experimental domain. These assumptions cannot be taken for granted. They must always be checked, e.g. by plotting the residuals in different ways. Such diagnostic tests will be discussed later on.

Assume that two estimates, s_1^2 and s_2^2, of the error variance have been obtained through two different and independent sets of experiments. Assume that $s_1^2 = 3.75$ has been determined by running the experiment seven times. This means that s_1^2 will have six degrees of freedom. Assume also that $s_2^2 = 2.43$ has been determined with eleven degrees of freedom. It is seen that s_1^2 and s_2^2 are not identical. Is this due to a pure chance since errors are random events, or is it a real difference? To answer this question, determine the F ratio.

$$F = s_1^2 / s_2^2$$

This ratio will have a known probability distribution for variances estimated from samples drawn from a normally distributed population. This probability distribution is called *the F distribution*. Fig. 3.9 shows a typical F distribution. The actual shape of the curve depends on the numbers of degrees of freedom of the sample variance estimates. Tabulated critical F ratios are given in Appendix; Statistical Tables at the end of this book.

To determine the probability that the two estimates, s_1^2 and s_2^2, are actually measures of the same variance, σ^2, the F distribution can be used as follows. Assume that we are willing to accept a 95 % probability that an observed F-ratio actually refers to estimation of the same variance. From the F distribution we can determine a critical F value, F^{Crit}, such that the surface under the curve in the interval $[0 - F^{Crit}]$ covers 95 % of the total surface under the curve. This is equivalent to saying that the probability that the observed F by pure chance exceeds the value of F^{Crit} should be less than 5 %. The probability distribution of F, and hence the critical F, values depend on the number of degrees of freedom of the variance estimates.

In the present case $F = \sigma_1^2 / \sigma_2^2 = 3.75/2.43 \approx 1.54$ with 6 respectively 11 degrees of freedom, and the corresponding critical F value for 95 % probability is 3.09.

60

$$F^{\text{Obs}} = 1.54 < F^{\text{Crit}} = 3.09$$

The observed F is less than F^{Crit} and there is no indication, at the 95 % level, that the two estimated variances are significantly different.

This technique can be used to analyse the *lack of fit* of response surface models. If a fitted response surface model is a good approximation of the unknown response function f, then the residuals, $e_i = y_i^{\text{Exp}} - y_i^{\text{Pred}}$, should depend *only* on the experimental error. From the series of n experiments which has been used in the modelling of the response surface, we can calculate the sum of squared residuals, Σe_i^2. The whole set of experiments has already been used once to estimate the model parameters. If we assume that there are p parameters in the model, there will be only $(n - p)$ degrees of freedom left to describe the independent errors. *If the model is good*, then $s_1^2 = \Sigma e_i^2 / (n - p)$ would be an estimate of the experimental error variance. Another, and independent, estimate of the experimental variance, s_2^2, can be obtained by replication of an experiment. This latter estimate can then be compared to the estimate from the residuals. A poor model fit would give $F = s_1^2 / s_2^2 > F^{\text{Crit}}$. If this is true, the model cannot be regarded as a reliable approximation of the response function.

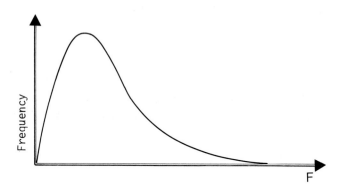

Fig.3.9:   An F distribution. The actual shape of the curve depends on the degrees of freedom of the estimated variances.

Let us have another look at the response surface model obtained in the elimination experiment.

$$y = 85.84 + 8.43\,x_1 + 0.93\,x_2 + 1.08\,x_1x_2 + e$$

We can predict the response by entering the values of the experimental variables in the different experiments, and then compute the residuals. The result is summarized in Table 3.3

Table 3.3: Calculation of residuals in the elimination experiment.

Exp. no	x_1	x_2	y^{Obs}	y^{Pred}	Residuals, e
1	−1	−1	78.1	77.56	0.54
2	1	−1	92.8	92.26	0.54
3	−1	1	77.8	77.26	0.54
4	1	1	96.8	96.28	0.52
5	0	0	83.7	85.84	−2.14

The sum of squared residuals is:

$$\Sigma\,e_i^2 = (0.54^2 + 0.54^2 + 0.54^2 + 0.52^2 + 2.14^2) = 5.7248 \approx 5.72$$

Five experiments were used to estimate four model parameters. There is only $(5 - 4) = 1$ degree of freedom left for computing the residual variance, $s_1^2 = \Sigma e_i^2 / (5 - 4) = 5.72$.

To determine the reproducibility of the experimental procedure, a pilot experiment $(T = 25\ °C, [Base] = 1.2\ M)$ was run five times before the response surface model was determined. These experiments were conducted within the experimental domain explored for the response surface model. They can therefore be used to obtain an independent estimate of the experimental error variance. The yields obtained in these experiments were: 79.1, 77.9, 78.7, 78.7, 81.0 %. This gives an average yield of 79.08 %. The calculated variance, s_2^2, will thus be:

$$s_2^2 =$$
$$[(79.1 - 79.08)^2 + (77.9 - 79.08)^2 + (78.7 - 79.08)^2 + (78.7 - 79.08)^2 + (81.0 - 79.08)^2]/(5 - 1)$$
$$= 5.368 / 4 = 1.342 \approx 1.34$$

An F−ratio for testing the lack of fit will thus be

$$F^{Lack\ of\ fit} = s_1^2 / s_2^2 = 5.72 / 1.34 \approx 4.27$$

The 95 % probability value of F_{Crit} with (1, 4) degrees of freedom is 7.71.

$$F^{\text{Lack-of-fit}} = 4.27 < F_{\text{Crit}} = 7.71$$

This result does not indicate a significant lack of fit. We can therefore use the model to evaluate the roles of the experimental variables.

5. Significance of estimated model parameters

5.1. Standard error

When we use experimental data to estimate parameters in response surface models, we will carry on the experimental error to the estimated model parameters. This will be manifested in such a way that the values of the model parameters will not be precisely known, i.e. they will have a probability distribution. To illustrate this, a simple example with only one experimental variable will first be discussed. It will then be obvious how the principles can be applied to models with several variables.

Example: A common routine task in the laboratory is to establish a calibration curve, e.g. to calibrate the signal from a chromatographic detector to known concentrations of the sample. When the concentration range is not too wide, often a straight line will give a good calibration, i.e.

$$y = \beta_0 + \beta_1 x_1 + e$$

It is a common experience, however, that if we repeat the calibration experiment, the estimated values of the intercept, b_0, and the slope, b_1, will not be identical. If we then repeat the calibration experiment several times, we will find that the estimated values of the parameters will be distributed around an average value, $\bar{b}_i$, sometimes denoted $\hat{\beta}$ (beta hat).[6] The situation is depicted in Fig. 3.10.

---

[6] The circumflex sign, ^, ("hat") on a parameter always denotes an estimated value. When a parameter is shown without the hat symbol it will denote the "true" value.

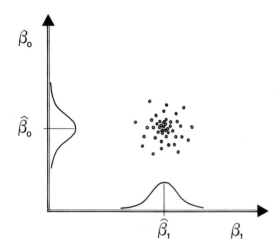

Fig.3.10: Experimentally determined parameters will have a probability distribution around an average value.

From a series of n calibration experiments it will be possible to compute both the average

$$\overline{b}_i = \Sigma b_i / n$$

and the corresponding standard deviation

$$s_i = \sqrt{[\Sigma(b_i - \overline{b}_i)^2 / (n - 1)]}$$

This standard deviation is called the *standard error* of the estimated parameter. If the estimated average value of the parameter and its standard error are known it will be possible to assess the probability that *new* estimates of the model parameters will measure the *same* properties, i.e. they should not be outside the expected range of random variation of the previously known estimates. This can be tested as follows:

Let d be the difference between the new estimate and the previously known average.

64

$$d_i = b_i - \overline{b}_i$$

Determine the ratio t of this difference to the corresponding standard error, s_i.

$$t_i = d_i \,/\, s_i$$

This t ratio will have a known probability distribution if the experimental errors are approximately normally distributed.

The probability distribution of t is a bell-shaped curve, similar to the normal distribution, see Fig. 3.11. However, the shape of the curve depends on the number of the degrees of freedom used to estimate the standard error. When this number increases, the t distribution approaches the normal distribution.

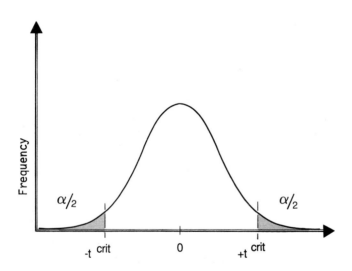

Fig.3.11: A t distribution. The actual shape of the curve depends on the number of degrees of freedom of the estimated standard deviation.

The t distribution can be used to assess the probability that an observed difference d_i is actually outside the expected range of random variation for the model parameter. To do so, we must decide on a significance level α. Often a significance level $\alpha = 5\ \%$ is used. This means that we accept that there is a 5 % risk that the observed difference *by pure chance* is outside the expected range. For the given significance level α we can use the t-distribution to determine a critical value, t^{Crit}, such that the surface under the t-distribution curve outside the limits $\pm t^{Crit}$ is α of the total surface under the curve. This is equivalent to saying that the surface under the curve in the interval $(-t^{Crit} - +t^{Crit})$ should be $(1 - \alpha)$ of the total surface. Critical t-values are tabulated at the end of the book.

With the assumption that the fitted model is correct, we can use the critical t-values, the estimated values of the model parameter and its standard error to assign *confidence limits* to the "true" model parameter. These limits are obtained by multiplying the standard error by the critical t-value.

$$\beta_i = \bar{b}_i \pm t^{Crit} \cdot s_i$$

This means that the probability is $(1 - \alpha)$ that the *true* value of β_i is within the confidence limits.

5.2 Standard errors of parameters in response surface models

It would not be practical if one had to run several independent determinations of a response surface model, in order to establish the standard errors of the model parameters. Fortunately, this is not necessary. The standard errors can be determined from an estimate, s^2, of the experimental error variance, σ^2, *and* from the experimental design used to establish the response surface model. The dispersion matrix, $(\mathbf{X'X})^{-1}$, is used. When this matrix is multiplied by the experimental error variance σ^2, the *variance–covariance matrix* is obtained. This matrix will contain the variances, $V(b_i)$, and the covariances[7], $Cov(b_i,b_j)$ of the estimated parameters.

Assume that the response surface model is:

$$y = \beta_0 + \Sigma \beta_i x_i + \Sigma\Sigma \beta_{ij} x_i x_j + e$$

The variance–covariance matrix of the estimated parameters will be as shown on the next page.

It is seen, that the diagonal elements consist of the variances of the estimated model parameters. The square roots of these variances are the standard errors.

---

[7] The covariance is defined $Cov(b_i,b_j) = [\Sigma(b_i - \bar{b}_i)(b_j - \bar{b}_j)]/N$. If b_i and b_j are independently estimated, the covariance will be zero. If they are not independently estimated, the value of b_i will depend on the value of b_j.

$$(\mathbf{X'X})^{-1}\sigma^2 = \begin{bmatrix} V(b_0) & Cov(b_0,b_1) & Cov(b_0,b_2) & \cdots & Cov(b_0,b_{ij}) \\ & V(b_1) & Cov(b_1,b_2) & \cdots & Cov(b_1,b_{ij}) \\ & & V(b_2) \cdots & \cdots & Cov(b_2,b_{ij}) \\ \text{Symmetric} & & & \cdots & \cdots \\ & & & & \cdots \\ & & & & V(b_{ij}) \end{bmatrix}$$

Let us now go back to the elimination experiment. The estimated error variance with four degrees of freedom determined by the pilot experiment was $s^2 = 1.34$ The variance−covariance matrix $(\mathbf{X'X})\,s^2$ will therefore be

$$\begin{bmatrix} V(b_0) & 0 & 0 & 0 \\ 0 & V(b_1) & 0 & 0 \\ 0 & 0 & V(b_2) & 0 \\ 0 & 0 & 0 & V(b_{12}) \end{bmatrix} = \begin{bmatrix} 1.34/5 & 0 & 0 & 0 \\ 0 & 1.34/4 & 0 & 0 \\ 0 & 0 & 1.34/4 & 0 \\ 0 & 0 & 0 & 1.34/4 \end{bmatrix} =$$

$$\begin{bmatrix} 0.268 & 0 & 0 & 0 \\ 0 & 0.335 & 0 & 0 \\ 0 & 0 & 0.335 & 0 \\ 0 & 0 & 0 & 0.335 \end{bmatrix}$$

The standard errors will thus be

Parameter	Standard error
b_0	$\sqrt{0.268} \approx 0.52$
b_1	$\sqrt{0.355} \approx 0.58$
b_2	$\sqrt{0.355} \approx 0.58$
b_{12}	$\sqrt{0.355} \approx 0.58$

The t distribution for four degrees of freedom gives $t^{\text{Crit}} = 2.78$, (at the 95 % probability level). We can therefore determine 95 % confidence limits for the estimated parameters.

$$
\begin{aligned}
b_0 &= 85.84 \pm 2.78 \cdot 0.52 \approx 1.45 \\
b_1 &= 8.43 \pm 2.78 \cdot 0.58 \approx 1.61 \\
b_2 &= 0.93 \pm 2.78 \cdot 0.58 \approx 1.61 \\
b_{12} &= 1.08 \pm 2.78 \cdot 0.58 \approx 1.61
\end{aligned}
$$

5.3. Significant model parameters

The response surface model is used to determine the influence of the experimental variables on the response. The model parameters *measure* this influence. A value of a model parameter equal to *zero* would imply that the corresponding term is not influencing the response. We can therefore assess the probability that a variable has a real influence by determining the probability that the corresponding model parameter is significantly different from zero. This means that the value zero should not belong to the confidence interval of the model parameter. Another way of saying this is that the value of the model parameter shall be outside the noise level given by the confidence limits. This can be determined from the ratio of the estimated parameter to its standard error. To be significant, this ratio should exceed the critical t-value.

From this it can be concluded that it is only the temperature variation, x_1, which has a significant (95 % level) influence on the yield in the elimination experiment.

5.4. Analysis of variance as a tool for determination of significance of models

From a series of n experiments used to determine a response surface model we can compute the sum of the squared responses. This is called the *total sum of squares, SST*. If y is the vector of responses over the series of experiments it is seen that

$$
SST = \Sigma \, y_i^2 = y'y
$$

An average response, y, is computed by

$$
y = (\Sigma \, y_i) \, / \, n
$$

To determine how much of the total sum of squares is a contribution from the average response, compute the difference

$$\Sigma \, y_i^2 - n \cdot [(\Sigma \, y_i) \, / \, n]^2 = \Sigma \, y_i^2 - (\Sigma \, y_i)^2 \, / \, n$$

This difference is called *sum of squares corrected for the mean, SSC.* *SSC* describes the squared deviations from the mean which should be accounted for by the regression parameters associated with the experimental variables. If the response surface should be totally flat (no slopes, no twists, no curvatures), the average response would be representative for all experiments, i.e. the estimated response surface model would be

$$y = b_0 + e$$

In that case it is seen that *SSC* would be a measure of the squared error terms. In most cases there is a variation in *y* which is related also to the variation of the experimental variables. It is possible to use the model and compute to what extent the model terms contribute to the sum of squares. This means that it is possible to determine how much *each* term in the model contributes. As we shall see, this offers another means of assessing the significance of the model parameters.

If we put the values of the variables in experiment *i* in the model we can calculate a predicted value of the response, y_i^{Pred}. For the series of experiments it is thus possible to compute the sum of squares of the predicted responses. This is called the *sum of squares due to regression, SSR.* If y^{Pred} is the vector of predicted responses it is seen that

$$y^{Pred} = \mathbf{X}b$$

where **b** is the vector of estimated model parameters. From this it follows that

$$SSR = \Sigma \, (y_i^{Pred})^2 = y^{Pred}, y^{Pred} = (\mathbf{X}b)'\mathbf{X}b = b'\mathbf{X}'\mathbf{X}b$$

The contribution to the sum of squares from the individual terms in the model is determined from the corresponding column vector, x_j, in the model matrix **X** and the estimated model parameter b_j by computing the sum over all experiments, $i = 1$ to n.

$$x_j'x_j b_j^2 = [\, \Sigma x_{ji}^2 \,] \cdot b_j^2$$

What is left when the sum of squares due to the regression is subtracted from the total sum of squares is the sum of squared residuals $\Sigma(y_i - y_i^{\text{Pred}})^2 = \Sigma e_i^2$. This is called the *residual sum of squares* or the error sum of squares, *RSS*. If the model is adequate, *RSS* should be a manifestation of the experimental error.

When we compute the model parameters we use all the observed responses, once for each model parameter. This means that for each estimation of the model parameters one degree of freedom is consumed. With n experimental observations and with p parameters in the model there are $(n - p)$ degrees of freedom left, over which the error sum of squares could be distributed. If the model is adequate, the *error mean square, MSE = RSS / (n − p)* will be an estimate of the experimental error variance.

We can also compute the *mean square due to regression, MSR*, by dividing the sum of squares due to regression by the corresponding degrees of freedom.

$$MSR = SSR \,/\, p$$

It is, of course, also possible to split up *MSR* to show the contributions from each model parameter to *MSR*.

Thus, we can use the F distribution to establish whether the contribution from the model terms is above the noise level given by the experimental error. If an independent estimate of the experimental error variance, s^2 (with r degrees of freedom), is available, this can be used to sharpen the analysis, and also to allow for a test of the lack of fit.

If the model is adequate, the ratio

$$F = MSE \,/\, s^2$$

should not exceed the critical F ratio for $(n - p)$ and r degrees of freedom.

For the model to describe a systematic variation above the noise level, the ratio

$$F = MSR \,/\, s^2$$

should be larger than the critical F ratio for p and r degrees of freedom.

For the individual terms in the model to be significant their contribution to the sum of squares divided by s^2

$$F = x_i' x_i \cdot b_i^2 \,/\, s^2$$

should be larger than the critical F ratio for 1 and r degrees of freedom. Often, an independent estimate of the error variance is not available and in such cases the error mean square, *MSE*, is used as an estimate of the error variance to assess the significance of the model terms.

This technique to apportion the total sum of squares over the different sources of contribution and to compare the estimated mean squares thus obtained to estimates of the error variance is called *analysis of variance*. It is often abbreviated as ANOVA. The analysis of variance is usually presented as a table showing: *(a)* the total sum of squares, the sum of squares due to regression (sometimes divided into the contribution of the individual terms in the model), the error sum of squares, *(b)* the degrees of freedom associated with the sums of squares, *(c)* the mean squares, *(d)* the F ratios, and sometimes also *(e)* the significance of the F ratios.

Table 3.4: Analysis of variance

Source of variation	Sum of squares	Degrees of freedom	Mean square	F ratio
Total	$SST =$ $\Sigma\, y_i^2 = y'y$	n		
Regression	$SSR =$ $b'X'Xb$	p	$MSR =$ $SSR\,/\,p$	$MSR\,/\,MSE$
Residuals	$RSS =$ $SST - SSR$	$n - p$	$MSE =$ $SSE\,/\,(n - p)$	

Analysis of variance applied to the elimination experiment, Table 3.1, yields the following result. From the pilot experiment an independent estimate of the experimental error variance was availabe, $s^2 = 1.34$, with four degrees of freedom. This estimate has been used to calculate the F ratios in Table 3.5. Thus, an estimate of the lack of fit is also obtained. The critical F ratio (95 % level) for 1 and 4 degrees of freedom is $F^{\text{Crit.}} = 7.71$. It is seen from Table 3.5 that the model does not show a significant lack of fit, and that it is only variable x_1 which has a significant influence. This was also found by analyzing the confidence limits of the model parameter using the t distibution.

Table 3.5: ANOVA in the elimination experiment

Source of variation	Sum of squares	Degrees of freedom	Mean square	F ratio
Total	37 140.22	5		
Regression	37 134.4995	4	9 283.6249	6 928.0783
on b_0	36 842.528	1	36 842.528	27 494.4239
on b_1	283.9225	1	283.9225	211.88
on b_2	3.4225	1	3.4225	2.55
on b_{12}	4.625	1	4.625	3.45
Residuals	5.7245	$5 - 4 = 1$	5.7245	4.27

References

1. C.J. Suckling, K.E. Sucklingand C.W. Suckling
   Chemistry Through Models
   Cambridge University Press, Cambridge 1978.

2. H. Eyring
   J. Chem. Phys. 3 (1935) 107.

3. L.P. Hammett
   Physical Organic Chemistry, 2nd Ed.
   McGraw-Hill, New York 1970.

4. G.E.P. Box and N.R. Draper
   Empirical Model-Building and Response Surfaces
   Wiley, New York 1987.

5. B. Ostle
   Statistics in Research
   The Iowa State University Press, Ames 1963, p. 72.

Appendix 3A: Least squares fit of response surface models by multiple linear regression

If the number of experiments, n, exceed the number of model parameters, p, there will always be some aberrations between the model and the experimental data, i.e. there will be at least some residuals which are different from zero. Here is shown that the relation

$$(X'X)^{-1}X'y = b$$

will always give the least squares estimate of the model parameters.

Let X be the model matrix for a series of n experiments; let b be the vector of model parameters to be estimated; let y be the vector of measured responses, such that $y = [y_1\, y_2\, ... \, y_n]$; and let e be the vector of unknown residuals, $e' = [e_1\, e_2\, ... \, e_n]$. For the series of experiments, the following relation applies

$$y = Xb + e \quad \text{i.e.}$$

$$e = y - Xb \qquad \qquad \qquad \qquad \text{(Equation A)}$$

$$\Sigma e_i^2 = e'e = e\,e \quad \text{(a scalar product)}$$

Develop the right part of (Eq. A) as a scalar product. This gives

$$\Sigma e_i^2 = (y - Xb) \cdot (y - Xb)$$

$$\Sigma e_i^2 = y \cdot y - y \cdot Xb - Xb \cdot y + Xb \cdot Xb$$

$$\Sigma e_i^2 = y \cdot y - 2y \cdot Xb + Xb \cdot Xb$$

In matrix notation the scalar products can be written

$$\Sigma e_i^2 = y'y - 2y'Xb + (Xb)'Xb$$

$$\Sigma e_i^2 = y'y - 2y'Xb + b'X'Xb$$

We wish to determine which set of model parameters minimizes the sum of squared residuals. This minimum can be found by solving the following systems of equations

$$\partial(\Sigma\, e_i^2)/\partial\beta_j = 0 \quad \text{for all parameters } \beta_j.$$

In matrix notation (see Appendix: Matrix calculus), the above expression can be written

$$\partial(\Sigma\, e_i^2)/\partial\beta = o$$

which corresponds to

$$\partial(y'y)/\partial\beta - \partial(2\ y'X\beta)/\partial\beta + \partial(\beta'X'X\beta)/\partial\beta = o$$

This gives

$$2(y'X)' + \{X'X + (X'X)'\}b = 0$$

The matrix $X'X$ is symmetric and $X'X = (X'X)'$ and the above expression is equivalent to

$$X'y = X'Xb$$

Provided that $(X'X)^{-1}$ exists, a multiplication of both sides by $(X'X)^{-1}$ affords the least squares relation

$$(X'X)^{-1}X'y = b$$

Q.E.D.

Chapter 4

General outline for screening experiments

1. Some initial questions and comments

A screening experiment is a series of experiments run with a view to finding an answer to the question:

Which experimental variables have a real influence on the result?

1.1. When is a screening experiment appropriate?

The screening problem is often encountered in organic synthesis. Some examples will illustrate this:

* With newly discovered reactions, experimental observations are scarce. If the reaction seems promising, it might possibly be developed into a *synthetic method*. To achieve a better understanding of the performance of the reaction, it will be necessary to identify those factors which may have an influence on yield and selectivity. This will also be an impetus to probe on pertinent details to gain an understanding at the mechanistic level. Any suggested reaction mechanism must be able to account for the observation that some variables are important, while others are not.

* A known procedure is to be elaborated to improve the performance. The most profitable way to do this will obviously be to adjust the most influential variables. There would be no point in trying to adjust factors which have a negligible influence on the result.

* A lab-scale procedure is to be transferred to the pilot-plant. Will the factors known to be important at the lab-scale also be important after scale-up?

1.2. Principles of screening experiments

In explorative research, the purpose of experiments is done to furnish *new* information. An experimental design is a detailed plan of a series of experiments suitably arranged so that the observed result will give the desired information. It is therefore necessary that the experimenter, prior to any experiment, clearly states *what* the desired information should be. For a screening experiment, this implies that the experimenter should first specify all the potentially important variables, and then decide upon a suitable set of experimental runs so that the influence of *each* variable can be established. Then, the significant variables can be identified as those which have an influence above the noise level of the experimental error.

In principle, this is a straightforward strategy. *In practice*, however, this is not trivial. *Before* any screening experiment can be laid out, an imaginative and creative process is required. *During* this process, a number of questions can hopefully be answered without any experiments. *After* this process, the problem will be more clearly stated than before. A clearly defined question is a necessary requirement for good experimental design.

It is, of course, impossible to give an algorithm for creativity, but some advice on how to approach the screening problem is appropriate:

Consider a problem with an attitude of *humble audacity*, meaning that we should acknowledge that we do not *know* everything beforehand, and that even our *assumptions* of what is going on in the reactor can be wrong. At the same time we must not let this state of partial ignorance prevent us from trying to do things.

Adopt an operational attitude towards the problem, i.e. to acknowledge that we can learn things by doing. After an experiment we know more than before, and this will help us to ask more detailed questions for the next experiment, see Fig. 2.1.

2. Steps to be taken in a screening experiment

Any synthetic procedure consists of a series of events where the experimenter may intervene at the various steps to determine the detailed settings of the experimental variables:

Substrate and reagents must be introduced into the reactor; the reaction must proceed for a certain period of time; during the reaction time additional reagents may be introduced; upon completion of the desired reaction, the product is isolated by some work-up procedure, which in turn consists of another series of events (extraction, washings, precipitation etc.)

It is evident, that even a simple synthesis can be modified in a very large number of ways if all possible experimental factors are to be considered. To reduce this number and identify those variables which might be assumed to be critical so that these can be studied in a manageable number of experiments, a strategy for the selection is helpful.

It should be remembered that every new problem is a unique problem in certain respects : It is not *identical* to previously studied synthetic procedures. It has its own back-ground history, e.g. there may be previous observations made on the reaction. Often there are also at least some assumptions on the mechanistic level.

It may be non-unique in that respect that it resembles other, well-known, reactions. In that case, certain features of the procedure may be analogous to known methods, and this may simplify the problem.

As there are often many aspects to consider, it will not be possible to envisage a set of clear-cut rules for the definition of an infallible strategy. Below, some advices are given on how the problem can be simplified.

2.1. Summary of the steps in a screening experiment

A sequence of steps to be taken is summarized in the points below. These points are discussed in detail in the next section.

(1) Analyze the synthetic procedure and determine the critical steps.

(2) Determine which response to measure

(3) Analyze the experimental procedure and determine the experimental variables.

(4) Rank the variables in priority categories as variables which are: (i) known to influence; (ii) suspected to influence; (iii) suspected not to influence; (iv) known not to influence.

(5) Proceed with variables in the categories (i), (ii), and if it is convenient also (iii), and determine the experimental domain.

(6) Can some variables be removed due to the constraints imposed by the experimental domain?

(7) Define the variables as economically as possible.

(8) Identify possible interaction effects.

(9) Repeat step (1) to (8).

(10) Identify variables which are difficult to vary. Consider blocking for these variables.

(11) Suggest a model.

(12) Choose an experimental design.

(13) Run the experiment and evaluate the result.

2.2. Discussion of the steps to be taken

(1) Go through the overall experimental procedure and answer the following questions:
(a) Are there several distinguishable *discrete* steps? If so, is there any step which can be considered as the *critical* step?
(b) Is the reaction run in specialized equipment, or can it be run in standard glassware?
(c) How much of the various chemicals will be needed? Will there be a sufficient supply, or will there be a limited amount available of some compound? Will a limited supply reduce the number of possible experimental runs?

(2) Determine a suitable response (or suitable responses) to measure. Measure this response *as close to the interesting event as possible*.
If possible, consider the error of measurement when the response is determined. If a systematic variation which is induced by changing the experimental conditions is to be detected, it must, at least, be above the noise level of the measurement error.

Often, a chromatographic or spectroscopic method, the precision of which is generally known, can be used to measure the response. If the precision is not known, it can easily be determined by replicated analyses of known samples. The analytical error is rarely an obstacle in screening experiments.

Quite often, a lot of time can be saved at this step. A response measured directly on the reaction mixture after the critical step may give the desired information without the necessity to go through the subsequent steps. For instance, a poor conversion as measured on the reaction mixture will also lead to a poor yield of isolated product.

There is another reason why measuring the response close to the interesting event is beneficial, viz. that any unnecessary manipulation will increase the experimental error, thus lowering the probability of detecting influences of the experimental variables on the response.

If the response of interest, e.g. the yield, can be measured directly on the reaction mixture, it will not be necessary to use the *time of reaction* as an experimental variable. Instead, monitoring yield over time will provide an extra response which can be used to obtain *additional* information on the influence of the experimental variables on rate phenomena.

(3) Go through the experimental procedure, up to the step where the response(s) is/are measured, and note *all* experimental variables which could be manipulated during the reaction.
Consider both continuous and discrete variations.

This will usually result in quite a long list containing all perceivable perturbations of the experimental conditions. The following steps suggest to how this list can be shortened.

(4) Go through the list and answer the questions below, and rank the variables into priority categories *(i)* - *(iv)*:

(i) Which variables are *known* to be important, either by prior experience or by analogy to known reactions? The question is not whether these variables will influence, but what magnitude their influence will be, and whether the experimental domain is suitably chosen. Such variables must always be included in the study.

(ii) Which variables can be *suspected* to have an influence? Such variables must be taken into account in a screening.

(iii) Which variables can be suspected to have a negligible influence? Often, such variables can be ignored, but sometimes they are included in the design as a safeguard to assure that they really are negligible.

(iv) Which variables are *known* not to have any influence in the experimental domain? Such variables can safely be removed from further considerations.

It is more economical to have one or more extra variables in a first screening, than to introduce an overlooked variable at later stage.

When ranking the variables in priority classes, use all available sources of information. There is no point in redetermining already known facts: *Two weeks in the lab can save one hour in the library.*

* What is already known from the literature?

* What previous observations of the reaction are available and which conclusions can be drawn from these?

* Are details of the reaction mechanism known with some degree of confidence and which suggestions will emerge from these?

A rule of thumb is:

Variables which can influence the kinetics of reactions must always be considered as potentially important.

Examples of such variables are: (Continuous variables): *Reaction temperature, concentration of reagents, ratios of reagent concentrations, reaction times of intermediary steps*; (Discrete variables): *type of solvent, type of catalyst.*

(5) Determine the experimental domain.

This implies assigning the upper and lower bounds to all continuous variables, and specifying the alternatives for discrete variables.

At this stage, practical constraints will intervene.

For example, the *concentration range* of a chemical can vary from almost infinite dilution up to the neat compound. High dilution may sometimes be beneficial, but is obviously not the best choice if a high-capacity process is the desired goal.

Other constraints will be imposed by physical limitations. For instance, the *reaction temperature* in open systems can be varied between the melting point and the boiling point of the solvent.

Yet other constraints will be determined by the technical performance of the equipment, e.g. safety specifications.

Remember, that it is you as the experimenter who decide, what experimental domain is the domain of interest.

Some advices that may help you to avoid mistakes:

* Do not turn continuous concentration-related variables into discrete variation by assigning a low limit of *zero* concentration to any constituent in the reaction mixture. The absence of a species will increase the risk that a totally different reaction mechanism begins to operate.

* Do not span too large a (or too narrow a) range of variation of the experimental variables. It is obvious, that a lot of different events may occur when the temperature is allowed to vary between -78°C and room temperature. Such a span would be too large a variation to explore in a screening experiment. Use your common sense and assign ranges of variation which *a priori* can be supposed to produce a *similar* change in the response. This can be difficult, especially with new reactions.

There is a simple initial set of experiments which can be run to make sure that a sufficiently large experimental domain has been chosen, allowing influences of the experimental variables to be detected:

A: Run two experiments: One with all variables set to their lower bounds, an the other with all variables set to the upper bound. These experiments will span a maximum in variation.

(a) If the results of these experiments are different it is highly probable that a screening experiment will reveal the important experimental variables.

Eventually, move the bounds to embrace the levels of the "better" experiment.

(b) If the results are approximately the same, it is an indication that the domain is too small. To confirm that nothing surprising will happen between these two extreme experiments it is advisable to run a third experiment, see below.

B: Run a third experiment with all continuous variables at their average values (at the center of the experimental domain). The discrete variables can be set at any level.

(a) If this experiment also gives approximately the same result as the extreme experiment, it is a clear indication that the experimental domain should be enlarged.

(b) If this result is different from the results of the extreme experiments, it is an indication that the response surface is curved. The experimental domain should be shrunk.

From these experiments it is not possible to identify any individual experimental variable responsible for the observed change. To do so, more experiments are needed.

(6) Can some variables be removed from further considerations, due to the constraints imposed by the experimental domain?

An example will illustrate this: Any chemical reaction is influenced by the temperature. In an enzyme-catalyzed reaction in water solution, the temperature should be maintained at the optimum temperature[1] for that particular enzyme. In this case *the temperature* is not an experimental variable which should be changed.

(7) Define the variables as economically as possible.

If we have a set of k experimental variables at least $(k + 1)$ individual experiments must be run to make an evaluation of each individual variable possible. This means that to keep the number of experiments acceptably low, redundant variables should be removed.

Examples:

* It will not be necessary to consider *both* the concentration of an added reagent and the rate of its addition. Either of these would be sufficient.

* If we are interested in knowing whether the concentration of the *substrate* and two reagents A and B influence the result, then of course the three concentration variables: [*Substrate*], [*A*], and [*B*] should be used. If it is the *relative proportions* of the reagents to the substrate which are of interest (how large an excess should be used?), then the concentration ratios [*A*] / [*Substrate*] and [*B*] / [*Substrate*] should be used as variables. This gives only two variables to consider.

(8) Consider possible interaction effects.

Examples:

* It is reasonable to believe that there will be interaction effects between variables related to the concentration of species, if other than first order kinetics are involved.

* Other consequences of the kinetics are, that variation in temperature often interacts with variables related to concentrations, as well as with variations in the reaction time of intermediary steps.

There are also variables which can be assumed *not* to interact, e.g. variables influencing *different* steps in multi-step reactions.

(9) Reconsider steps *(1)* to *(8)*!

---

[1] Literature data of "optimum temperatures" of enzymes has often been determined by studying one variable at a time. As interaction effects between other variables and the reaction temperature are to be expected, the concept of "optimum temperature" is dubious. With enzymes from warm-blooded animals, a reaction temperature close to the normal body temperature of the animal, will probably be a good approximation in a *screening experiment*. For a detailed *optimization* the temperature variation must be considered.

After going through the steps *(1)* to *(8)*, the experimenter will have determined: *(a)* a suitable response (or responses) to measure, *(b)* a set of potentially important variables, and *(c)* a set of suspected interactions.

Do not rush into the lab with these, without a critical re-examination!

Important points may have been overlooked. Do not fall victim to the Heureka syndrome, i.e. do not be so satisfied with the first, sudden stream of thoughts, that you cannot change your point of view. Discuss your ideas with a colleague whose judgement you trust, or distract yourself from the ideas for a while. Go to the pub, to the cinema, play bridge with your friend, meditate, or do whatever you like to relax, and then go back to reconsider the problem the next day. I'm not joking. Distracting allows the problem to be twisted subconciously and this may induce the creative process, which will hopefully give an intuitive comprehension of the problem.

> *(10)* Rank the variables according to how easily they can be varied
> between different experimental runs.

Most experimental variables can be changed without difficulty in the different runs. This is important for *randomizing* the order of the experimental runs. Why randomization is important is discussed in the next chapter.

There may, however, be some variables which would it be very inconvenient to change in every run. Examples of such "difficult" variables are:

* Two different kinds of equipment are used. We wish to know if they produce a different result in connection with variation of other experimental variables. A change between these alternatives would imply cumbersome reconstructions of the apparatus. It would be more convenient to make the reconstruction only once, and run the experiments in two series, one with each type of equipment.

* Two different catalysts, *A* and *B*, are to be tested. The preparation of the catalysts is time-consuming. Unfortunately, the catalysts are not stable over time and it is advisable to use them shortly after their preparation. Also in this case, will it be more convenient to run the experiments in two series, one series with each catalyst.

Variables which are difficult to vary between runs can be handled by dividing the total set of experiments into smaller separate *blocks* of experiment. *Block designs* are discussed in Chapter 6.

(11) Suggest a response model for the screening experiment.

We wish to discern the influence of each experimental variable, x_i, on the response, y.

$$y = f(x_1,.....,x_k) + e$$

In the preceeding chapter it was shown that a model approximated by a plynomial in the experimental variables can describe f in a limited experimental domain. Such models useful for screening experiments are:

The linear model

$$y = \beta_0 + \Sigma\beta_i x_i + e$$

The second-order interaction model

$$y = \beta_0 + \Sigma\beta_i x_i + \Sigma\Sigma\beta_{ij} x_i x_j + e$$

These models will rarely give a perfect description of the response surface (curvature is not described), but they will give estimates of the *slopes, b_i,* along each variable axis, and the *twists, b_{ij},* to describe interactions. *This is exactly what is needed to assess the influence of the variables.*

If it is known or strongly suspected that interaction effects will be present, then, of course, a model with interaction terms should be used as the first choice.

If nothing is known and/or if there are a large number of experimental variables to consider, it is advisable to start with a linear model. It is always possible to run a second, complementary set of experiments to estimate also the interaction coefficients. The advantage of starting with a linear model is, that it may be possible to remove certain variables from further consideration after a first evaluation of the model. If it is still necessary to determine interaction effects, the complementary set of experiment will be smaller after the elimination of insignificant variables. Variables which do not have a significant linear influence rarely show strong interaction effects.

(12) Choose an experimental design to estimate the model.

To estimate the model parameters, it will be sufficient to study each variable at two levels only. This means that the number of necessary experiments will be of the same order as the number of parameters in the model, e.g. to fit a linear model with seven variables, eight experiments would be sufficient.

There are several types of designs which can be used in screening experiments. These designs are discussed in the chapters to follow:

Two-level full factorial designs (Chapter 5)

Two-level fractional factorial designs (Chapter 6)

Plackett-Burman designs (Chapter 7)

D-Optimal designs (Chapter 7)

(13) Run the experiment and fit the model. Assess the significance of the model parameters.

Significance of the model parameters can be assessed by t statistics, see Chapter 2. To do so, an estimate of the experimental error variance, s^2, is needed. This can be obtained by replication of one or more experimental runs.

By another technique, *plotting on normal probability paper*, the significant parameters can be detected without prior knowledge of the experimental error variance. This very useful technique is described in Chapter 6.

3. Example: Synthesis of 1,4-dibromobenzene

The procedure below has been used as a laboratory exercise at the author's university during the first semester undergraduate course in organic chemistry.

Assume that you wish to run the synthesis of 1,4-dibromebenzene on a larger scale and that this particular procedure is a candidate for scale-up. To allow for an optimization of the yield it will be necessary to identify the important experimental variables.

Synthesis of 1,4-dibromobenzene

All operations should be carried out in a hood.

Bromination: To 5 ml of benzene in a 100 ml Erlenmeyer flask is added one small piece of iron filings. Cool the flask in ice-water and add cautiously 6.0 ml of bromine. Swirl the flask and allow it to stand at room temperature until the next day.

Work-up: Add 50 ml of water to the flask and a small magnetic stirring bar. Heat the content to boiling with vigorous magnetic stirring. Cool the mixture until the dibromobenzene solidifies, and decant the water. Repeat the water treatment once, followed by a third washing with 50 ml of 2 % NaOH. Place the flask on the steam-bath and add *ca.* 80 ml of ethyl alcohol (95 %). When the crude product is dissolved, filter the hot solution through a fluted filter. Heat the filtered solution to boiling and carefully add hot water until the solution becomes *slightly* turbid. Place the hot flask in a large beaker with cotton wool, cover the mouth of the flask with a small watch-glass, and let the solution cool slowly. Collect the precipitated crystals by filtration, dry in air on a large filter paper. Weigh the product to determine the yield, and determine the melting point. Compare the observed melting point with literature data.

The procedure is divided into two steps, bromination and work-up. The critical step is the bromination. A maximum conversion in this step is necessary for a maximum final yield. Hence, variables involved in the first step will be the most important. A first screening can therefore be limited to these. An overall optimization can possibly be achieved through a step-wise approach: First optimize the yield of the bromination step, then adjust the work-up procedure to fit an optimum bromination. This is not always true, sometimes by-products interfere with the work-up procedure.

The yield of dibromobenzene can be determined directly from the reaction mixture by gas chromatography after addition of an internal standard. With this technique, the work-up step can be omitted in the screening experiments.

Although the bromination procedure is very simple, several variables can be varied. A first list may look as follows.

Variables

A:	Amount of bromine/benzene: (stoiciometric or excess)
B:	Amount of Fe/benzene: (catalytic amounts, but how much is that?)
C:	Type of iron: (filings, wool, powder...)
D:	Rate of bromine addition (slow, rapid)
E:	Reaction temperature (room temperature or higher)
F:	Stirring rate
G:	Time of reaction
H:	Presence of a co-solvent: (yes, no)
I:	Type of solvent if *H* is *yes*

A few moments of contemplation may lead to the following considerations:

A: The reaction is run with approximately stoichiometric amounts of the reactants. There are two bromination steps. The first step is fairly rapid and leads to bromobenzene, the second bromination step is slow and converts the bromobenzene to 1,4-dibromobenzene. An excess of benzene would lead to an insufficient amount of bromine for the second step. An excess of bromine may lead to tribrominated products, especially at elevated temperatures. Thus, a stoichiometric amount of bromine should be used: *A* shall not be varied.

B: The role of the iron is to form ferric bromide by reaction with bromine. Ferric bromide is a Lewis acid catalyst. The catalytic action is probably carried out by dissolved catalyst. The solubility of ferric bromide in the medium is limited. A large amount of iron would lead to a solid precipitate of inactive catalyst. The amount of iron should be varied, but the variation should not be too large, presumably in the range $2 - 4\%$.

C: To react efficiently with bromine the iron should have a large surface. Therefore, use finely divided iron powder and do not vary the type of iron.

D: The reaction is rather slow at room temperature, at least in the second bromination step. Run a pilot experiment to see if the reaction can be kept under control (exothermic?) if the bromine is added rapidly. If the reaction is not exothermic, use rapid addition in all experiments and do not vary D.

E: Try to run the reaction at an elevated temperature ($\approx 40 - 50\ ^\circ C$) to increase the rate. Maybe it is possible to rise the final temperature to 80 $^\circ C$ to avoid solidification of the final product.

F: A synthetic reaction is only rarely improved by inefficient stirring. As there is a solid catalyst present which is in equilibrium with the dissolved catalyst, stirring is necessary. For lab-scale application of the method, it will, however, not be necessary to evaluate the stirring *rate* (rpm). Use efficient but fairly constant stirring. If the procedure later on should be applied to pilot-plant synthesis, reconsider stirring rate as a potential variable.

G: If it is possible to measure the consumption of the substrate or a limiting reagent, do not vary the reaction time. Monitoring the reaction gives access to an extra response related to the kinetics. In this case, unreacted bromine can be detected by a blue starch−potassium iodide reaction. For convenience, fix a limiting time (e.g. 12 h) after which the mixture is quenched with bisulfite if there is still bromine left.

H, I: Do not involve a solvent in the first screening. This leads to unnecessary complication of the problem. Reconsider the possibility of adding a solvent, only if the present method should give unsatisfactory results.

The above reasoning suggest that

B, *E* and possibly *D*, shall be studied in a screening experiment;

A, *C*, and *G* shall not be varied at all;

F may need consideration after very large scale-up;

H, *I* shall not be included in the first run.

Admittedly, this example with such a simple experimental procedure is highly artificial. However, the idea was to show the approach. With more complex experimental procedures, the number of potential variables will be much larger.

Chapter 5

Two-level factorial designs

1.   Introductory remarks

1.1   The problem

Often there are quite a few experimental variables which are believed to have an influence on the experimental result and we wish to know how these variables exert their influence. Interaction effects are common when the reaction temperature is varied together with other factors which may influence the kinetics of a reaction. To clarify such effects it will be necessary both to determine the direct influence of the experimental variables as well as their possible interaction effects. For this, two-level factorial designs are the appropriate tools.

1.2.   Why use a design?

It is a common belief that valuable information can only be obtained from experiments which have given "good" results, e.g. high yields. *This is a misconception.* To understand why certain experimental conditions give good yields, it is essential to know also why other conditions do not. The *information* which can be obtained in a single experimental run will depend on the results obtained in other runs. The quality of information is therefore dependent on the ensemble of all experimental runs, i.e. the experimental design.

1.3.   What is obtained through a factorial design?

The idea of a factorial design is to arrange the experiments in such a way that the variation in response obtained with different settings of the experimental variables, *factors*, can be traced back to the variations of the factors. By proper arrangement of the factor settings it will be possible to determine influence of the variation of each factor on the response in the presence of a simultaneous variation of all other factors. It will also be possible to determine what effects changing two or more factors simultaneously and independently will have on the variations of the response. In fact, it is possible to determine all interaction effects, i.e. interaction effects of two factors, three factors, four factors ..., all factors.

All these effects can be determined *independently* of each other. This means, that the estimated value of any effect does not depend on the estimated value of any other effects.

1.4. What is a factorial design?

In a factorial design, each variable (factor) is investigated at fixed levels:
In a two-level factorial design, each variable can take two values. For continuous variables, this will signify a *low level* and a *high level*. For discrete variables, this will signify the alternatives. In a three-level factorial design, three different levels of each variable is investigated, etc.

A complete factorial design contains all possible settings of the experimental variables. A factorial design in which k factors are studied at r levels will thus contain r^k different experimental runs: A two-level factorial design with k factors contains 2^k runs.

It is seen that the number of individual experimental runs increases rapidly with an increasing number of factors, and even more rapidly when more than two levels are used.

The arrangement of experimental runs in a factorial design makes it possible to use each observed response more than once. Each result is actually used in combinations with the other results to compute: *an average response, all main effects, all two-factor interaction effects, ... all multi-factor interaction effects*. This means that each individual experimental run is *equally important* to the overall result.

1.5. Randomize the order of runs!

The consequences of an experimental error on the conclusions drawn from experimental data were discussed in Chapter 3. To allow this error to be analyzed by known statistical probability distributions, such as the *normal distribution*, the *t-distribution*, the *F-distribution*, certain assumptions as to the experimental error must be made. These assumptions are:

(a) The experimental errors should be *independent*, i.e. the disturbances leading to the error should occur independently of each other between experimental runs.

(b) The experimental error variance, σ^2, should be constant in the experimental domain.

However, chemical systems can be influenced by disturbances which are not random and which will produce a systematic error, e.g.:

* If a series of Friedel-Crafts alkylations is run using the same package of aluminum chloride, the yield will drop over time. This occurs because aluminum chloride is sensitive to moisture and will lose activity over time.

* The experimental skill is improved over time and the experimental error will decrease.

* Repeated injections on a gas chromatograph will contaminate the detector, which gradually lowers the sensitivity.

There is always a risk that the experimental result may be influenced by non-random, time-dependent errors. Such risks may be counteracted by precautions taken by the experimenter, and the remedy is *randomization*. This means, that in any situation where the experimenter has a choice as to in which order he/she should do things, the choice should be a random choice. For example, the order of executing the experimental runs should be randomized; the order of analyzing samples drawn from the reaction mixture should also be randomized if several samples are analyzed on the same occasion.

These precautions will break time-dependent phenomena, and transform systematic errors into random errors.

A random order of experiments can be established in several ways, e.g. by writing an identification number of the experiments on separate pieces of paper, putting them in a hat, and then drawing them at random, by using tabulated random numbers or by using the random number generator of a computer. Personally, I use a well-shuffled pack of cards.

2. Different representations of factorial designs

Factorial designs were introduced in the pioneering works by Ronald Fisher in the 1920s [1]. Almost any textbook on statistics contains some description of factorial designs. At first glance, these descriptions may seem totally different. We shall, however, see that they are all equivalent. To show this, the same example will be treated by different approaches to evaluation. There is a reason for presenting different methods of analysis in this book: When results of experimental studies using factorial designs are presented in publications, the analyses look different, depending on which method of analysis the authors are familiar with. It is therefore desirable that the reader should be able to understand the result, regardless of the method used. There is also another reason as well: It is my experience, that factorial designs are more easily understood if the classic method of evaluation is presented first.

2.1. Example to be treated: Lewis acid catalyzed rearrangement

A Lewis acid catalyzed rearrangement of an olefin was studied. The measured response, y, was the yield (%) of the rearranged product. Three experimental variables were considered:

A: the amount of catalyst
B: the reaction temperature
C: the reaction time

The amounts of substrate and solvent were maintained constant in all experiments.

It is a common practice to use uppercase letters to denote factors in factorial and fractional factorial designs.

2.2. Design and analysis

To set up a factorial design, the experimental domain must be specified. For a two-level design this means assigning a low level, (−), and a high level, (+), to each factor. In this case the following levels were chosen:

Table 5.1: Variables and experimental domain in the rearrangement study

Factors	Levels	
	(−) level	(+) level
A: Amount of catalyst/substrate		
(mol/mol)	0.01	0.05
B: Reaction temperature (°C)	40	60
C: Reaction time (h)	0.5	1.0

Three variables will give eight different combinations of the variable settings. These experiments were run and are summarized in the classic representation of a factorial design shown in Table 5.2.

The bottom line of Table 5.2 contains the experiment labels. These should be interpreted as follows: *1* is the experiment in which all factors are at the low level; *a* is the experiment in which all factors, *except A*, are at the low level; *ab* is the experiment in which all factors *except A and B* are at the low level etc. The experiments given in Table 5.2 are more easily surveyed if they are presented as the *sign table* shown in Table 5.3

Table 5.2: Factorial design in the rearrangement study

Factors	Levels							
C	0.5				1.0			
B	40		60		40		60	
A	0.01	0.05	0.01	0.05	0.01	0.05	0.01	0.05
Yield (%)	74	69	78	81	76	87	84	91
Label	1	a	b	ab	c	ac	bc	abc

Table 5.3: Sign table to show the factorial design

Label	Factors			Yield
	A	B	C	(%)
1	−	−	−	74
a	+	−	−	69
b	−	+	−	78
ab	+	+	−	81
c	−	−	+	76
ac	+	−	+	87
bc	−	+	+	84
abc	+	+	+	91

2.3. Geometric illustration

The factorial experiments above can be illustrated geometrically as shown in Fig. 5.1. The experimental domain is a cube spanned by the factor axes. The experiments are located at the "corners" of the cube.

2.4. Calculations of the effects of the factors

The geometric representation in Fig. 5.1 is used to illustrate what is meant by *effects* the factors.

Main effects: The main effect of factor A is an average of the observed difference of the response when A is varied from the low level to the high level. There are four such comparisons which can be made: Along the edge between *1* and *a*, only factor A has been changed, and $(a - 1) = (69 - 74) = -5$, is such an observed difference.

Three other differences can be determined along the remaining edges, parallel to the A-axis: $(ab - a)$, $(ac - c)$, $(abc - bc)$.

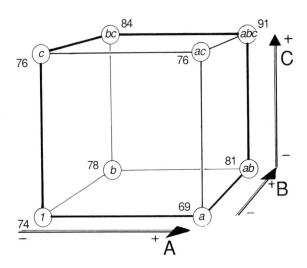

Fig.5.1: Distribution of the experimental points on the space defined by the factor.

The *main effect of A* is defined[1] as half of the average of the observed differences.

$$A = 1/2 \cdot [(a - 1) + (ab - b) + (ac - c) + (abc - bc)] / 4 =$$
$$= 1/8 \cdot [(a - 1) + (ab - b) + (ac - c) + (abc - bc)] =$$
$$= 1/8 \cdot [(69 - 74) + (81 - 78) + (87 - 76) + (91 - 84)] = 2.0$$

The main effects of B and C are calculated analogously by comparing experiments along the edges parallel to the B-axis and the C-axis, respectively.

---

[1] Some authors define the *main effect* as the average of the observed differences. Here half of this average is used. By this the calculated effect will be equal to the coefficients in polynomial approximations of the response functions, i.e. the parameters, b_i, b_{ij}, in response surface models.

$B \quad = 1/8 \cdot [(78 - 74) + (81 - 69) + (84 - 76) + (91 - 87)] = 3.5$

$C \quad = 1/8 \cdot [(76 - 74) + (87 - 69) + (84 - 78) + (91 - 81)] = 4.5$

Interaction effects: If there is an interaction effect AB, this means that the influence of changing factor A will depend on the setting of factor B. This can be analyzed by comparing the *effects* of A when B are at different levels. If these effects should be found to be equal, this means that there is no interaction effect AB. If, on the other hand, they should be found to be different there is an interaction effect. We can therefore use the observed *difference* as a measure of the interaction effect.

An interaction effect AB is defined as half of the difference of the effects of A when they are determined with factor B on its high, respectively its low level.

$$AB \quad = 1/2 \cdot [A_{B+} - A_{B-}]$$

The effect of A when B is at its high level

$$A_{B+} \quad = 1/2 \cdot [(ab - b) + (abc - bc)]/2 =$$
$$= 1/4 \cdot [(ab - b) + (abc - bc)] = 1/4 \cdot [(81 - 78) + (91 - 84)] = 2.5$$

The effect of A when B is on its low level

$$A_{B-} \quad = 1/2 \cdot [(a - 1) + (ac - c)]/2 =$$
$$= 1/4 \cdot [(a - 1) + (ac - c)] = 1/4 \cdot [(69 - 74) + (87 - 76)] = 1.5$$

These effects are different and there is an interaction effect AB which is calculated

$$AB = 1/2 \cdot (2.5 - 1.5) = 0.5$$

The other two-factor interaction effects are calculated analogously.

The three-factor interaction effect ABC is computed as half the average of the difference of the responses in experiments where all three factor have been changed between the runs. These experiments are joined by the large diagonals of the cube:

$$ABC \quad = 1/2 \cdot [(abc - 1) + (a - bc) + (b - ac) + (c - ab)]/4 =$$
$$= 1/8 \cdot [(91 - 74) + (69 - 84) + (78 - 87) + (76 - 81)] = -1.5$$

The calculated effects are

Average yield

80.0 %

Main effects	Two-factor interaction	Three-factor interaction
$A = 2.0$	$AB = 0.5$	$ABC = -1.5$
$B = 3.5$	$AC = 2.5$	
$C = 4.5$	$BC = -0.5$	

From the above example it is evident that there is nothing mysterious about the evaluation of factorial designs. The effects are obtained by direct comparisons of the experimental observations.

A tentative interpretation of these results is that all three factor have an influence on the yield. The average change in yield is 5, 7 and 9 %, respectively, when the factors A, B and C are changed from their low levels to their high levels. It is also possible that there is an important interaction effect between factors A and C. Other interaction effects are probaly small and insignificant. It is not possible, however, to draw any conclusions as to the *significance* of the estimated effects. To do so, they must be compared to an estimate of the experimental error variation.

3.   Generalization to any number of factors

It is difficult to visualize by geometric interpretation, the computation of effects with more than three factors. A "cube" in four dimensions has 16 "corners". How should one proceed in such cases?

3.1.1 Evaluation from sign tables

Table 5.3 showed a sign table for a factorial design 2^3. Sign tables are easy to construct for any number of factors, and examples are shown below for 2^2, 2^3, 2^4 factorial designs.   Look at these sign tables column by column:

Column A: alternating $(-)$ and $(+)$;
Column B: two $(-)$, two $(+)$, two $(-)$, two $(+)$ ... etc.
Column C: four $(-)$, four $(+)$, four $(-)$, four $(+)$ ... etc.
Column D: eight $(-)$, eight $(+)$ ... etc.

This can be generalized to any number of factors. These sign tables show the factor combination to be used in the individual experimental runs,i.e. the *experimental design*.

Exp no	Factors A	B
1	–	–
2	+	–
3	–	+
4	+	+

Exp no	Factors A	B	C
1	–	–	–
2	+	–	–
3	–	+	–
4	+	+	–
5	–	–	+
6	+	–	+
7	–	+	+
8	+	+	+

Exp no	Factors A	B	C	D
1	–	–	–	–
2	+	–	–	–
3	–	+	–	–
4	+	+	–	–
5	–	–	+	–
6	+	–	+	–
7	–	+	+	–
8	+	+	+	–
9	–	–	–	+
10	+	–	–	+
11	–	+	–	+
12	+	+	–	+
13	–	–	+	+
14	+	–	+	+
15	–	+	+	+
16	+	+	+	+

3.2. Sign of interaction

To compute the interaction effects, a new concept, *the sign of an interaction* is to be introduced. The sign of an interaction is defined according to normal multiplication rules as the sign obtained by multiplication of $(-)$ and $(+)$ signs, i.e.

$$(-) \cdot (-) = (+)$$
$$(+) \cdot (-) = (-)$$
$$(-) \cdot (+) = (-)$$
$$(+) \cdot (+) = (+)$$

Through these rules it is possible to construct sign columns for the interactions in factorial designs:

2^2 Factorial design

A	B	AB
−	−	+
+	−	−
−	+	−
+	+	+

2^3 Factorial design

A	B	C	AB	AC	BC	ABC
−	−	−	+	+	+	−
+	−	−	−	−	+	+
−	+	−	−	+	−	+
+	+	−	+	−	−	−
−	−	+	+	−	−	+
+	−	+	−	+	−	−
−	+	+	−	−	+	−
+	+	+	+	+	+	+

2^4 Factorial design

A	B	C	D	AB	AC	AD	BC	BD	CD	ABC	ABD	ACD	BCD	ABCD
−	−	−	−	+	+	+	+	+	+	−	−	−	−	+
+	−	−	−	−	−	−	+	+	+	+	+	+	−	−
−	+	−	−	−	+	+	−	−	+	+	+	−	+	−
+	+	−	−	+	−	−	−	−	+	−	−	+	+	+
−	−	+	−	+	−	+	−	+	−	+	−	+	+	−
+	−	+	−	−	+	−	−	+	−	−	+	−	+	+
−	+	+	−	−	−	+	+	−	−	−	+	+	−	+
+	+	+	−	+	+	−	+	−	−	+	−	−	−	−
−	−	−	+	+	+	−	+	−	−	−	+	+	+	−
+	−	−	+	−	−	+	+	−	−	+	−	−	+	+
−	+	−	+	−	+	−	−	+	−	+	−	+	−	+
+	+	−	+	+	−	+	−	+	−	−	+	−	−	−
−	−	+	+	+	−	−	−	−	+	+	+	−	−	+
+	−	+	+	−	+	+	−	−	+	−	−	+	−	−
−	+	+	+	−	−	−	+	+	+	−	−	−	+	−
+	+	+	+	+	+	+	+	+	+	+	+	+	+	+

It is seen that all columns are different to each other. Now, if a column **I** consisting of only (+) signs is added, the complete sign table for the calculation of the effects is obtained. This column is used to calculate the average response.

In Chapter 6 it is shown how such complete sign tables can be used to construct and to analyze *fractional factorial designs*. To show their use for calculating the effects from factorial designs, the example given above will be used.

Exp no	I	A	B	C	AB	AC	BC	ABC	Response
1	+	−	−	−	+	+	+	−	74
2	+	+	−	−	−	−	+	+	69
3	+	−	+	−	−	+	−	+	78
4	+	+	+	−	+	−	−	−	81
5	+	−	−	+	+	−	−	+	76
6	+	+	−	+	−	+	−	−	87
7	+	−	+	+	−	−	+	−	84
8	+	+	+	+	+	+	+	+	91

The effects are computed from the responses and the signs in the corresponding columns, as shown by the following examples:

The effect of A is obtained by first assigning the response in experiment i the corresponding sign in column A, computing the sum of the signed responses and then dividing the sum obtained by the number of experiments.

$$A = (-74 + 69 - 78 + 81 - 76 + 87 - 84 + 91)/8 = 2.0$$
$$B = (-74 - 69 + 78 + 81 - 76 - 87 + 84 + 91)/8 = 3.5$$

.

.

.

$$AB = (+74 - 69 - 78 + 81 + 76 - 87 - 84 + 91)/8 = 0.5$$

.

.

.

$$ABC = (-74 + 69 + 78 - 81 + 76 - 87 - 84 + 91)/8 = -1.5$$

The column I is used to compute the average response, y

$$y = (74 + 69 + 78 + 81 + 76 + 87 + 84 + 91)/8 = 80.0$$

This type of calculations can easily be carried out on a pocket calculator. It is not necessary to have access to a computer and dedicated software to evaluate factorial designs. However, using hand-calculation in large factorial experiments, is of course a time-consuming process.

3.3. Drawbacks of using sign tables

The calculations involved in the evaluation of factorial designs by using sign tables rely on an unrealistic assumption, viz. that each experiments was conducted *exactly* as was specified by the design. For instance, that the temperature was adjusted exactly to its high and low levels. In practice, this is never obtained in synthetic chemistry. The variables can be adjusted fairly close to the specified levels, but over the series of experiments. there will always be small differences between the runs. It is therefore better, and more honest, to use the settings *actually used* in the evaluation of the experiments. For this, the appropriate tool is multiple linear regression which is used to fit response surface models to the experimental data. This technique is described in the next section.

A method often used for hand-calculation of effects from factorial designs is the Yates algorithm.[2] As this algorithm also assumes that the levels of the variables are exactly as specified by the design. It is therefore suggested that it should not be used to the evaluation of synthesis experiments.

3.4. Least squares fit of response surface models to experiments in factorial designs

A factorial design can be used to fit a response surface model to the experimental results. In this case, the effects will be the corresponding model parameters. To achieve this, the factors are scaled through a linear transformation to *design variables*, x_i, as was described in section 3.4.2, see also Fig. 5.2.

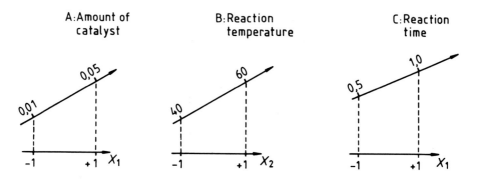

Fig.5.2: Linear transformation of the natural variables into design variables.

The experimental design will be described by a design matrix. As an example a 2^3 design is shown in Table 5.4.

Table 5.4: Design matrix of a 2^3 factorial design

Exp no	x_1	x_2	x_3
1	−1	−1	−1
2	1	−1	−1
3	−1	1	−1
4	1	1	−1
5	−1	−1	1
6	1	−1	1
7	−1	1	1
8	1	1	1

A response surface model which accounts for the average, all main effects, all two-factor interactions, and the three-factor interaction will thus be

$$y = \beta_0 + \beta_1 x_1 + \beta_2 x_2 + \beta_3 x_3 + \beta_{12} x_1 x_2 + \beta_{13} x_1 x_3 + \beta_{23} x_2 x_3 + \beta_{123} x_1 x_2 x_3 + e$$

To compute the model parameters we must first construct the corresponding *model matrix*, **X**, by augmenting the design matrix with a column I for the constant term in the model, and columns for all interaction terms in the model. The model matrix is given in Table 5.5.

Table 5.5: Model matrix for a 2^3 factorial design

Exp no	I	x_1	x_2	x_3	$x_1 x_2$	$x_1 x_3$	$x_2 x_3$	$x_1 x_2 x_3$
1	1	−1	−1	−1	1	1	1	−1
2	1	1	−1	−1	−1	−1	1	1
3	1	−1	1	−1	−1	1	−1	1
4	1	1	1	−1	1	−1	−1	−1
5	1	−1	−1	1	1	−1	−1	1
6	1	1	−1	1	−1	1	−1	−1
7	1	−1	1	1	−1	−1	1	−1
8	1	1	1	1	1	1	1	1

The same example as above is used to illustrate the computations. A brief summary of matrix calculus is given in Appendix: Matrix calculus at the end of this book. The whole series of experiments can be summarized by the now well-known matrix relation

$y = \mathbf{X}\beta + e$

which in the present case corresponds to

$$
\begin{bmatrix} 74 \\ 69 \\ 78 \\ 81 \\ 76 \\ 87 \\ 84 \\ 91 \end{bmatrix}
=
\begin{bmatrix}
1 & -1 & -1 & -1 & 1 & 1 & 1 & -1 \\
1 & 1 & -1 & -1 & -1 & -1 & 1 & 1 \\
1 & -1 & 1 & -1 & -1 & 1 & -1 & 1 \\
1 & 1 & 1 & -1 & 1 & -1 & -1 & -1 \\
1 & -1 & -1 & 1 & 1 & -1 & -1 & 1 \\
1 & 1 & -1 & 1 & -1 & 1 & -1 & -1 \\
1 & -1 & 1 & 1 & -1 & -1 & 1 & -1 \\
1 & 1 & 1 & 1 & 1 & 1 & 1 & 1
\end{bmatrix}
\begin{bmatrix} \beta_0 \\ \beta_1 \\ \beta_2 \\ \beta_3 \\ \beta_{12} \\ \beta_{13} \\ \beta_{23} \\ \beta_{123} \end{bmatrix}
+
\begin{bmatrix} e_1 \\ e_2 \\ e_3 \\ e_4 \\ e_5 \\ e_6 \\ e_7 \\ e_8 \end{bmatrix}
$$

A least squares fit of the model is obtained by

$(\mathbf{X}'\mathbf{X})^{-1}\mathbf{X}'y = b$

For this, first compute $\mathbf{X}'\mathbf{X}$

$$
\mathbf{X}'\mathbf{X} =
\begin{bmatrix}
1 & 1 & 1 & 1 & 1 & 1 & 1 & 1 \\
-1 & 1 & -1 & 1 & -1 & 1 & -1 & 1 \\
-1 & -1 & 1 & 1 & -1 & -1 & 1 & 1 \\
-1 & -1 & -1 & -1 & 1 & 1 & 1 & 1 \\
1 & -1 & -1 & 1 & 1 & -1 & -1 & 1 \\
1 & -1 & 1 & -1 & -1 & 1 & 1 & -1 \\
1 & 1 & -1 & -1 & -1 & -1 & 1 & 1 \\
-1 & 1 & 1 & -1 & 1 & -1 & -1 & 1
\end{bmatrix}
\begin{bmatrix}
1 & -1 & -1 & -1 & 1 & 1 & 1 & -1 \\
1 & 1 & -1 & -1 & -1 & -1 & 1 & 1 \\
1 & -1 & 1 & -1 & -1 & 1 & -1 & 1 \\
1 & 1 & 1 & -1 & 1 & -1 & -1 & -1 \\
1 & -1 & -1 & 1 & 1 & -1 & -1 & 1 \\
1 & 1 & -1 & 1 & -1 & 1 & -1 & -1 \\
1 & -1 & 1 & 1 & -1 & -1 & 1 & -1 \\
1 & 1 & 1 & 1 & 1 & 1 & 1 & 1
\end{bmatrix}
$$

$$= \begin{bmatrix} 8 & 0 & 0 & 0 & 0 & 0 & 0 & 0 \\ 0 & 8 & 0 & 0 & 0 & 0 & 0 & 0 \\ 0 & 0 & 8 & 0 & 0 & 0 & 0 & 0 \\ 0 & 0 & 0 & 8 & 0 & 0 & 0 & 0 \\ 0 & 0 & 0 & 0 & 8 & 0 & 0 & 0 \\ 0 & 0 & 0 & 0 & 0 & 8 & 0 & 0 \\ 0 & 0 & 0 & 0 & 0 & 0 & 8 & 0 \\ 0 & 0 & 0 & 0 & 0 & 0 & 0 & 8 \end{bmatrix}$$

A diagonal matrix is obtained in which the diagonal elements are equal to the number of experiments. This is the case for all two-level factorial designs. Then, compute the dispersion matrix $(\mathbf{X'X})^{-1}$ which in this case will be

$$(\mathbf{X'X})^{-1} = \begin{bmatrix} 1/8 & 0 & 0 & 0 & 0 & 0 & 0 & 0 \\ 0 & 1/8 & 0 & 0 & 0 & 0 & 0 & 0 \\ 0 & 0 & 1/8 & 0 & 0 & 0 & 0 & 0 \\ 0 & 0 & 0 & 1/8 & 0 & 0 & 0 & 0 \\ 0 & 0 & 0 & 0 & 1/8 & 0 & 0 & 0 \\ 0 & 0 & 0 & 0 & 0 & 1/8 & 0 & 0 \\ 0 & 0 & 0 & 0 & 0 & 0 & 1/8 & 0 \\ 0 & 0 & 0 & 0 & 0 & 0 & 0 & 1/8 \end{bmatrix} =$$

$$= 1/8 \cdot \begin{bmatrix} 1 & 0 & 0 & 0 & 0 & 0 & 0 & 0 \\ 0 & 1 & 0 & 0 & 0 & 0 & 0 & 0 \\ 0 & 0 & 1 & 0 & 0 & 0 & 0 & 0 \\ 0 & 0 & 0 & 1 & 0 & 0 & 0 & 0 \\ 0 & 0 & 0 & 0 & 1 & 0 & 0 & 0 \\ 0 & 0 & 0 & 0 & 0 & 1 & 0 & 0 \\ 0 & 0 & 0 & 0 & 0 & 0 & 1 & 0 \\ 0 & 0 & 0 & 0 & 0 & 0 & 0 & 1 \end{bmatrix} = 1/8 \cdot I_8$$

The least squares fit

$$(X'X)^{-1}X'y = b$$

can be written

$$1/8 \cdot I_8 X'y = 1/8 X'y = b$$

which corresponds to the following

$$1/8 \cdot \begin{bmatrix} 1 & 1 & 1 & 1 & 1 & 1 & 1 & 1 \\ -1 & 1 & -1 & 1 & -1 & 1 & -1 & 1 \\ -1 & -1 & 1 & 1 & -1 & -1 & 1 & 1 \\ -1 & -1 & -1 & -1 & 1 & 1 & 1 & 1 \\ 1 & -1 & -1 & 1 & 1 & -1 & -1 & 1 \\ 1 & -1 & 1 & -1 & -1 & 1 & -1 & 1 \\ 1 & 1 & -1 & -1 & -1 & -1 & 1 & 1 \\ -1 & 1 & 1 & -1 & 1 & -1 & -1 & 1 \end{bmatrix} \begin{bmatrix} 74 \\ 69 \\ 78 \\ 81 \\ 76 \\ 87 \\ 84 \\ 91 \end{bmatrix} = \begin{bmatrix} b_0 \\ b_1 \\ b_2 \\ b_3 \\ b_{12} \\ b_{13} \\ b_{23} \\ b_{123} \end{bmatrix} = \begin{bmatrix} 80.0 \\ 2.0 \\ 3.5 \\ 4.5 \\ 0.5 \\ 2.5 \\ -0.5 \\ -1.5 \end{bmatrix}$$

This is exactly the same computations as were carried out using the sign tables. The computed effects can be interpreted geometrically as properties of the fitted surface (intercept, slopes along the variable axes, twist of the surface to describe interactions etc.) as was discussed in Chapter 3. This leads to the following important conclusion:

> *Running and evaluating a two-level factorial design by any of the methods described above is equivalent to determining a response surface model which includes terms to describe all interaction effects.*

From this conclusion follows, that a factorial design can be used to fit a response surface model to account for main effects and interaction effects. In the concluding section of this chapter is discussed how the properties of the model matrix X influence the quality of the estimated parameters in multiple regression. It is shown that factorial design have optimum qualities.

Fitting a response surface model by means of multiple regression can be carried out even if the settings of the experimental variables should not be exactly as were specified by the design. In such cases, the dispersion matrix, $(X'X)^{-1}$, would probably not be a perfect diagonal matrix and it would not be possible to fit the response surface model by hand-calculation. Computer programs for multiple linear regression are abundant. Some useful programs for response surface modelling are given in the reference list.

The two-level factorial designs are the most convenient and efficient way of laying out a screening experiment if the number of experimental variables does not exceed four. With more than four variables, it is more convenient to use a *fractional factorial design*. These designs are discussed in the next chapter.

4. Examples of two-level factorial designs in exploring synthetic procedures

4.1. 2^2-Design: Reduction of an enamine

Enamines can be reduced to the corresponding saturated amine by treatment with formic acid.[3] To determine suitable experimental conditions for the reduction of enamines derived from camphor, a factorial design was used to explore the reaction of the morpholine enamine. The reaction was conducted by adding formic acid dropwise to the neat enamine at such a rate that the foaming caused by the evolution of carbon dioxide could be kept under control. The reaction is rapid and completed within a few minutes. The main product in the reaction was the desired bornylmorpholine (mixture of *endo* and *exo* isomers. There were also varying amounts of unreacted enamine and some camphor formed by hydrolysis of the enamine.

Enamine   Amine   Camphor

The experimental procedure is very simple and there are only two variables to consider: x_1, the amount of formic acid, and x_2, the reaction temperature. The levels of the variables are shown in Table 5.6.

Table 5.6: Variables and experimental domain in the enamine reduction

Variables	Levels of the scaled variables	
	-1	$+1$
x_1: Amount of formic acid/enamine (mol/mol)	1.0	1.5
x_2: Reaction temperature (°C)	25	100

When the reaction was complete, the resulting mixture was treated with aqueous sodium hydroxide. Then the composition of the organic layer was analyzed by gas chromatography. The recovery of organic material was quantitative.
The measured responses were:

y_1:   The yield of bornylmorpholine (%)
y_2:   The amount of unreacted enamine (%)
y_3:   The amount of camphor (%)

The experimental design and the measured responses are summarized in Table 5.7.

Table 5.7: Experimental design and responses in the enamine reduction

Exp no	x_1	x_2	y_1	y_2	y_3
1	-1	-1	80.4	12.5	6.7
2	1	-1	72.4	14.0	10.5
3	-1	1	94.4	0	5.5
4	1	1	90.6	0	7.7

From these experiments a response surface model

$$y = \beta_0 + \beta_1 x_1 + \beta_2 x_2 + \beta_{12} x_1 x_2 + e$$

was determined for each response. For this purpose, the model matrix **X** is constructed.

I	x_1	x_2	$x_1 x_2$
1	−1	−1	1
1	1	−1	−1
1	−1	1	−1
1	1	1	1

This corresponds to the dispersion matrix $(\mathbf{X'X})^{-1}$

$$\begin{bmatrix} 1/4 & 0 & 0 & 0 \\ 0 & 1/4 & 0 & 0 \\ 0 & 0 & 1/4 & 0 \\ 0 & 0 & 0 & 1/4 \end{bmatrix}$$

Estimates of the model parameters are obtained by computing

$$b = (\mathbf{X'X})^{-1}\mathbf{Xy}$$

The calculations with y_1 are shown on the next page. Calculations for the other responses are carried out analogously.

After rounding-off to significant figures, the response surface model for y_1 will thus be

$$y_1 = 84.5 - 3.0\, x_1 + 8.1\, x_2 + 1.1\, x_1 x_2 + e \qquad \text{and for the other responses}$$

$$y_2 = 6.6 + 0.4\, x_1 - 6.6\, x_2 - 0.4\, x_1 x_2 + e$$

$$y_3 = 7.6 + 1.5\, x_1 - 1.0\, x_2 - 0.4\, x_1 x_2 + e$$

$$
\begin{bmatrix} b_0 \\ b_1 \\ b_2 \\ b_{12} \end{bmatrix} = \begin{bmatrix} 1/4 & 0 & 0 & 0 \\ 0 & 1/4 & 0 & 0 \\ 0 & 0 & 1/4 & 0 \\ 0 & 0 & 0 & 1/4 \end{bmatrix} \begin{bmatrix} 1 & 1 & 1 & 1 \\ -1 & 1 & -1 & 1 \\ -1 & -1 & 1 & 1 \\ 1 & -1 & -1 & 1 \end{bmatrix} \begin{bmatrix} 80.4 \\ 72.4 \\ 94.4 \\ 90.6 \end{bmatrix} =
$$

$$
1/4 \cdot \begin{bmatrix} 80.4 + 72.4 + 94.4 + 90.6 \\ -80.4 + 72.4 - 94.4 + 90.6 \\ -89.4 - 72.4 + 94.4 + 90.6 \\ 80.4 - 72.4 - 94.4 + 90.6 \end{bmatrix} = 1/4 \cdot \begin{bmatrix} 337.8 \\ -11.8 \\ 32.2 \\ 4.2 \end{bmatrix} = \begin{bmatrix} 84.45 \\ -2.95 \\ 8.05 \\ 1.05 \end{bmatrix}
$$

It is not possible to make any statistical analysis of the model parameters since no estimate of the experimental error is available.

Three-dimensional plots of the fitted surfaces are shown in Fig. 5.3

These plots show that suitable experimental conditions are one equivalent of formic acid (x_1 at low level) at high temperature (x_2 at high level). It is not possible to use less than stoichiometric amounts of formic acid, and as the boiling point of formic acid is 101 °C it is not likely that the temperature could be increased much further. To further improve the result it might, however, be possible to increase slightly the temperature of the enamine prior to introducing the formic acid. For the sake of convenience, heating on a boiling water-bath was employed in preparative scale (1 mol) runs, which afforded yields in the range of 84 − 87 % of pure isolated distilled bornylamine.

In this case, the experiment is very simple and the information can be obtained directly from the experimental design. It is not necessary to make any calculations at all. The information obtained from the three-dimensional plots in Fig. 5.3 can also be obtained by plotting the experimental data shown in Fig. 5.4. Such plots can always be obtained directly from the experimental data.

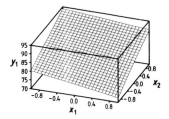

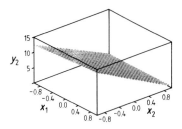

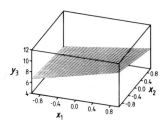

Fig.5.3: Three dimensional plots of the response surfaces: y_1, bornylamine; y_2, unreacted enamine; y_3, camphor.

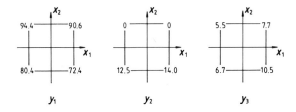

Fig.5.4: Approximate description of the response surfaces by plotting the experimental data directly. (a) Bornylamine; (b) unreacted enamine; (c) camphor.

4.2. 2^3-Design: Bromination of an enamine

Synthesis of bromomethyl ketones can be accomplished by treating the morpholine enamine from the parent ketone with elemental bromine followed by hydrolysis.

When this procedure was applied to the enamine derived from methyl neopentyl ketone, poor yields (< 60 %) were obtained. This was unexpected, since a variety of other ketones had afforded good yields (> 85 %) by the same procedure. The problem was to determine:*(1)* Why the enamine from methyl neopentyl ketone did not work in this procedure, and *(2)* what to do about it.

The procedure involved two steps, *bromination* and *hydrolysis*. The poor yield might depend on either or both of these steps: Due to steric hindrance, the bromination reaction might not have gone to completion when water was introduced, and/or the hydrolysis was slow and was not completed when the mixture was worked up. To clarify this, the important variables must be identified. It was known from other similar reaction, that the bromination must be effected at low temperature to suppress side reactions, and that the hydrolysis should be carried out not above room temperature. During the hydrolysis step, bromomethyl ketones are susceptible to acid catalyzed rearrangements to the isomeric 3-bromoketone. With these limitations taken into account, the following experimental variables were considered to be able to influence the result.

x_1:   the concentration of bromine might possibly have an influence on the rate of the first step.

x_2:   the time of reaction for the bromination step.

x_3:   the time of reaction of the hydrolysis step.

The experimental domain is given in Table 5.8.

Table 5.8: Experimental domain in the bromination of an enamine

Factors	Levels	
	(−)	(+)
x_1: Bromine concentration (mol/dm^3)	0.25	0.50
x_2: Bromination time (min)	2	5
x_3: Hydrolysis time (min)	5	30

Table 5.9: Factorial design in the enamine bromination

Exp no	x_1	x_2	x_3	Yield (%)
1	−1	−1	−1	45.0
2	1	−1	−1	46.0
3	−1	1	−1	42.2
4	1	1	−1	43.0
5	−1	−1	1	72.2
6	1	−1	1	66.0
7	−1	1	1	66.0
8	1	1	1	63.0

The low levels of the variables correspond to the conditions used with other enamine substrates. Under these conditions, the yield of bromomethyl neopentyl ketone was 45 %. To see whether or not the experimental domain is sufficiently large to give an observable variation in the experimental result, the next experiment to be run was with all variables set to their high levels. This afforded a 63 % yield, which is a considerable improvement. To be able to detect *which* variable(s) was/were critical, the remaining experiments aimed at giving a complete two-level factorial design were run. The results are summarized in Table 5.9.

From these experiment a second-order interaction response surface model

$$y = \beta_0 + \beta_1 x_1 + \beta_2 x_2 + \beta_3 x_3 + \beta_{12} x_1 x_2 + \beta_{13} x_1 x_3 + \beta_{23} x_2 x_3 + e$$

was determined.

The estimated model parameters are

$$b_0 = 55.4 \qquad b_{12} = 0.4$$
$$b_1 = -0.9 \qquad b_{13} = 1.4$$
$$b_2 = -1.9 \qquad b_{23} = -0.4$$
$$b_3 = 11.4$$

There is no need for a statistical analysis to understand that variable x_3 is important. This result identifies unequivocally the hydrolysis step as the bottleneck. To increase the yield of bromethyl neopentyl ketone, the time of hydrolysis should be at the high level. Possibly it should be increased The conditions for the bromination step is not critical and this step can be carried out under the previously used conditions.

A final yield of 90% of bromethyl neopentyl ketone could be achieved when the time of hydrolysis was increased to 1h.

4.3. 2^4-Design: Catalytic hydrogenation

The experiments were run to determine how four experimental variables influence the yield of tetrahydrofuran in catalytic hydrogenation of furan over a palladium catalyst. The variables and the experimental domain are shown in Table 5.10.

Table 5.10: Variables and experimental domain in catalytic hydrogenation of furan

Variables	Levels	
	(−)	(+)
x_1: Amount of catalyst / substrate (g / mol)	0.7	1.0
x_2: Hydrogen pressure (bar)	45	55
x_3: Reaction temperature (°C)	75	100
x_4: Stirring rate (rpm)	340	475

The experimental design and the yields, y, obtained are given in Table 5.11.

If the objective is to determine the experimental condition for maximum yield, this is all that is required. All variables at the high level give a quantitative yield.

However, this was not the only objective. It was also desired to determine the influence of the variables, as well as their interactions. Thus, a second order interaction model was assumed to give a satisfactory description, i.e.

$$y = \beta_0 + \Sigma \, \beta_i \, x_i + \Sigma\Sigma \, \beta_{ij} \, x_i x_j + e$$

This means that three- and four-factor interaction effects were assumed to be small compared to the experimental error. The model contains 11 parameters and the experimental design contains 16 runs. The excluded higher order interaction effects allow an estimate of the residual sum of squares, SSE, which will give an estimate of the experimental error variance, s^2, with $16 - 11 = 5$ degrees of freedom. This estimate can then be used to compute confidence limits for the estimated model parameters so that their significance can be evaluated.

A least squares fit of the model to the experimental data in Table 5.11 afforded the estimates of the model parameters given below:

$$
\begin{aligned}
b_0 &= 91.65 \pm 0.28 & b_{12} &= 0.19 \pm 0.28 \\
b_1 &= 2.10 \pm 0.28 & b_{13} &= -0.36 \pm 0.28 \\
b_2 &= 2.76 \pm 0.28 & b_{14} &= -0.52 \pm 0.28 \\
b_3 &= 0.79 \pm 0.28 & b_{23} &= 0.32 \pm 0.28 \\
b_4 &= 5.75 \pm 0.28 & b_{24} &= -2.54 \pm 0.28 \\
& & b_{34} &= 0.39 \pm 0.28
\end{aligned}
$$

It is seen that all variables have significant linear influences on the yield, although the temperature, x_3, is not that important. One highly significant interaction effect, β_{24}, is found and this is to be expected. There is a compensation effect between the hydrogen pressure, x_2, and the stirring rate, x_4. A low pressure can be compensated by a high stirring rate, while this is less pronounced when the pressure is high. The twist of the response surface plane due to the interaction effect is clearly seen in the projection of the response surface model down to the space spanned by $[y, x_2, x_4]$, Fig.5.5. In this projection x_1 and x_3 have been set to their high levels. These linear coefficients for the variables are positive, and they should therefore be set to their high levels.

Table 5.11: Experimental design and yields obtained in the catalytic hydrogenation of furan

Exp no	Variables				Yield
	x_1	x_2	x_3	x_4	y
1	−1	−1	−1	−1	77.5
2	1	−1	−1	−1	83.8
3	−1	1	−1	−1	87.8
4	1	1	−1	−1	92.9
5	−1	−1	1	−1	77.8
6	1	−1	1	−1	83.3
7	−1	1	1	−1	90.0
8	1	1	1	−1	94.1
9	−1	−1	−1	1	94.1
10	1	−1	−1	1	98.3
11	−1	1	−1	1	94.2
12	1	1	−1	1	98.3
13	−1	−1	1	1	97.0
14	1	−1	1	1	99.3
15	−1	1	1	1	98.0
16	1	1	1	1	100.0

The confidence limits of the estimated parameters have been determined as follows:

An estimate of the experimental error variance with five degrees of freedom is obtained from the residual sum of squares, *RSS*

$$s^2 = RSS \ /(n - p) = RSS \ / \ 5$$

RSS is computed as the difference between the total sum of squares, *SST*, and the sum of squares due to regression, *SSR*. These are computed as follows

$$SST = \Sigma \ y_i^2 = (77.5^2 + 83.8^2 + \ ......+ 98.0^2 + 100.0^2) = 135 \ 242.2$$

$$SSR = b'X'Xb = 16 \cdot (b_0^2 + b_1^2 + \ ......+ b_{34}^2) =$$
$$= 16 \cdot (91.65^2 + 2.10^2 + \ .....+ 0.39^2) = 16 \cdot (8452.5789) = 135 \ 241.2624$$

$$RSS = SST - SSR = 135 \ 242.2 - 135 \ 241.2624 = 0.9376$$

$$s^2 = 0.9376 \ / \ 5 = 0.18752$$

Since a factorial design was used, the model parameters are estimated with the same precision and their variances, $V(b_i)$, are

$$V(b_i) = s^2 / 16 = 0.18752 / 16 = 0.01172$$

The standard error of the parameter will therefore be

Standard error $= \sqrt{0.01172} \approx 0.108$

The critical t−value at the 95 % level for five degrees of freedom is $t^{\text{Crit.}} = 2.571$. The confidence limits of the estimated parameters will thus be

Confidence limits $= \pm\, t^{\text{Crit.}} \cdot$ *Standard error* $= 2.571 \cdot 0.108 \approx 0.28$

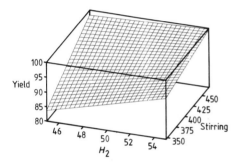

Fig.5.5:   Three-dimensional plot showing the variation in yield when x_2 (hydrogen pressure) and x_4 (strirring rate) are varied. The twist of the response surface due to the interaction is clearly seen.

5.   Some remarks on the quality of model parameters estimated by least squares fit

5.1.   General aspects

The "quality" of estimated model parameters is determined by the experimental error variance, σ^2, and the properties of the dispersion matrix, $(\mathbf{X'X})^{-1}$, which in turn is determined by the experimental design. Let us have a closer look at the concept "quality".

It was shown in Chapter 3 that it is possible to assign confidence limits to each model parameter. This gives the precision of the estimate of each parameter

116

separately. As the response surface model contains more than one model parameter, a more severe criterion would be to assign probability limits for the joint variation of all the model parameters, i.e. how much they can vary in relation to each other for a given probability level. We shall therefore define a *joint confidence region*[6] which takes all model parameters into account simultaneously. The principles can be illustrated geometrically as in Fig. 5.6.

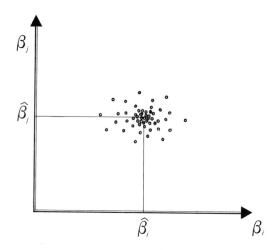

Fig.5.6:   The individual estimated model parameters will have a probability distribution around an average value.

Let each model parameter define a coordinate axis along which possible values of the parameters can be plotted, and assume that $\hat{\beta}_0$, $\hat{\beta}_1$,..,$\hat{\beta}_i$,..,$\hat{\beta}_k$,...,$\hat{\beta}_{ij}$ are the estimated value of these parameters. These estimates will define a point in the space spanned by the coordinate axes. As each estimated value is affected by the experimental error it will have a probability distribution. For two model parameters the situation can be depicted as in Fig.5.6. It is evident that values of the parameters close to the center point are more probable estimates than values which define a point far away from the center point.

It is possible to use the F distribution to draw boundaries for the probable variation of all parameters simultaneously for any given probability level. These boundaries will define a closed space which confines probable estimates of the model parameters.

Let s^2 be an estimate of the experimental error variance with r degrees of freedom; let **B** be the vector of estimated model parameters; let p be the number of estimated model parameters; and let $F_{\alpha(p,r)}$ be the critical F value with (p,r) degrees of freedom and the significance level α. The boundaries of the joint confidence region is defined by the equation

$$(\boldsymbol{\beta} - \boldsymbol{\hat{\beta}})'(\mathbf{X'X})(\boldsymbol{\beta} - \boldsymbol{\hat{\beta}}) = p \cdot s^2 \cdot F_{\alpha(p,r)}$$

In two dimensions this will define concentric elliptic contours for different levels of significance, see Fig. 5.7a. In three dimensions the boundaries will be ellipsoidic shells which confine the model parameters, see Fig. 5.7b. In higher dimensions the boundaries will be hyperellipsoides.

The joint probability region can have different orientations and extensions in the parameter space. This will correspond to different "quality" aspects of the estimated values of the model parameters. These quality aspects will depend on the properties of the dispersion matrix $(\mathbf{X'X})^{-1}$, see Fig. 5.8.

The "volume" of the joint confidence region is proportional to the experimental error variance σ^2 and to the square root of the determinant of the dispersion matrix.

$$\text{"Volume"} \propto |(\mathbf{X'X})^{-1}| \cdot \sigma^2$$

To obtain a high precision in the estimates of all model parameters the experiments should be laid out in such a way that $|(\mathbf{X'X})^{-1}|$ is as small as possible.

The confidence limits for individual model parameters are proportional to the main axes of the ellipsoids and these are related to the eigenvalues of the dispersion matrix. To obtain the same precision in all model parameters, the confidence region should be spherical. Thus, the experiment should be laid out so that the eigenvalues of $(\mathbf{X'X})^{-1}$ are equal.

Independent estimates of the model parameters are obtained when the main axes of the confidence (hyper)ellipsoid are *parallel* to the parameter axes. This is not the case in Fig. 5.8d. Such situations occur if the covariances of the model parameters are not equal to zero. This will happen if the two or more variables have been varied in a correlated way over the series of experiments. In such cases it will not be possible to unequivocally discern *which* of these variables is responsible for an observed variation of the response. To obtain independent estimates of the model parameters, the experiments should be laid out in such a way that the dispersion matrix is a diagonal matrix. This implies that the settings of the variables of the model are uncorrelated over the set of experiments, which is equivalent to saying

118

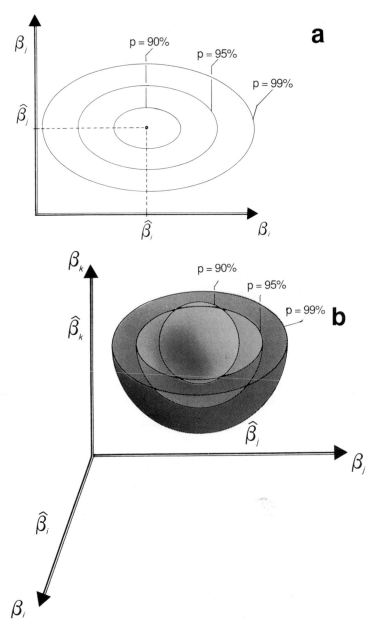

Fig.5.7: Joint confidence regions: With two model parameters the confidence limits are defined by elliptic contours. With three parameters these limits are defined by ellipsoidic shells. With many parameters, these limits are defined by hyperellipsoids.

that the column vectors of the model matrix should be orthogonal to each other. An experimental design which fulfills this, is said to be an *orthogonal design*.

The reasoning above is general and applies to all least squares estimations. In the example below is shown how these criteria are fulfilled when a two-level factorial design is used to estimate the parameters of a response surface model.

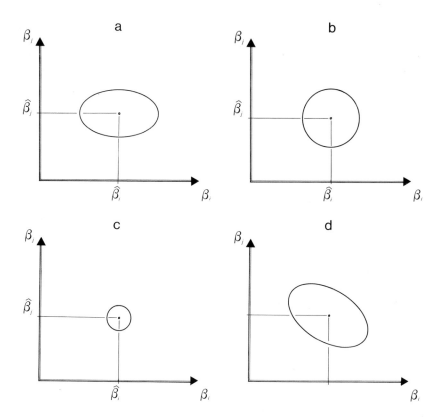

Fig.5.8: Different shapes of the joint confidence region: *(a)* The parameters are independenly estimated since the main axes of ellipses are parallel to the parameters axes. The main axes of the ellipses have unequal length and the estimates do not have the same precision.*(b)* The estimates are independent and are determined with equal precision. *(c)* The same as *(b)* but the precision is better. *(d)* The parameters are not independently estimated. The main axes of the confidence ellipsoids are not parallel to the parameter axes.

5.2. Quality of estimated parameters from a factorial design

The application of the above principles is illustrated by the previously discussed 2^3 design. The dispersion matrix $(\mathbf{X'X})^{-1}$ is shown below.

The determinant $|(\mathbf{X'X})^{-1}| = 8^{-8}$ which is the minimum value. The eigenvalues of the dispersion matrix are all equal and the variance of all estimated model parameters are $\sigma^2 / 8$. The parameters are independently estimated and the dispersion matrix is a diagonal matrix (the covariances of the models parameters are zero). This means that parameters estimated from a two-level factorial design are *independently estimated*, with *equal* and *maximum* precision.

$$
\begin{bmatrix}
1/8 & 0 & 0 & 0 & 0 & 0 & 0 & 0 \\
0 & 1/8 & 0 & 0 & 0 & 0 & 0 & 0 \\
0 & 0 & 1/8 & 0 & 0 & 0 & 0 & 0 \\
0 & 0 & 0 & 1/8 & 0 & 0 & 0 & 0 \\
0 & 0 & 0 & 0 & 1/8 & 0 & 0 & 0 \\
0 & 0 & 0 & 0 & 0 & 1/8 & 0 & 0 \\
0 & 0 & 0 & 0 & 0 & 0 & 1/8 & 0 \\
0 & 0 & 0 & 0 & 0 & 0 & 0 & 1/8
\end{bmatrix}
$$

There are innumberable ways by which the settings of experimental variables can be varied to determine their influence on the measured response. To obtain good and reliable estimates the lay-out of the experimental runs in the experimental domain is very important. Different designs in that respect will correspond to different dispersion matrices, and hence to the quality of the estimated parameters. It is seen that two-level factorial designs have excellent statistical properties; we cannot do better.

In the chapters to follow, we shall see how the criteria discussed above can be satisfied with other types of experimental designs.

References

1. R.A. Fisher
   J. Minist. Agric. 33 (1926) 503.

2. F. Yates
   Techn. Commun. No 35
   Imperial Bureau of Soil Science, Harpenden 1937.

3. R.Carlson and Å. Nilsson
   Acta Chem. Scand. B 39 (1985) 181.

4. R. Carlson, unpublished.

5. C. Godawa
   Etude de l'hydrogenation catalytique du furanne: Optimisation du procede
   Diss. L'Institut National Polytechnique de Toulouse, Toulouse, 1984.

6. A.C. Atkinson and J.S. Hunter
   Technometrics 10 (1968) 271.

Suggestions for further reading

G.E.P. Box, W.G. Hunter and J.S. Hunter
Statistics for Experimenters
Wiley,New York 1978.

G.E.P. Box and N.R. Draper
Empirical Model-Building and Response Surfaces
Wiley, New York 1987.

W.G. Cochran and G.M. Cox
Experimental Designs
Wiley, New York 1957

Computer programs

Some available programs for response surface modelling are listed below.

For personal computers (IBM PC XT/AT and compatibles):

MODDE
Umetri AB,
P.O.Box 1456, S-901 24 Umeå, Sweden

NEMROD
LPRAI
Att. Prof R. Phan-Tan-Luu
Université d'Aix-Marseille III
Av. Escadrille Normandie Niemen
F-13397 Marseille Cedex 13, France

STATGRAPHICS
STSC, Inc.
2115 East Jefferson Street
Rockville, MD 20852, USA

ECHIP
Expert in a Chip, Inc.
RD! Box 384Q
Hockesin, DE 19707, USA

XSTAT
Wiley Professional Software
John Wiley and Sons, Inc.
605 Third Avenue
New York, NY 10158, USA

122

For mini and mainframe computers:

RS1/Discover
BBN Sofware Products Corp.
10 Fawcett Street
Cambridge, MA 02238, USA

SAS
SAS Institute, Inc.
SAS Circle
P.O.Box 800
Cary, NC 27512, USA.

For a critical review of programs for experimental design, see

C.J. Nachtheim
J. Qual. Technol. 19 (1987) 132.

Chapter 6

Two-level fractional factorial design

1. Introductory remarks

1.1. Screening experiments with many variables

When an experimental procedure is analyzed as was discussed in Chapter 4, it is rather common that more than a handful of variables need to be considered. Determination of the influence of k variables through a complete two-level factorial design calls for 2^k individual experimental runs. Seven variables give 128 runs ; ten variables give 1024 runs; fifteen variables give 32,768 runs. It is evident that many variables would result in a prohibitively large number of runs if analyzed by a factorial design.

For chemical systems in a limited experimental domain, it is reasonable to assume that interaction effects between three or more variables are small compared to main effects and two-variable interaction effects. To run a screening experiment with seven variables, it would be satisfactory to determine: the average response, seven main effects, and 21 two-variable interaction effect, i.e. 29 different parameters. To do so, a minimum of 29 experimental runs is necessary. In fact, exactly 29 runs would be sufficient. It would therefore be a wasteful approach to use a complete factorial design, which in addition to allowing for an estimation of the desired parameters, also affords estimates of all the remaining higher order interaction effcts. As these are assumed to be negligible they will afford an estimate of the experimental error *with 99 degrees of freedom*. This is a little too much.

In this chapter, it is discussed how to select a subset of experiments from a complete factorial design in such a way that it will be possible to estimate the desired parameters through a limited number of experimental runs. We shall see that it is very easy to construct designs which are 1/2, 1/4, 1/8, 1/16,... $1/2^p$ fractions of a complete factorial design. This will give a total of 2^{k-p} experimental runs, where k is the number of variables, and p is the size of the fraction.

1.2. A step-wise strategy is possible

At the outset of an exploratory study of a new synthetic procedure, the roles played by the various variables are not known. Under these conditions we are not primarily interested in very *precise* measures of the influence of the variables, but rather in obtaining information *whether or not* they are influencing, and for the influencing variables, the *magnitude* and the *direction* of their influence. In such cases, an approximation of the response surface by a plane will give sufficient information, i.e.

$$y = \beta_0 + \Sigma \, \beta_i \, x_i + e$$

The parameters in this model measure the slope of the plane and, hence, the sensitivity of the response to variations in the variable settings.
In such cases it is possible to analyze three variables in four runs, up to seven variables in eight runs, up to fifteen variables in sixteen runs etc.

This will provide us with a first check of which variables are influencing the response. It is then always possible to run a *complementary set* of new experiments to augment the model by cross-product terms to allow for an analysis of interaction effects, i.e.

$$y = \beta_0 + \Sigma \, \beta_i \, x_i + \Sigma\Sigma \, \beta_{ij} \, x_i x_j + e$$

Fortunately, in many cases it will not be necessary to investigate interaction effects between *all* variables *initially* considered. After the first series of experiments and evaluation of the linear model, some variables often turn out to have a minor or even negligible linear influence. Such variables rarely show strong interaction effects with other variables. Interaction effects are more likely to be encountered between variables which also have strong linear influence. Of course, this is not fool-proof, but it offers a means of simplifying the problem, and will therfore lead to a reduction of the number of additional experimental runs.

From the above is seen that it is advantageous to include *all* potentially interesting variables at the outset of an experimental study and in an initial series of screening experiments. This can be done without an excessive number of experimental runs. Then, analysis of these experiments will identify *which* variables should be included in a subsequent more detailed study.

2. How to construct a fractional factorial design?

It was seen in the preceeding chapter that the model matrix $\mathbf{X}$ of a factorial design has orthogonal columns, i.e. $\mathbf{X'X}$ is a diagonal matrix. It is the model matrices from complete factorial designs that are used to define the *design matrices* of fractional factorial designs. This means that we let the columns in the model matrix, $\mathbf{X}$, of a factorial design define the settings of the variables over the series of experiments. We can include as many variables as there are columns in $\mathbf{X}$. In this chapter we shall use bold-face arabic numeral characters to denote the variable columns in the matrices. This has practical consequences as we shall see later on.

Example: A two-variable factorial design has the following complete model matrix

I	**1**	**2**	**12**
1	−1	−1	1
1	1	−1	−1
1	−1	1	−1
1	1	1	1

The column **12** is to be read as the product of $\mathbf{1} \cdot \mathbf{2}$. We can use the colums **1, 2, 12** to define the variation of three variables, x_1, x_2, x_3 in four runs and this gives the design matrix

Exp no	x_1	x_2	x_3
1	−1	−1	1
2	1	−1	−1
3	−1	1	−1
4	1	1	1

This is one half-fraction of a complete three-variable factorial design and it is seen that the experiments correspond to Exp no 5, 2, 3, 8 in the complete factorial design. Fig. 6.1 shows how the experiments of the half-fraction are distributed in the space spanned by the three variables.

Exp no	x_1	x_2	x_3
1	−1	−1	−1
2	1	−1	−1
3	−1	1	−1
4	1	1	−1
5	−1	−1	1
6	1	−1	1
7	−1	1	1
8	1	1	1

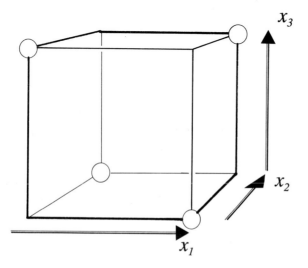

Fig. 6.1: Distribution of the experimental points in 2^{3-1} fractional factorial design.

It is seen that the experimental points describe a tetrahedron, which is the largest volume of the space which can be spanned by only four points. This shows an important property of fractional factorial designs: the experiments cover as large a part as possible of the space spanned by the variables. We cannot cover the whole space by a limited number of points, but a fractional factorial design selects experiments which are as widely distributed as possible and which hence do their best to cover a maximum variation over the experimental space.

Another example: We can study seven variables in a 2^{7-4} fractional factorial design. This design is defined from the model matrix of a 2^3 factorial design, see Fig. 6.2.

Design matrix 2^{7-4}

Exp no	I	x_1 1	x_2 2	x_3 3	x_4 12	x_5 13	x_6 23	x_7 123
1	1	−1	−1	−1	1	1	1	−1
2	1	1	−1	−1	−1	−1	1	1
3	1	−1	1	−1	−1	1	−1	1
4	1	1	1	−1	1	−1	−1	−1
5	1	−1	−1	1	1	−1	−1	1
6	1	1	−1	1	−1	1	−1	−1
7	1	−1	1	1	1	−1	1	−1
8	1	1	1	1	1	1	1	1

A complete seven-variable factorial design contains 128 runs. The design points define the "corners" of a hypercube in the seven-dimensional space spanned by the variable axes. The fractional design 2^{7-4} is 1/16 of the complete design and selects eight of the 128 cornes in such a way that they are as far apart from each other as possible and hence cover a maximum of the experimental space.

By the same principles the columns of a 2^4 factorial design can be used to define the variations of (up to) 15 variables in 16 runs, i.e. a 2^{15-11} fractional factorial which is a 1/2048 of a complete 2^{15} factorial design.

These designs can be used to fit linear models

$$y = \beta_0 + \Sigma\, \beta_i\, x_i + e$$

to the experimental data. The computations for obtaining the models are made analogously as was described for the factorial designs.

Of course, the information on each variable which can be obtained by a fractional factorial design is less than what can be obtained from a complete factorial design. There is a price to pay for being lazy, and the price is confounding.

3.   What is lost by using fractional factorial designs instead of complete factorial designs?

To illustrate what is meant by *confounding*, we shall first treat a simple case: three variables in a 2^{3-1} fractional factorial design. It is seen that over the set of experiments, each variable always varies in the same way as the product of the remaing two variables, and that the constant column $\mathbf{I}$ is the same as the three-variable product $x_1x_2x_3$.

	$x_1x_2x_3$	x_2x_3	x_1x_3	x_1x_2
Exp no	$\mathbf{I}$	x_1	x_2	x_3
1	1	-1	-1	1
2	1	1	-1	-1
3	1	-1	1	-1
4	1	1	1	1

If a complete factorial design with three variables was chosen, it would be possible to establish the following model:

$$y = \beta_0 + \beta_1 \, x_1 + \beta_2 \, x_2 + \beta_3 \, x_3 + \beta_{12} \, x_1x_2 + \beta_{13} \, x_1x_3 + \beta_{23} \, x_2x_3 + \beta_{123} \, x_1x_2x_3 + e$$

Now only four experiments have been run, and it is possible to estimate four parameters. In the fractional design the following relations apply:

$$x_1 = x_2x_3$$
$$x_2 = x_1x_3$$
$$x_3 = x_1x_2$$
$$\mathbf{I} = x_1x_2x_3$$

If we put these equalities into the model above, we obtain

$$y = \beta_0 + \beta_{123} + \beta_1 \, x_1 + \beta_2 \, x_2 + \beta_3 \, x_3 + \beta_{12} \, x_3 + \beta_{13} \, x_2 + \beta_{23} \, x_1 + e$$

This can be reduced to

$$y = (\beta_0 + \beta_{123}) + (\beta_1 + \beta_{23}) \, x_1 + (\beta_2 + \beta_{13}) \, x_2 + (\beta_3 + \beta_{12}) x_3 + e$$

which is a linear model with four parameters

$$y = b_0 + b_1 \, x_1 + b_2 \, x_2 + b_3 \, x_3 + e$$

where the parameters are linear combinations of the "true" effects

$$b_0 = \beta_0 + \beta_{123}$$
$$b_1 = \beta_1 + \beta_{23}$$
$$b_2 = \beta_2 + \beta_{13}$$
$$b_3 = \beta_3 + \beta_{12}$$

The parameter b_1 is an estimator of the "true" parameter β_1, but this estimate is contaminated by a contribution from the "true" two-variable interaction β_{23}. It is said that β_1 is *confounded* with β_{23}, and that b_1 is an *alias* of the confounded effects.

This is the price to pay when fractional designs are used. It is possible to obtain estimates of the effects but these estimates will be sums of the true effects. We shall see that this is not a serious drawback. It is possible to construct fractional designs so that main effects and two-variable interaction effects are confounded with higher order interaction effects. As higher order interaction effects quite safely can be assumed to be negligible, the fractional designs can be efficiently used to obtain good estimates of the "true" main effects and the "true" two-variable interaction effects.

3.1. Generator of a fractional factorial design

To analyze which effects will be confounded in a fractional design, we shall introduce a new concept, the *generator* of a fractional factorial design.[1] As an example to illustrate the principles, we shall use a fractional factorial design 2^{4-1}. To understand why the generators are practical to use we shall write down the complete variable matrix of a 2^4 factorial design.

I	1	2	3	4	12	13	14	23	24	34	123	124	134	234	1234
1	-1	-1	-1	-1	1	1	1	1	1	1	-1	-1	-1	-1	1
1	1	-1	-1	-1	-1	-1	-1	1	1	1	1	1	1	-1	-1
1	-1	1	-1	-1	-1	1	1	-1	-1	1	1	1	-1	1	-1
1	1	1	-1	-1	1	-1	-1	-1	-1	1	-1	-1	1	1	1
1	-1	-1	1	-1	1	-1	1	-1	1	-1	1	-1	1	1	-1
1	-1	1	1	-1	-1	-1	1	1	-1	-1	-1	1	1	-1	1
1	1	1	1	-1	1	1	-1	1	-1	-1	1	-1	-1	-1	-1
1	-1	-1	-1	1	1	1	-1	1	-1	-1	-1	1	1	1	-1
1	1	-1	-1	1	-1	-1	1	1	-1	-1	1	-1	-1	1	1
1	-1	1	-1	1	-1	1	-1	-1	1	-1	1	-1	1	-1	1
1	1	1	-1	1	1	-1	1	-1	1	-1	-1	1	-1	-1	-1
1	-1	-1	1	1	1	-1	-1	-1	-1	1	1	1	-1	-1	1
1	1	-1	1	1	-1	1	1	-1	-1	1	-1	-1	1	-1	-1
1	-1	1	1	1	-1	-1	-1	1	1	1	-1	-1	-1	1	-1
1	1	-1	1	-1	-1	1	-1	-1	1	-1	-1	1	-1	1	1
1	1	1	1	1	1	1	1	1	1	1	1	1	1	1	1

Among these experiments there are exactly eight in which variable **4** varies in the same way as the product between **1**, **2** and **3**, i.e. **4** = **123**. These experiments have been shadowed. In these experiments it is also found that **1** = **234**, **2** = **134**, **3** = **124**, and that **12** = **34**, **13** = **24**, **14** = **23**, and **I**= **1234**. It is rather cumbersome to identify such relations by inspection of the variable matrices from complete factorial designs, especially with large matrices as the one above. The eight experiments selected by fulfilling the relation **4** = **123** is one half-fraction of the complete factorial design. Eight experiments correspond to a complete factorial design 2^3. Hence, we can use the complete model matrix from the 2^3-design to define the variation of the *extra* variable **4**, and obtain the fractional factorial design 2^{4-1}. The half-fraction in which **4** = **123** is obtained by letting the column **123** in the model matrix of the 2^3-design define the variations of variable **4**.

							4
I	**1**	**2**	**3**	**12**	**13**	**23**	**123**
1	−1	−1	−1	1	1	1	−1
1	1	−1	−1	−1	−1	1	1
1	−1	1	−1	−1	1	−1	1
1	1	1	−1	1	−1	−1	−1
1	−1	−1	1	1	−1	−1	1
1	1	−1	1	−1	1	−1	−1
1	−1	1	1	−1	−1	1	−1
1	1	1	1	1	1	1	1

Confirm that these are the same experiments as were selected from the complete design.

The matrix above has rather peculiar mathematical properties. If the columns are multiplied by each other, the result will always be another column in the matrix.[1]

---

[1] Mathematically this means that the columns of a two-level fractional factorial design is a *group*. A group is a set of elements on which a composition rule is defined (in this case a multiplication) and the result obtained by joining the elements from the set always gives another element in the set. In a group there is also a *neutral element* which when joined to another element gives the same element as result. In this case column I is the neutral element.

Other examples of mathematical groups are the *integer numbers* on which multiplication and addidion is defined. Multiplication of two integer numbers will produce another integer number. The neutral element in multiplication is the number *(1)*. The same holds for addition, where zero *(0)* is the neutral element.

It is seen that

$$I \cdot 1 = 1 \quad \text{i.e.} \quad I1 = 1$$
$$I \cdot 2 = 2 \quad\quad\quad I2 = 2$$
$$I \cdot 3 = 3 \quad\quad\quad I3 = 3$$
$$I \cdot 12 = 12 \quad\quad I12 = 12$$

etc.

Multiplication by **I** does not change anything since we are multiplying the elements of the other colum by $(+1)$.

Multiplying any column by itself implies that (-1) is multiplied by (-1), and $(+1)$ is multiplied by $(+1)$. This will alawys give the column **I**.

$$1 \cdot 1 = 1^2 = I$$
$$2 \cdot 2 = 2^2 = I$$
$$3 \cdot 3 = 3^2 = I$$
$$12 \cdot 12 = 1^2 \cdot 2^2 = I$$

etc.

In the factorial design above, we have chosen to vary **4** as the product **123**. In doing so, we have deliberately confounded β_4 with β_{123}. However, other effects will also be confounded.

We have *generated* the fractional factorial design from the smaller complete factorial design by selecting **4 = 123**. The *generator* of the design is

$$4 = 123$$

The generator can be used to identify the confounding pattern. To this end, we perform a transformation as follows:
Multiply both sides of the relation **4 = 123** by **4**.

$$4 \cdot 4 = 123 \cdot 4 \qquad \text{which gives}$$

$$I = 1234$$

This important relation which presents the generator in a form which shows how *different* columns can be multiplied together to give the colum **I**. Hereafter, the generators of a fractional design will be given in this form. The generator can be used to identify the confounding patterns as follows:

The basis for the fractional design is a complete factorial design. If we multiply the column labels in the variable matrix of the fractional design by the generator, we can identify which multiplications of different column will produce the columns of factorial design matrix, i.e. multiplying $I = 1234$ by

I	1	2	3	12	13	23	123

gives

$I \cdot (I = 1234)$	i.e.	$I = 1234$
$1 \cdot (I = 1234)$		$1 = 234$
$2 \cdot (I = 1234)$		$2 = 134$
$3 \cdot (I = 1234)$		$3 = 124$
$12 \cdot (I = 1234)$		$12 = 34$
$13 \cdot (I = 1234)$		$13 = 24$
$23 \cdot (I = 1234)$		$23 = 14$
$123 \cdot (I = 1234)$		$123 = 4$

These results are more lucid when given at the column labels of the variable matrix

1234	234	134	124	34	24	14	4
I	1	2	3	12	13	23	123
1	−1	−1	−1	1	1	1	−1
1	1	−1	−1	−1	−1	1	1
1	−1	1	−1	−1	1	−1	1
1	1	1	−1	1	−1	−1	−1
1	−1	−1	1	1	−1	−1	1
1	1	−1	1	−1	1	−1	−1
1	−1	1	1	−1	−1	1	−1
1	1	1	1	1	1	1	1

When these columns are used to calculate the effects, the following results are obtained:

From column **I**   the estimate $\beta_0 + \beta_{1234}$
   1 $\beta_1 + \beta_{234}$
   2 $\beta_2 + \beta_{134}$
   3 $\beta_3 + \beta_{124}$
   12 $\beta_{12} + \beta_{34}$
   13 $\beta_{13} + \beta_{24}$
   23 $\beta_{23} + \beta_{14}$
   123 $\beta_{123} + \beta_4$

3.2. More on generators: Highly fractionated factorial designs

Suppose that we wish to screen six variables. For this, a complete factorial design will give 64 runs. To have a rough estimate whether or not the variables influence the result, a fractional factorial design with eight runs would be sufficient. To define such a design, the now well-known(?) three-variable matrix from a 2^3 full factorial design can be used. This matrix will contain the following columns

I   1   2   3   12   13   23   123

To define the "extra" variables, **4, 5** , and **6**, any three of the four interaction columns can be used.

If we choose to define these variables as

4 = 12   i.e. $4^2 = 124$
5 = 13 $5^2 = 135$
6 = 23 $6^2 = 236$

we obtain the following *independent generators*:

I = 124 = 135 = 236

It is also true that all multiplications of the independent generators will also give the column **I** as result and that

$$I = 124 \cdot 135 = 124 \cdot 236 = 135 \cdot 236 = 124 \cdot 135 \cdot 236$$

which can be simplified to

I = 2345 = 1346 = 1256 = 456

We must therefore append these *derived* generators to the independent generators to obtain the complete set of generators

I = 124 = 135 = 236 = 2345 = 1346 = 1256 = 456

This set of generators contains eight "words". The number of such "words" can be rationalized if we remember that the 2^{6-3} fractional design is $1/8$ of the complete design. Eight rows and 64 columns of the complete matrix have to be plied together to give an 8 x 8 matrix and as a result of this, eight of the 64 columns will coincide on each column in the 8 x 8 matrix.

From the set of generators, the following confounding pattern is derived:

I	1	2	3	12	13	23	123
124	24	14	1234	4	234	134	34
135	35	1235	15	235	5	125	25
236	1236	36	26	136	126	6	16
2345	12345	345	245	1345	1245	45	145
1346	346	12346	146	2346	46	1246	246
1256	256	156	12356	56	2356	1356	356
456	1456	2456	3456	12456	13456	23456	123456

With the assumptions that interaction effects between three or more variables are small compared to main effects and two-variable interaction effects, it is seen that the experiments above afford estimates of the following model parameters:

From	I	β_0
	1	$\beta_1 + \beta_{24} + \beta_{35}$
	2	$\beta_2 + \beta_{14} + \beta_{36}$
	3	$\beta_3 + \beta_{15} + \beta_{26}$
	12	$\beta_4 + \beta_{12} + \beta_{56}$
	13	$\beta_5 + \beta_{13} + \beta_{46}$
	23	$\beta_6 + \beta_{23} + \beta_{45}$
	123	$\beta_{34} + \beta_{25} + \beta_{16}$

We could have defined the variables differently. It is always the experimenter who decides which columns to use for defining the variation of the variables. We could have chosen, e.g. **4 = 123, 5 = 12** and **6 = 23**, which would have given the following set of generators

$$I = 1234 = 125 = 236 = 345 = 146 = 1356 = 2456$$

and hence, a confounding pattern different from that of the previous design would result.

3.3. Resolution of a fractional factorial design

Degree of resolution, or shortly, the *resolution* of a fractional factorial design is defined by the length of the shortest "word" in the set of generators.[1] The resolution is commonly specified by roman numeral characters.

In a design of

Resolution III,

> main effects are confounded with two-variable interaction effects.

Resolution IV,

> main effects are confounded with three-variable interaction effects, and two-variable interaction effects are confounded with each other.

Resolution V, main effects are confounded with four-variable interaction effects, two-variable interaction effects are confounded with three-variable interaction effects.

Fractional factorial designs with higher resolution than V are rarely used in sceening experiments.

4. Example: 2^{4-1} fractional factorial design. Synthesis of a semicarbazone.

A procedure for the synthesis of semicarbazone from phenylglyoxalic acid was studied.[2]

The reaction is one of the steps in a multi-step synthesis of azauracil, which is a anti-leucemic cytostatic agent. The experiments were run to determine suitable conditions for scale-up synthesis in the pilot plant.

Three responses were measured:

y_1: The yield (%) of crude semicarbazone, determined gravimetrically;
y_2: The purity of the crude product, determined by titration;
y_3: The ease of filtration, subjectively judged and assigned a value in the interval [−5 to 5]; −5 (difficult), 5 (easy).

The desired result was, of course, to establish experimental conditions for obtaining a high yield of pure semicarbazone which could be easily filtered.

The experimental variables and the domain explored are given in Table 6.1. The experimental design and the responses obtained are given in Table 6.2. The design was a fractional factorial design 2^{4-1} with the generator $I = 1234$.

Table 6.1: Variables and experimental domain in the synthesis of semicarbazone.

Variable	Experimental domain	
	−1	1
x_1: Time of addition of the glyoxalic acid (h)	1.0	2.0
x_2:Time of stirring after the addition of glyoxalic acid (h)	0.5	2.0
x_3:Reaction temperature (° C)	20	60
x_4:Amount of water added to the semicarbazone (ml/mol)	75	200

Table 6.2: Experimental design and responses in the synthesis of semicarbazone

Exp no	x_1	x_2	x_3	x_4	y_1	y_2	y_3
1	−1	−1	−1	−1	88.7	94.8	−5
2	1	−1	−1	1	88.8	95.7	0
3	−1	1	−1	1	88.2	96.7	−5
4	1	1	−1	−1	88.6	94.8	−5
5	−1	−1	1	1	86.1	98.2	5
6	1	−1	1	−1	89.4	97.9	5
7	−1	1	1	−1	86.5	97.8	5
8	1	1	1	1	88.6	98.3	5

By a visual inspection of the responses in Table 6.2 it is immediately seen (without any calculations) that the variations in yield, y_1, are not large. Therefore no large effects of the variables on this response can be expected. The variation of the purity, y_2, is more pronounced and there may be an influence of the variables on this response. It is also seen that y_2 and y_3 are associated and vary in the same direction. With pure crystals the filtration is easy, which is to be expected. There does not seem to be any obvious relation between the yield and the purity. A plot of y_2 against y_1 , Fig. 6.2, confirms that they are not correlated.

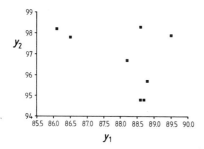

Fig 6.2: The yields of crude semicarbazone (y_1) and its purity (y_2) are not correlated.

To evaluate the influence of the variables, the complete variable matrix from a 2^3 design is used. It is seen that the columns of x_1, x_2 and x_3 have been varied according to a factorial design 2^3, and it is seen that $x_4 = x_1x_2x_3$, ($\mathbf{I} = \mathbf{1234}$). Multiplying the column labels of the complete variable matrix by the generator gives the confounding pattern.

1234	234	134	124	34	24	14	4			
I	1	2	3	12	13	23	123	y_1	y_2	y_3
1	−1	−1	−1	1	1	1	−1	88.7	94.8	−5
1	1	−1	−1	−1	−1	1	1	88.8	95.7	0
1	−1	1	−1	−1	1	−1	1	88.2	96.7	−5
1	1	1	−1	1	−1	−1	−1	88.6	94.8	−5
1	−1	−1	1	1	−1	−1	1	86.1	98.2	5
1	1	−1	1	−1	1	−1	−1	89.4	97.9	5
1	−1	1	1	−1	−1	1	−1	86.5	97.8	5
1	1	1	1	1	1	1	1	88.6	98.3	5

The responses must be evaluated one at a time. Assuming that interaction effects between three or more variables are small as compared to main effects and two-variable interaction effects, the following estimates are obtained:

Estimates	y_1 (Yield)	y_2 (Purity)	y_3 (Ease of filtration)
b_0	88.11	96.78	0.6
b_1	0.74	−0.10	0.6
b_2	−0.14	0.13	−0.6
b_3	−0.46	1.28	4.4
b_4	−0.10	0.45	0.6
$b_{12} + b_{34}$	−0.11	−0.25	−0.6
$b_{13} + b_{24}$	0.61	0.15	−0.6
$b_{14} + b_{23}$	0.04	−0.13	0.6

For the responses y_2 and y_3 the interpretation is clear. The reaction temperature, x_3, is the only variable which has an influence. The effect is positive, which means that an increase in temperature will increase the purity as well as the ease of filtration. The reaction should be run with x_3 at its high level (60 ° C). It is possible that an improvement can be obtained if the temperature is further increased.

For the response y_1, the interpretation of the result is not evident by mere regarding the estimated effects. They are all rather small. Two main effects, b_1 and b_3, are slightly larger than the others. This might indicate that x_1 and x_3 have a small influence on the yield. Among the confounded interaction effects, ($b_{13} + b_{24}$) is also slighly larger than the other confounded effects. It is not possible to determine which of these confounded effects contributes to the slightly higher value of the estimate. A guess is that b_{13} is the dominating term, since x_1 and x_3 are suspected to have a small linear influence, while x_2 and x_4 have not. However, b_3 is negative which might indicate that an increase in temperature lowers the yield. To see which role the suspected interaction effect may play, a projection of the response surface over the plane spanned by x_1 and x_3 can be made from the experimental design and the responses obtained. The projection is shown in Fig. 6.3.

The projection in Fig. 6.3 has been obtained as follows. The experimental domain is marked as a square in the plane spanned by x_1 and x_3. The average response, 88.11, is given at the origin. The corner in the lower left quadrant corresponds to the variable settings $x_1 = -1$, and $x_3 = -1$. There are exactly two experiments in the design which have been run with these seetings of x_1 and x_3, i.e. experiments number 1 and 3. The responses in these experiments are 88.7 and 88.2. The average of these responses is 88.45. This value is written in the lower left corner. The same procedure is then used for the other combinations of the variable settings $(x_1, x_3) = (+1, -1)$;

$(-1, +1)$; $(+1, +1)$. It is seen in Fig. 6.3 that the yield drops slightly in the upper left quadrant, but it is fairly constant in the other parts of the domain. The conclusions that might be drawn from this is that the reaction temperature, x_3, can be increased without a drop in yield, if the time of addition of glyoxalic acid, x_1, is concomitantly increased.

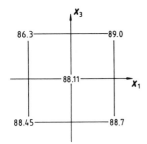

Fig.6.3:   A two-dimensional projection showing the variation of yield of semicarbazone when the time of addition of glyoxalic acid (x_1) and the reaction temperature (x_3) are varied.

Analysis of variance is used to see if the assumption that x_1 and x_3 have a significant influence on the yield while x_2 and x_4 are insignificant, can be supported from the experimental data. If x_2 and x_4 are insignificant, their settings over the series of experiments should not have any influence on the variation of y_1. The observed variation for different settings of these variables should therefore be nothing but experimental error noise. If x_1 and x_3 are significant and a model is fitted to the data to model y_1 as a function of only x_1 and x_3, the sum of squares due to regression should be significantly greater than the residual sum of squares. This model would be

$$y_1 = 88.11 + 0.74 \, x_1 - 0.46 \, x_3 + 0.61 \, x_1x_3 + e$$

Since fractional factorial design give *independent* estimates of the parameters, the value of the coefficients will not change if terms are removed from the model. The ANOVA is summarized in Table 6.3. It is, however, not possible to use the critical *F*-ratios given in at the end of this book for a significance test in this case. The degrees of freedom of the mean squares are *not* the true degrees of freedom of independent estimates. We have already used all data once to fit a model with all variables included. From this model we have selected those variables which have the largest coefficients. Using the tabulated critical *F* ratios for $n - p$, and p degrees of freedom in a significance test of the reduced model involves the assumption that it is the first tested model, and this is not true. We can, however, use the F ratios in the ANOVA table as hints as to the significance of the variables. For a discussion on regression analysis with several tentative models, see the book by Draper and Smith.[3]

The ANOVA table indicates that the variation of x_1 and x_3, produces a systematic variation of y_1, and that this variation is probably significantly above the noise level given by the residual sum of squares.. A three-dimensional plot of the response surface is shown in Fig. 6.4. The twist of the plane caused by the interaction is clearly seen.

Table 6.3: Analysis of variance in the synthesis of semicarbazone

Source	SS	df	MS	F
Total	62,120.11	8		
Regression				
due to b_0	62,119.56	1		
to b_1, b_3, b_{13}	9.06	3	3.02	20.95
to b_1	4.35	1	4.35	30.17
to b_3	1.71	1	1.71	11.87
to b_{13}	3.00	1	3.00	20.81
Residual	0.5467	4	0.1442	

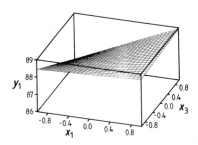

Fig. 6.4: Three-dimensional plot of the response surface of the yield of semicarbazone.

From the above analysis, the following conclusions can be drawn as to which experimental conditions should be used in the pilot plant experiment.

x_1: The time of addition of glyoxalic acid should be at its high level. Due to the interaction effect it could be slightly increased to avoid a drop in yield if the temperature is raised.

x_2: The time of reaction after addition of glyoxalic acid is not critical. A short reaction time is convenient and x_2 could be kept at its low level.

x_3: The reaction temperature is important for all three responses. A slight increase of the temperature from its high level mights further improve the result.

x_4: The amount of added water is not critical. A small amount of water increases the capacity of the reactor. This may be of economic importance if the procedure is to be used in large production scale.

The experimental results obtained in the pilot plant confirmed the above conclusions.

The above example is a nice illustration that substantial amounts of information can be obtained from few experiments with a proper experimental design.

5.   How to separate confounded effects?

5.1.   Separation of specific effects

Let us start with a simple case. We shall use the design above as an example. It was suspected that the alias $(b_{13} + b_{24})$ was important. It is not possible to know if either or both of the terms are important. Everywhere in the design above, the variation of **13** is always the same as the variation of **24**. To be able to separate the confounded effects, we must therefore run experiments in which the variation of **13** is *different* from the variation of **24**. This can be achieved in two experimental runs for which the following relation applies

13	24
1	−1
1	1

There are many possible settings for which the above relation applies. One example of such variable settings would be as shown on the next page.

1	2	3	4	13	24	
1	1	1	-1	1	-1	*New experiment*
1	1	1	1	1	1	*Exp. no 8*

The experiment with all variables at the (+1)-level has already been run, *Exp. no 8*. It will be necessary to run one complementary experiment. These two experiments define the smallest possible fractional factorial design, 2^{4-3}. It is seen that the following confounding pattern is obtained

13	
3	
2	
1	4
I	24
1	-1
1	1

From this small design the following estimates can be obtained

$$b_0 + b_1 + b_2 + b_3 + b_{13} \text{ and}$$

$$b_{24} + b_4$$

As estimates of b_0, b_1, b_2, and b_4 are already known from the previous design, the two interaction effects can now be estimated.

This illustrates a great advantage of fractional factorial designs, namely, that confounded effects can be separated by doing *complementary runs*. For more details, see Box, Hunter and Hunter.[4]

5.2. Complementary fractions to separate confounded effects

To illustrate the principles, a 2^{5-2} fractional factorial design is discussed. The complete variable matrix from a 2^3 design is used as a base. Assume that the "extra" variables **4** and **5** have been defined as **4** = **12** and **5** = **13** which give the independent generators

$$I = 124 = 135$$

The design is a Resolution III design and main effects are confounded with two-variable interaction effects. The design is one quarter of a full, 2^5, factorial design.

Another quarter is obtained by switching the signs of column $4 = -12$. Yet another quarter of the full design is obtained by switching the signs of column $5 = -13$. The remaining quarter is obtained by switching the signs of both columns, $4 = -12$ and $5 = -13$. These different ways of defining the "extra" variables correspond to different sets of generators:

Design A: $I = 124 = 135 = 2345$
Design B: $I = -124 = 135 = -2345$
Design C: $I = 124 = -135 = -2345$
Design D: $I = -124 = -135 = 2345$

The first two are the independent generators, and ± 2345 has been obtained by multiplying the independent generators. As the sets of generators are different, the confounding patterns will be different. For clarity, only main effect and two−variable interaction effects are shown.

Design A:	Design B:	Design C:	Design D:
β_0	β_0	β_0	β_0
$\beta_1 + \beta_{24} + \beta_{35}$	$\beta_1 - \beta_{12} + \beta_{35}$	$\beta_1 + \beta_{24} - \beta_{35}$	$\beta_1 - \beta_{24} - \beta_{35}$
$\beta_2 + \beta_{14}$	$\beta_2 + \beta_{13}$	$\beta_2 + \beta_{14}$	$\beta_2 - \beta_{14}$
$\beta_3 + \beta_{15}$	$\beta_3 + \beta_{15}$	$\beta_3 - \beta_{15}$	$\beta_3 - \beta_{15}$
$\beta_4 + \beta_{12}$	$\beta_4 - \beta_{12}$	$\beta_4 + \beta_{12}$	$\beta_4 - \beta_{12}$
$\beta_5 + \beta_{13}$	$\beta_5 + \beta_{13}$	$\beta_5 - \beta_{13}$	$\beta_5 - \beta_{13}$
$\beta_{23} + \beta_{45}$	$\beta_{23} - \beta_{45}$	$\beta_{23} - \beta_{45}$	$\beta_{23} + \beta_{45}$
$\beta_{34} + \beta_{25}$	$\beta_{34} - \beta_{25}$	$\beta_{25} - \beta_{34}$	$-\beta_{34} - \beta_{25}$

Assume that a series of experiments has been run according to fraction A. We can then run a second series of experiments as specified by another fraction. It is seen that fraction D is complementary to fraction A in such a way that all main effect can be separated from confoundings with two-variable interaction effect. The two-variable intraction effects will, however be confounded with each other.

When a fractional factorial design has been employed, there is always a possibility to run complementary fractions to resolve confounded effects.

5.3. Fold-over design to resolve main effects from confoundings with two-variable interactions

This section presents a technique, by which a complemetary fraction can be found which can resolve all main effects from confoundings with two-variable interaction effects.

Assume that a 2^{7-4} fractional factorial screening experiment has been run with seven variables, and where the "extra" variables have been defined as **4 = 12**, **5 = 13**, **6 = 23**, and **7 = 123**. The independent generators will thus be

$$I = 124 = 135 = 236 = 1237$$

By making all possible multiplications of these, the following rather large complete set of generators is obtained

$$I = 124 = 135 = 236 = 1237 = 2345 = 1346 = 347 = 1256 = 257 = 167 =$$
$$= 2467 = 456 = 1457 = 3567 = 1234567$$

Each estimate from this design will be an alias of 16 confounded effects. If we consider only main effects and two-variable interaction effects and ignore all higher interactions, the following estimates can be obtained:

β_0

$\beta_1 + \beta_{24} + \beta_{35} + \beta_{67}$

$\beta_2 + \beta_{14} + \beta_{36} + \beta_{57}$

$\beta_3 + \beta_{15} + \beta_{26} + \beta_{47}$

$\beta_4 + \beta_{12} + \beta_{37} + \beta_{56}$

$\beta_5 + \beta_{13} + \beta_{27} + \beta_{46}$

$\beta_6 + \beta_{23} + \beta_{17} + \beta_{45}$

$\beta_7 + \beta_{34} + \beta_{25} + \beta_{13}$

All main effects are confounded with three two-variable interaction effects. Any significant estimate can be due either to the main effect or to any of the confounded interaction effects. Therefore, it can be dangerous to draw far-reaching conclusions as to the significance of the *variables* from such experiments. It would be better if the main effects could be separated from confounding with the two-variable interactions. There is a simple trick by which this can be accomplished, namely to run a complementary fraction which is obtained by *fold-over*. This fraction is obtained by switching the signs of all experimental settings of the variables in the first design. The result of this is more easily surveyed if it is done in the following way.

Mirror the first design in its bottom line. Switch the signs of the last row and write it as the first row in the fold-over design. Do the same with the last but one row and write it as the second row in the fold-over design etc. This gives the result shown on the next page.

In the fold-over design, $4 = -12$, $5 = -13$, $6 = -23$, and $4 = 123$, which corresponds to the following independent generators

$$I = -124 = -135 = -236 = 1237$$

As these generators are different from the generators of the parent design, they will give a different confounding pattern. The following estimates can be obtained.

β_0

$\beta_1 - \beta_{24} - \beta_{35} - \beta_{67}$

$\beta_2 - \beta_{14} - \beta_{36} - \beta_{57}$

$\beta_3 - \beta_{15} - \beta_{26} - \beta_{47}$

$\beta_4 - \beta_{12} - \beta_{37} - \beta_{56}$

$\beta_5 - \beta_{13} - \beta_{27} - \beta_{46}$

$\beta_6 - \beta_{23} - \beta_{17} - \beta_{46}$

$\beta_7 - \beta_{34} - \beta_{25} - \beta_{13}$

First design

| | | | 4 | 5 | 6 | 7 |
1	2	3	12	13	23	123
-1	-1	-1	1	1	1	-1
1	-1	-1	-1	-1	1	1
-1	1	-1	-1	1	-1	1
1	1	-1	1	-1	-1	-1
-1	-1	1	1	-1	-1	1
1	-1	1	-1	1	-1	-1
-1	1	1	-1	-1	1	-1
1	1	1	1	1	1	1

Fold-over

| | | | 4 | 5 | 5 | 7 |
1	2	3	-12	-13	-23	123
-1	-1	-1	-1	-1	-1	-1
-1	1	-1	1	-1	1	1
1	1	-1	-1	1	1	-1
-1	-1	1	-1	1	1	1
1	-1	1	1	-1	1	-1
-1	1	1	1	1	-1	-1
1	1	1	-1	-1	-1	1

When the estimates from these two fractions are taken together, it is seen that the average sums of the estimates from corresponding columns will be the main effects, free from confounding with the two-variable interactions. The average differences will be estimated aliases of the confounded two-variable interactions. It is seen that in each alias, the same variable is encountered only once. This sometimes offers an opportunity to make an *educated guess* of which term may dominate in the aliases. Significant interaction effects are generally found for variables which also have significant main effects, (see the example with the semicarbazone synthesis).

5.4. Example: Fold-over design. Rearrangement of n-pentane in superacid medium

The example is taken from a study of the reactions which occur when *n*-pentane is treated with superacids, $R_F\text{-}SO_3H + SbF_5$. [5]

The initiating reaction is a hydride transfer from n-pentane to form a carbocation. This ion then initiates a series of events: It rearranges to isopentyl carbocation, which is then attacking other hydrocarbon molecules to produce a number of both cracked and polymeric products.

The original paper [5] reports several responses. To show the principles of a fold-over design, only one response, the reported initial rate of formation of isopentane, $k^{Exp.}$, is shown here. The experimental variables are summarized in Table 6.4.

Table 6.4: Variables and experimental domain in the rearrangement of pentane

Variables		Domain	
		-1	1
x_1:	Ratio of SbF_5/R_FSO_3H (mol/mol)	0.20	1.50
x_2:	Ratio of R_FSO_3H/n-pentane (mol/mol)	0.015	0.030
x_3:	The total pressure in the reactor (bar)	5	10
x_4:	The hydrogen pressure (bar)	1	4
x_5:	The reaction temperature (° C)	25	30
x_6:	The stirring rate (rpm)	350	500
x_7:	The type of sulfonic acid	CF_3SO_3H	$C_4F_9SO_3H$

The experiments are summarized in Table 6.5. Experiments number $1 - 8$ form the first design, and experiments number $9 - 16$ form the fold-over design.

The following estimates are obtained from Experiments $1 - 8$:

$$b_0 = 0.216$$
$$b_1 + (b_{24} + b_{36} + b_{57}) = 0.187$$
$$b_2 + (b_{14} + b_{35} + b_{67}) = 0.044$$
$$b_3 + (b_{25} + b_{16} + b_{47}) = 0.041$$
$$b_4 + (b_{12} + b_{56} + b_{34}) = 0.030$$
$$b_5 + (b_{23} + b_{46} + b_{17}) = 0.108$$
$$b_6 + (b_{13} + b_{45} + b_{27}) = 0.039$$
$$b_7 + (b_{34} + b_{15} + b_{26}) = 0.099$$

and from Experiment $9 - 16$

$$b_0 = 0.166$$
$$b_1 - (b_{24} + b_{36} + b_{57}) = 0.125$$
$$b_2 - (b_{14} + b_{35} + b_{57}) = 0.033$$
$$b_3 - (b_{25} + b_{16} + b_{47}) = 0.002$$
$$b_4 - (b_{12} + b_{56} + b_{34}) = 0.040$$
$$b_5 - (b_{23} + b_{46} + b_{17}) = 0.024$$
$$b_6 - (b_{13} + b_{45} + b_{27}) = 0.002$$
$$b_7 - (b_{34} + b_{15} + b_{26}) = 0.032$$

From these results, estimates of the main effect and aliases of confounded two-variable interaction effects can be computed. These values are shown on the next page.

Table 6.5: Experimental design and observed initial rate, k^{Exp}, in the rearrangement of pentane

Exp no	x_1	x_2	x_3	x_4	x_5	x_6	x_7	k^{Exp} /(h^{-1})
1	−1	−1	−1	1	1	1	−1	0.023
2	1	−1	−1	−1	1	−1	1	0.455
3	−1	1	−1	−1	−1	1	1	0.032
4	1	1	−1	1	−1	−1	−1	0.190
5	−1	−1	1	1	−1	−1	1	0.009
6	1	−1	1	−1	−1	1	−1	0.203
7	−1	1	1	−1	1	−1	−1	0.053
8	1	1	1	1	1	1	1	0.763
9	−1	−1	−1	−1	−1	−1	−1	0.039
10	1	−1	−1	1	−1	1	1	0.270
11	−1	1	−1	1	1	−1	1	0.041
12	1	1	−1	−1	1	1	−1	0.304
13	−1	−1	1	−1	1	1	1	0.055
14	1	−1	1	1	1	−1	−1	0.165
15	−1	1	1	1	−1	1	−1	0.026
16	1	1	1	−1	−1	−1	1	0.424

$$b_0 = 0.19 \qquad Block\ difference = 0.050$$
$$b_1 = 0.156 \qquad b_{24} + b_{36} + b_{57} = 0.031$$
$$b_2 = 0.039 \qquad b_{14} + b_{35} + b_{67} = 0.006$$
$$b_3 = 0.022 \qquad b_{25} + b_{16} + b_{47} = 0.020$$
$$b_4 = -0.005 \qquad b_{12} + b_{56} + b_{34} = -0.005$$
$$b_5 = 0.042 \qquad b_{23} + b_{46} + b_{17} = 0.042$$
$$b_6 = 0.019 \qquad b_{13} + b_{45} + b_{27} = 0.019$$
$$b_7 = 0.066 \qquad b_{34} + b_{15} + b_{26} = 0.034$$

The estimates of the average b_0 differ between the two fractions. The difference, 0.050, is a measure of systematic variation of the average between the two sets of experiments. The two fractions were run on two different occasions, and obviously it was not possible to create identical conditions for the two series. Such effects occur rather frequently, (see the section: *Running experiments in blocks*, at the end of this chapter). The conclusions drawn from this experiment were that variable x_1, and to some extent, variable x_7 influence the rate. The alias of confounded two-variable interactions containing b_{17} is the largest. It vas therefore suspected that b_{17} dominates in this alias.

6. Normal probability plots to discern significant effects

6.1. The problem

An experimental design defines the settings of the experimental variables over a series of experiments. The objective is to determine the systematic variation in the presence of a random experimental error. With a good design, it is possible to estimate the influence of each variable on the systematic variation. However, these estimates will have a probability distribution due to the experimental error. It was discussed in Chapter 3 how significant variables can be identified by a t-test in which the estimated effects are compared to their standard error. It is also possible to use Analysis of variance and assess the significance of the regression by the F distribution. These methods call for an estimate of the experimental error variance. If a factorial design is used to fit a model with linear and cross-product terms, there will be degrees of freedom left from the omitted higher order interaction effects. In a screening experiment, these higher order interaction effects can be assumed to be negligible. An estimate of the error variance can then be obtained from the residual sum of squares.

However, in many cases a *useful* estimate of the experimental error is not available when a screening experiment is run. To be useful, an estimate of the experimental error must have sufficient degrees of freedom to permit clear conclusions. Personally, I am not prepared to make a series of replicate runs of a new synthetic procedure, *until* I know for certain that the procedure gives interesting results. To realize whether or not the reaction will do so it is often necessary to check the reaction under varying experimental conditions. An early step in the exploration of a new reaction is therefore to run a highly fractionated screening experiment which can accomodate all potentially important variables. This ensures a large variation of the experimental conditions, and it can rapidly be established whether the reaction seems promising. By such an experiment, it will also be possible to see which experimental variables are important. However, no useful estimate of the experimental error can be obtained from highly fractionated screening experiments. How should the significant effects be assessed in such cases?

There is a method by which this can be achieved without prior knowledge of the experimental error variance, viz. through using *normal probability plots*. This very

150

useful technique was introduced as an aid to experimental design evaluation by C. Daniel.[6] To understand the technique, a short background is needed.

6.2. Normal probability plots

If we have a set of normally distributed measured data, $[y_1, y_2, ..., y_n]$, their frequency is described by the bell-shaped normal distribution, Fig. 6.5.

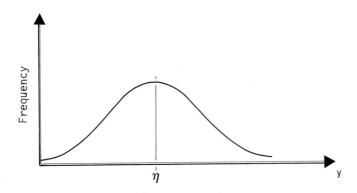

Fig.6.5: Normal probability frequency distribution.

The surface under the curve is normalized to unity

$$\int_{-\infty}^{\infty} \frac{1}{\sigma\sqrt{2\pi}}\ e^{\frac{(y-\mu)^2}{2\sigma^2}}\ dy = 1 = 100\ \%$$

The probability, $P(y \le a)$, that a measured value randomly drawn from the set is less or equal to a given value a, is proportional to the surface under the curve in the interval $[-\infty, a]$, see Fig. 6.6.

$$P(y \le a) = \int_{-\infty}^{a} \frac{1}{\sigma\sqrt{2\pi}}\ e^{\frac{(y-\mu)^2}{2\sigma^2}}\ dy$$

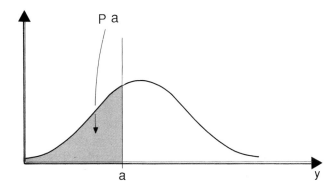

Fig.6.6: The probability $P(y \leq a)$ is proportional to the surface under the frequency distribution.

If $P(y)$ is plotted against y, a sigmoid curve is obtained, Fig. 6.7a.

The relation shown in Fig. 6.7a is called the *cumulative probability distribution*. It is now possible to adjust the graduation of the ordinate $(P(y))$ axis in such a way that the S-shaped curve is straightened out that the relation can be described by a straight line. Graph paper with such graduation, *normal probability paper*, is commercially available.

A plot of $P(y)$ against y on normal probability paper will give a result as shown in Fig. 6.7b.

Straight lines have two advantages: *(1)* Data which fit to a straight line are easily distinguished in a plot. *(2)* The line is easily drawn by using a ruler.

152

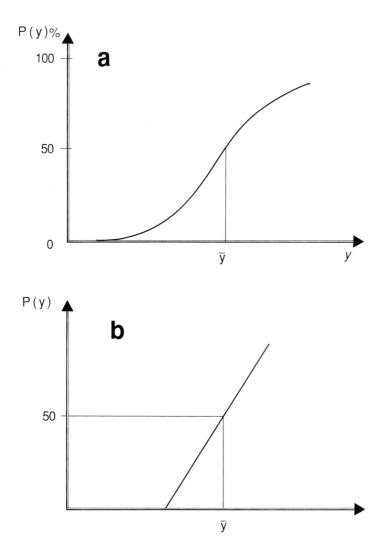

Fig.6.7:   Cumulative normal probability distributions. *(a)* Plot on linear graph paper; *(b)* Plot on normal probability paper.

If we have a set of data which are supposed to be normally distributed, we can construct cumulative probability distribution and plot this on normal probabilty paper. The result may be as in Fig. 6.8.

The majority of the data fit well to the line, but there are three aberrant points. Two are too large and one is too small to fit on the line. The conclusion which can be drawn from this is, that the aberrant data do not belong to the same distribution as the remainder of the data. *Nota bene*, the outliers should appear as they do in the plot above. Data too large to fit should appear to the right of the line in the upper right-hand part of the plot, and data too small to fit should appear to the left of the line in the lower left-hand part of the plot.

It can sometimes be that a straight line cannot be fitted to the data, or that the aberrant data appear differently. Such cases indicatate that the data are not normally distributed.

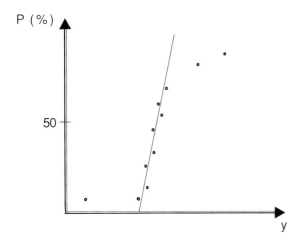

Fig. 6.8: Plot of a data on normal probability paper.

6.3. Significant effects in screening experiments

Assume that two variables, x_1 and x_2, have been studied in a 2^2 factorial design. Assume also that the "true" response is constant, $y = a$, in the experimental domain, i.e. the response surface is completely flat and the variables do not influence the response at all. In the presence of a random experimental error, the observed responses may appear to be different due to the presence of the error term in the response surface model.

$$y = B_0 + B_1 x_1 + B_2 x_2 + B_{12} x_1 x_2 + e$$

If the parameters of the model are computed using the model matrix and the observed responses, the following estimates are obtained.

I	x_1	x_2	$x_1 x_2$	y
1	−1	−1	1	$a + e_1$
1	1	−1	−1	$a + e_2$
1	−1	1	−1	$a + e_3$
1	1	1	1	$a + e_4$

$$\begin{aligned}
b_0 &= [(a + e_1) + (a + e_2) + (a + e_3) + (a + e_4)] / 4 = \\
&= a + (e_1 + e_2 + e_3 + e_4) / 4 \\
b_1 &= [-(a + e_1) + (a + e_2) - (a + e_3) + (a + e_4)] / 4 \\
&= (-e_1 + e_2 - e_3 + e_4) / 4 \\
b_2 &= [-(a + e_1) - (a + e_2) + (a + e_3) + (a + e_4)] / 4 = \\
&= (-e_1 - e_2 + e_3 + e_4) / 4 \\
b_{12} &= [(a + e_1) - (a + e_2) - (a + e_3) + (a + e_4)] / 4 = \\
&= (e_1 - e_2 - e_3 + e_4) / 4
\end{aligned}$$

Factorial and fractional factorial designs are balanced; each column contains an equal number of (−) and (+) settings of the variables. Due to this, a constant response will be cancelled out when the effects are computed. The estimated effects in such cases will be nothing but different averaged summations of the experimental error. If we assume that the experimental error is normally distributed, the averaged sums will also be normally distributed.

Let us now have a look at general screening experiments with many variables. Assume that k variables $(x_1, x_2,..., x_k)$ have been studied by a fractional factorial design and that a response surface model with linear and cross-product interaction terms has been determined.

$$y = B + \Sigma B_i x_i + \Sigma \Sigma B_{ij} x_i x_j + e$$

If the variables do not have any influence whatsoever on the response, the response surface is completely flat and the *true* value of all β_i:s and β_{ij}:s is *zero*. The estimated effects in such cases would be nothing but different average summations of the experimental error. If we have randomized the order of execution of the experiments in the design, and have done all what we can do to avoid systematic errors, the set of estimated model parameters, $[b_1, b_2,...b_k, b_{12},...,b_{ij}]$, would be a *random sample*, drawn from a population of normally distributed random experimental errors. Since a random sample is representative for the population from which it has been drawn, we can expect that also the *random sample* will be approximately normally distributed. Hence, we can make a normal probability plot of the estimated effects. Those estimated effects which fall on the straight line of the normal distribution, can reasonably be assumed to be nothing but various summations of error terms, and hence, they do not represent any significant effects of the corresponding variables. On the other hand, effects which deviate from the straight line in such a way that they are either too large to fit (upper right) or too small to fit (lower left) can be assumed to represent real effects.

How such plots are constructed is shown in the example below.

6.4. Screening of variables in a Willgerodt-Kindler synthesis [7]

The rather peculiar oxidation/reduction/rearrangement of aryl alkyl ketones into ω-arylthiocarboxamides which occurs when the ketone is heated in the presence of elemental sulfur and amines, is known as the Willgerodt-Kindler reaction.[8] The reaction has been described in almost 2000 published papers. Despite the large effort spent to study the reaction, a generally accepted reaction mechanism has not yet been found. The reaction is mentioned in almost every text-book on organic chemistry, sometimes as an embarrasing foot-note. The reaction has a reputation for being rather useless for preparative purposes due to the poor to moderate yields ($< 70\%$) usually obtained from the foul-smelling, black, and messy reaction mixtures.

As a reaction mechanism is not known, it is not possible to deduce how the experimental conditions should be controlled to maximize the yield.

The following reaction was studied

Evaluation of the massive amount of literature data available, indicated that the variables shown in Table 6.6 could influence the yield.

Table 6.6: Variables and experimental domain in the Willgerodt-Kindler synthesis

Variables	Domain	
	−1	1
x_1: The amount of sulfur/ketone (mol/mol)	5	11
x_2: The amount of morpholine/ketone (mol/mol)	6	10
x_3: The reaction temperature (° C)	100	140
x_4: The particle size of sulfur (mesh)	240	120
x_5: The stirring rate (rpm)	300	700

The reaction occurrs in a heterogeneous system in which flowers of sulfur is suspended. The variables x_4 and x_5 were included to take the heterogeneity of the system into acount.

As the reaction mechanism is not known, it is not possible to exclude any two-variable interaction effect from consideration. A complete second order interaction response surface model was therefore attempted:

$$y = \beta_0 + \Sigma \beta_i x_i + \Sigma\Sigma \beta_{ij} x_i x_j + e$$

A 2^{5-1} fractional factorial design ($\mathbf{I} = \mathbf{12345}$) was used to estimate the model parameters. The design has a Resolution V and the desired parameters can be estimated free from confoundings with each other. The design matrix and the yields (%), y, obtained are given in Table 6.7.

The yields vary from 11.8 to 91.4 %, which is outside the range of error variation. It is therefore evident that there are significant effects. The estimated effects are

$$b_0 = \beta_0 + \beta_{12345} = 64.48$$
$$b_1 = \beta_1 + \beta_{2345} = 9.09$$
$$b_2 = \beta_2 + \beta_{1245} = 6.56$$
$$b_3 = \beta_3 + \beta_{1245} = 18.62$$
$$b_4 = \beta_4 + \beta_{1235} = -1.17$$
$$b_5 = \beta_5 + \beta_{1234} = -2.87$$

$$b_{12} = \beta_{12} + \beta_{235} = -2.43$$
$$b_{13} = \beta_{13} + \beta_{245} = -6.43$$
$$b_{14} = \beta_{14} + \beta_{235} = 0.84$$
$$b_{15} = \beta_{15} + \beta_{234} = 0.87$$
$$b_{23} = \beta_{23} + \beta_{145} = -7.93$$
$$b_{24} = \beta_{24} + \beta_{135} = -1.01$$
$$b_{25} = \beta_{25} + \beta_{124} = -0.84$$
$$b_{34} = \beta_{34} + \beta_{125} = 1.99$$
$$b_{35} = \beta_{35} + \beta_{124} = 0.92$$
$$b_{45} = \beta_{45} + \beta_{123} = -0.49$$

Table 6.7: Experimental design and yields obtained in the Willgerodt-Kindler synthesis

Exp no	x_1	x_2	x_3	x_4	x_5	y	Run order
1	−1	−1	−1	−1	1	11.5	9
2	1	−1	−1	−1	−1	55.8	1
3	−1	1	−1	−1	−1	55.7	2
4	1	1	−1	−1	1	75.1	3
5	−1	−1	1	−1	−1	78.1	10
6	1	−1	1	−1	1	88.9	11
7	−1	1	1	−1	1	77.6	4
8	1	1	1	−1	−1	84.5	12
9	−1	−1	−1	1	−1	16.5	14
10	1	−1	−1	1	1	43.7	13
11	−1	1	−1	1	1	38.0	5
12	1	1	−1	1	−1	72.6	6
13	−1	−1	1	1	1	79.5	16
14	1	−1	1	1	−1	91.4	7
15	−1	1	1	1	−1	86.2	15
16	1	1	1	1	1	78.6	8

If it is assumed that interaction effects between three or more variables can be neglected, it is seen that the estimated effect can be used as estimators of the model parameters. The 16 model parameters have been estimated from 16 runs. There are no degrees of freedom left to give an estimate of the residual variance which might have been used as an estimate of the error variance. Neither was any previous error estimate available. The significance of the estimated parameters must be assessed from a Normal probability plot.

To construct such a plot, the following steps are taken:

(1): Arrange the estimated effects in the order

smallest < next to smallest < ... < next to largest < largest.

Do not include the average response (b_0) in this set. This parameter cannot be a sum of error terms.

(2): Count the number, r, of estimated effects. In this case $r = 15$.

(3): Divide the interval [0 – 100 %] on the P-axis into r equally large intervals. Each interval will be 100 $/r$. In the present case

100 $/$ 15 = 6.666... = 6 ≈ 6.67

(4): Plot the mid-point of the first interval $[0 - 6.67] \approx 3.3$ % against the smallest effect. Then plot the mid-point in the second interval $[6.67 - 13.3] \approx 10.0$ % against the second next to smallest effect, etc.

The data to be plotted are summarized in Table 6.8, and the corresponding plot is shown in Fig. 6.9.

If a normal probability plot is to show the distribution of the experimental error, it should pass the point, (0, 50 %) The expected mean of the experimental error is zero. It should thus be possible to draw the error distribution line by pivoting a straight line in the point (0, 50 %) to obtain a maximum fit.

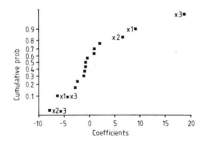

Fig.6.9: Normal probability plot of estimated coefficient in the Willgerodt-Kindler screening.

Three linear coefficients, b_1, b_2, b_3, and two cross-product coefficients, b_{13} and b_{23}, deviate from the normal probability line in Fig. 6.9. The line passes the point (0, 50 %) which supports the assumption that it represents the error distribution. All coefficients associated with variables x_4 and x_5 are projected on the line. It can therefore be concluded that the range of variation of these variables in the experiments do not have any influence on the yield. To further optimize the yield it will only be necessary to consider the variables x_1, x_2, and x_3. The screening experiment has therefore sorted out these variables as being the significant ones.

Table 6.8: Ranking order of estimates and the corresponding probability coordinate, P

Ranking order	Estimate	Parameter	P /(%)
1	−7.93	b_{23}	3.3
2	−6.34	b_{13}	10.0
3	−2.87	b_5	16.7
4	−2.43	b_{12}	23.3
5	−1.17	b_4	30.0
6	−1.01	b_{24}	36.7
7	−0.84	b_{25}	43.3
8	−0.49	b_{45}	50.0
9	0.84	b_{14}	56.7
10	0.87	b_{15}	63.3
11	0.92	b_{35}	70.0
12	1.99	b_{34}	76.7
13	6.56	b_2	83.3
14	9.09	b_1	90.0
15	18.62	b_3	96.7

To check these conclusions, the next step is to adopt a response surface model which contains only the significant coefficients and to see if such a model can give a satisfactory description of the observed response. This model is

$$y = 64.48 + 9.09\,x_1 + 6.56\,x_2 + 18.62\,x_3 - 6.34\,x_1x_3 - 7.93\,x_2x_3 + e$$

The analysis of variance is given in Table 6.8.

As all data already have been used once, it would not be honest to use the tabulated critical F ratios to assess the significance of the F ratios defined by $MSR\,/\,MSE$ given in the ANOVA table. These ratios are quite large and probably significant.

6.5. Diagnostic test of the fitted model. Residual plots

That a fitted model has been shown to be significant according by means of an analysis of variance is not sufficient. Sometimes a rather sloppy attitude in this respect can be found in published works. To be trustworthy, a model must withstand all diagnostic tests which could jeopardize the model. It is important that the experimenter should not fool himself/herself and draw conclusions which are not

Table 6.9: Analysis of variance of the reduced model in the Willgerodt-Kindler synthesis

Source	SS	df	MS	F
Total	76,097,62	16		
Regression				
on b_0	66522.73	1		
on variable terms	9225.48	5	1845.10	52.81
on b_1	1322.05	1	1322.05	37.87
on b_2	688.54	1	688.54	19.72
on b_3	5547.20	1	5547.20	158.90
on b_{13}	661.53	1	661.53	18.94
on b_{23}	1006.16	1	1006.16	28.82
Residuals	349.41	10	34.94	

supported by experimental evidence. One obvious thing to do when a model is fitted, is to analyze the residuals $e_i = y_i^{Exp} - y_i^{Pred}$. If the model is good, the residuals should be nothing but the experimental error. These errors can be expected to be normally distributed. To check this, a normal probability plot of the residuals should be drawn. This will also allow for a detection of clear outliers, which could appear if something went wrong in the execution of the experiment. Such experiments should be carefully analyzed to determine why they are aberrant as there is another possibility: an aberrant result may appear if something unexpected happens under these conditions, and this may lead to new discoveries. Fig. 6.10 shows a normal probability plot obtained from the residuals after fitting the reduced model to the Willgerodt-Kindler data. The fit is adequate. Other uses of normal probability plots are discussed in the next section.

Another diagnostic test which always should be made is to plot the residuals in each experimental point against the predicted response from the corresponding experiment. An adequate model should give a plot in which the points show a random scatter such that the upper and lower bonds of the pattern in the plot form two parallel horizontal lines at approximately equal distances from the zero line. The corresponding plot of the residuals in the Willgerodt-Kindler study is shown in Fig. 6.12. The plot is satisfactory.

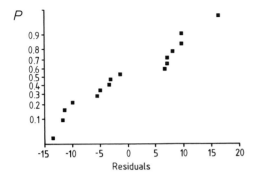

Fig.6.10:   Normal probability plot of residuals after fitting the reduced model to the data
in the Willgerodt-Kindler screening

However, plots of residuals can show different anomalous features, which indicate
that the model is poor:

If the upper and lower bonds diverge and the pattern of the residuals in the plot
is funnel-like, (Fig. 6.11a), it is an indication that the error is different in different
parts of the experimental domain. In such cases, the assumption of a constant error
variance is violated. To overcome this difficulty and to obtain a fairly constant error
distribution, the metric of the response variable should be changed through some
transformation, and the modell refitted to the transformed response. How this can
be done is discussed in Chapter 12.

If the upper and lower bonds are parallel but not horizontal, (Fig. 6.11b), there is
probably a systematic error. The most likely cause is that a linear term is missing in
the model.

If the pattern of the point is bent, and the upper and lower bounds form an arch
or a U, Fig. (6.11c), a second degree term is missing in the model.

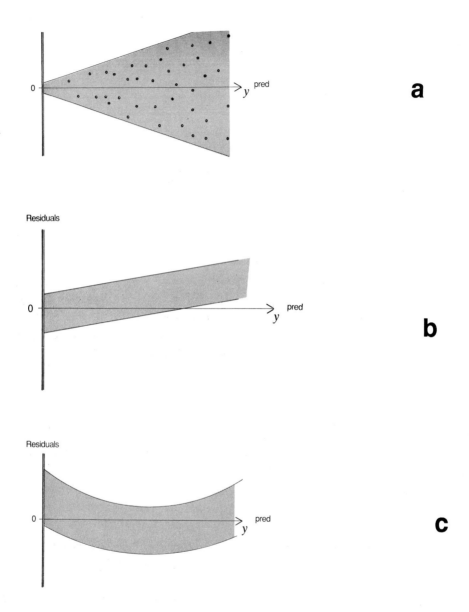

Fig.6.11: Different patterns which may appear when the residuals are plotted agains the predicted response.

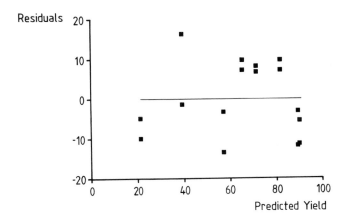

Fig.6.12: Plot of the residuals against predicted yield in the Willgerodt-Kindler screening.

Another informative plot is obtained by plotting the residuals in the run order of the experiments. The points should be evenly distributed around the zero line. When a pattern as in Fig.6.11b is obtained, it is an indication that a time-dependent phenomenon interferes to produce a systematic error. Often this is due to ageing of chemicals, or to contamination of chromatographic detectors.

For a discussion of the analysis of residuals, see Draper and Smith.[9]

7. Other uses of normal probability plots

In addition to the use of normal probability plots previously mentioned, i.e. to assess significant effects from screening experiments, and to check residuals after fitting models, there are some other useful applications which are briefly discussed in this section.

7.1. Estimates of mean and standard deviation directly from the plot

Normally distributed data form a straight line in the plot. The intersection of this line with the horizontal line $P(y) = 50 \%$, gives the abscisse coordinate which corresponds to the mean.

On commercially normal probability graph paper there are printed lines which will intersect the probability distribution line at $\pm \sigma$, $\pm 2\sigma$ and $\pm 3\sigma$. An estimate of the standard deviation can therefore be obtained directly from the plot.

7.2. Half-normal plots

Estimated effects, as well as residuals from eight-run designs, give plots in which the points are rather far apart. It may be difficult to see how a line should be drawn. A plot with more densely marked points can be obtained if the absolute values of the estimated effects, $|b_i|$ or the residuals, $|e_i|$, are plotted in the interval $P = [50 - 100 \%]$.

A word of caution on the use of half-normal plots is found in a book by Daniel.[10] Sometimes, the sign of an estimate can provide extra information, which will be lost if the absolute values are plotted. It is therefore advisable, not to use half-normal plots *routinely*.

7.3. Detection of erroneous experimental results in factorial and fractional factorial designs

A very useful technique by which erroneous experimental results can be detected has been given by Box and Draper.[11] The principles are illustrated by an example obtained from a study on Lewis acid catalyzed synthesis of carboxamides.[12]

Five variables were studied:

1: The amount of triethylamine.
2: The amount of boron trifluoride etherate
3: The amount of amine subtrate (benzylamine)
4: The order of introducing the reagents
5: Presence or absence of a drying agent

A fractional factorial design 2^{5-1} (**I = 12345**) was used to estimate the main effects and the two-variable interaction effects on the yield of carboxamide. The following estimates were obtained:

$$b_0 = 72.20 \qquad b_{12} = -2.31$$
$$b_1 = 17.30 \qquad b_{13} = -1.19$$
$$b_2 = -1.56 \qquad b_{14} = 15.81$$
$$b_3 = 11.60 \qquad b_{15} = -3.06$$
$$b_4 = 2.94 \qquad b_{23} = 3.06$$
$$b_5 = 2.56 \qquad b_{24} = -16.10$$
$$b_{25} = 1.81$$
$$b_{34} = -5.06$$
$$b_{35} = -2.44$$
$$b_{45} = 3.44$$

A normal probability plot of the estimated effects is shown in Fig. 6.13.

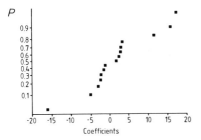

Fig.6.13: Normal probability plot of estimated model parameters in the synthesis of carboxamides.

The plot shows a strange feature. The normally distributed points are found along two parallel lines which seem to have been obtained by breaking the expected line at $P = 50\ \%$. The significant effects are discerned as outliers as usual.

The peculiar feature of the normal plot seen in Fig. 6.13 is obtained when there is one wild observation among the responses used to calculate the effects. An aberrant response can be due to a blunder or to an occasional large analytical error. It can also be due to *properties* of the studied system and to the fact that a different mechanism operates under the the conditions used in the "wild" experiment. An aberrant result is always interesting.

The strange feature of the plot occurs because factorial and fractional factorial designs are balanced designs; there is an equal number of (-1) and $(+1)$ in all columns used to calculate the effects. In all calculations, the erroneous response has been used once: it has either been added or substracted. This will result in half of

	−		+	+	+	−	−	−		−	−	+	+	
	17.31	−1.56	11.56	2.94	3.44	2.56	−2.31	−1.19	−2.44	15.81	−3.06	−5.08	3.08	1.81	−16.06
Exp no	1	2	3	4	5	12	13	14	15	23	24	25	34	35	45
1	−1	−1	−1	−1	1	1	1	1	−1	1	1	−1	1	−1	−1
2	1	−1	−1	−1	−1	−1	−1	−1	−1	1	1	1	1	1	1
3	−1	1	−1	−1	−1	−1	1	1	1	−1	−1	−1	1	1	1
4	1	1	−1	−1	1	1	−1	−1	1	−1	−1	1	1	−1	−1
5	−1	−1	1	−1	−1	1	−1	1	1	−1	1	1	−1	−1	1
6	1	−1	1	−1	1	−1	1	−1	1	−1	1	−1	−1	1	−1
7	−1	1	1	−1	1	−1	−1	1	−1	1	−1	1	−1	1	−1
8	1	1	1	−1	−1	1	1	−1	−1	1	−1	−1	−1	−1	1
9	−1	−1	−1	1	−1	1	1	−1	1	1	−1	1	−1	1	−1
10	1	−1	−1	1	1	−1	−1	1	1	1	−1	−1	−1	−1	1
11	−1	1	−1	1	1	−1	1	−1	−1	−1	1	1	−1	−1	1
12	1	1	−1	1	−1	1	−1	1	−1	−1	1	−1	−1	1	−1
13	−1	−1	1	1	1	1	−1	−1	−1	−1	−1	−1	1	1	1
14	1	−1	1	1	−1	−1	1	1	−1	−1	−1	1	1	−1	−1
15	−1	1	1	1	−1	−1	−1	−1	1	1	1	−1	1	−1	−1
16	1	1	1	1	1	1	1	1	1	1	1	1	1	1	1

the estimated effects having too high a value and the other half having too low a value. This is what can be seen in the plot.

It is possible to go backwards to trace *which* experiment is likely to be responsible for the observed effect:

(1) Write down the complete variable matrix which has been used to compute the effects (the model matrix **X** without the constant column **I**).

(2) Write the value of the estimated effects above the column labels.

(3) Write the *sign* of the estimated effect above its estimated value. Do this only with the *insignificant* effects. This gives a sequence of signs of the estimated effects. *(4)* Go through the variable matrix and look for the experiment in which the signs of the variable setting show the best correspondance to the sign sequence of the effects. This experiment is probably the aberrant one.

The procedure is illustrated by the variable matrix from the above example. The matrix is shown in the next page.

The conclusions to be drawn from this is that *Experiment no 13* is likely to have given an erroneous response. The next step would be to check this and repeat this experiment.

8. Running experiments in blocks

8.1. Discrete variations and systematic errors

In Chapter 3 it was discussed how the presence of a random error can be handled by statistical tools. The precautions which must be taken by the experimenter not to violate the assumption of independencies of the experimental error is *randomization*, which allows certain time-dependent systematic errors to be broken down and turned into random errors. There are, however, sources of error which can be suspected to produce *systematic* deviations which cannot be counteracted by randomization. In such cases, forseeable sources of systematic variation can be brought under control by dividing the whole set of experiments into smaller blocks which can be run under more homogeneous conditions. By a proper arrangement of these blocks, the systematic variation can be isolated through comparison of the between-block variation. Some examples where splitting the series of experiments into blocks is appropriate are:

* There is not time enough to run the whole set of experiments on one occasion. The experiments must be run during two different periods of time, maybe separated by several weeks. It is not granted that identical conditions can be obtained on these two occasions. In spite of this it is not likely that the influence of the *experimental variables* on the chemical reaction will be different at different times. It is other factors which are not controlled in the experiment that intervene to produce a systematic error. It would therefore be unfortunate if the systematic error should enter into the computation and bias the estimated effects of the *experimental variables*. This can be avoided by running the experiments divided into two blocks, one to be run on each occasion. The systematic error can then be isolated and removed to avoid confounding with the estimated effects.

* The experiments will be conducted by two different persons. Although they are both skilful and experienced, they are not identical. They will therefore behave slightly differently when they do the practical work. It is desirable that such personal bias can be separated from the influence of the experimental variables.

* The reaction will be run in different reactors and/or with different equipment. Although these are supposed to produce *similar* results, they are not identical, and thus there may be small differences which produce a systematic variation.

* The contents of one package of a chemical will not suffice for the whole series of experiment. It will be necessary to use two different packages. Their contents may differ slightly and this may produce a systematic variation.

* There are three or more varieties of a reagent which are to be tested. The objective is to select the one which seems most promising for further development of the procedure. In this case it is highly probable that the influence of experimental variables will be modified by the nature of the reagent, i.e. there are interaction effects between the type of reagent and the experimental variables.

Here, two examples will be discussed to illustrate some principles. The general problem of how to split large designs into smaller blocks has been the subject of considerable interest over the years. It is beyond the scope of this book to go into details on this. Readers who are interested in more thorough discussions on this subject should consult the specialized text-books given in the list of references.[13]

8.2. Example: Evaluation of three different reagents in a synthetic procedure

Assume that three different reagents A, B, and C, can be used to achieve a synthetic transformation. It is desired to select the most promising one for future development.

In addition to the type of reagent used, three experimental variables are assumed to influence the responses *yield* and *selectivity*. These variables are:

1: The reaction temperature
2: The amount of the reagent (A, B, C)
3: The time of addition of the reagent

The most common approach to this problem is to run only one experiment with each reagent under "standardized" conditions in which **1, 2** and **3** are maintained at fixed levels. A moment of reflexion reveals that this may be misleading. The *type* of reagent can interact with the variables and give compensatory effects; e.g. one reagent may be useful at a low temperature but may have a poor selectivity at ambient temperature, while another reagent may be insensitive to temperature variation.

To be absolutely sure is to know everything. In this case, this would require a knowledge of the roles played by all experimental perturbations of the reaction conditions. To make an absolutely fair comparison of the reagents, it would be necessary to establish the *optimum conditions* for each reagent. This would, however, lead to an excessive number of experiments, approximately 60 runs. It is obvious that a compromise is necessary, which implies that some information must be sacrificed.

One solution would be to run the experiments in three blocks where each block is a complete factorial design for each reagent. This would make it possible to estimate all main effects and interaction effects between the experimental variables, and also the effects caused by a change of reagent. This would give a total of 24 runs. The effects of the reagents is measured by the difference of the estimated effects of the variables obtained in the blocks.

* The *average effect* of changing the reagent from A to B is $b_0^B - b_0^A$.

* If the change of reagents modifies the influence of the variables, a *measure* of this is the difference of the effects of the variables determined in the different blocks, e.g. a difference in the sensitivity of variation in temperature is *measured* by the difference $b_1^B - b_1^A$.

A complete factorial design for each reagent would provide rather detailed information on main effects *and* interaction effects of the experimental variables. It would also provide information on to how a change of reagent modifies these effects. In a screening situation, this may not be necessary. With complete factorial designs, the number of experiments is still rather high.

A more convenient design in three blocks can be obtained by using a fractional factorial design 2^{3-1} ($\mathbf{I} = \mathbf{123}$) to vary the experimental conditions. In such design, main effects will be confounded with the two-variable interactions, $b_1 = (\beta_1 + \beta_{23})$... etc.

A comparison of the estimated effects between the blocks would anyhow reveal whether the influence of the experimental variables changes when the reagent is changed. Such an approach would give twelve runs.

Design matrices

Block A				Block B				Block C		
1	**2**	**3**		**1**	**2**	**3**		**1**	**2**	**3**
−1	−1	1		−1	−1	1		−1	−1	1
1	−1	−1		1	−1	−1		1	−1	−1
−1	1	−1		−1	1	−1		−1	1	−1
1	1	1		1	1	1		1	1	1

The following estimates are obtained from *Block A*

$$b_0^A = \beta_0^A + \beta_{123}^A$$
$$b_1^A = \beta_1^A + \beta_{23}^A$$
$$b_2^A = \beta_2^A + \beta_{13}^A$$
$$b_3^A = \beta_3^A + \beta_{12}^A$$

The estimates from the other blocks are obtained analogously. Since the same settings of the experimental variables are used in the different blocks, it is possible to compare the *runs* to each other. The effect of changing the reagent from A to B on the influence of the variables will thus be

Average response		$b_0^B - b_0^A$
Influence of	x_1	$b_1^B - b_1^A$
	x_2	$b_2^B - b_2^A$
	x_3	$b_3^B - b_3^A$

The other effects of the reagents are evaluated analogously.

The results obtained through this design allow for a rather safe selection of the most promising candidate for future work on the reaction. The block associated with the winning candidate can be augmented with a fold-over design to separate the confoundings. This would give good guidelines for the optimization of the procedure.

When there are a large number of possible reagents (or solvents etc.) selection of good candidates can be made through a strategy based on *principal properties*. This is discussed in Chapter 16.

8.3. Example: Discrete variables which are difficult to change

Assume that a reaction with the general formula

$$\text{Substrate} \xrightarrow{\text{Reagent A}} \xrightarrow{\text{Reagent B}} \text{Product}$$

is studied. In addition to three continuous variables, **1**, **2** and **3** given below, two discrete variations are also to be evaluated.

Variables

1: The reaction temperature
2: The amount of reagent A
3: The amount of reagent B
4: Type of reactor (-1 = Reactor I or $+1$ = Reactor II)
5: Type of solvent (-1 = THF or $+1$ = dioxane)

The solvents must be rigorously dried prior to use. It is therfore more convenient to run the experiments in dry THF in one context, and then do the drying procedure for dioxane and run the dioxane experiments.

It is not practical to change the apparatus randomly over the series of experiments.

If the aim is to determine all main effects as well as all two-variable interaction effects, a Resolution V design should be used. A fractional factorial design 2^{5-1} ($I = 12345$) meets these requirements. A total of 16 runs should be divided into blocks to cope with the practical constraints. The overall 2^{5-1} design is shown on the next page.Variable **4** specifies which type of reactor to use. This gives two blocks of eight runs. These are then further subdivided into two blocks of four experiments, which can be run with the selected solvents. Each block is a fractional factorial design in the continuous variables. After completion, the results obtained in the blocks can be combined to give the full model matrix by which estimates of all main effects and all two-variable interaction effects can be obtained.

It is of course possible to compute aliases of confounded effects from each block and then separate the confoundings by comparing the different aliases to each other. It will be a good exercise for the reader to do this. As a hint, the aliases of confounded effects which can be computed from the first block is given by the generators.

Exp no	1	2	3	4	5
1	−1	−1	−1	−1	1
2	1	−1	−1	−1	−1
3	−1	1	−1	−1	−1
4	1	1	−1	−1	1
5	−1	−1	1	−1	−1
6	1	−1	1	−1	1
7	−1	1	1	−1	1
8	1	1	1	−1	−1
9	−1	−1	−1	1	−1
10	1	−1	−1	1	1
11	−1	1	−1	1	1
12	1	1	−1	1	−1
13	−1	−1	1	1	1
14	1	−1	1	1	−1
15	−1	1	1	1	−1
16	1	1	1	1	1

It is seen that

−5			
−4			3
I	**1**	**2**	**12**
1	−1	−1	1
1	1	−1	−1
1	−1	1	−1
1	1	1	1

The independent generators are $(\mathbf{I} = -\mathbf{4} = -\mathbf{5} = \mathbf{123})$ and the complete set of generators is therefore

$$\mathbf{I} = -\mathbf{4} = -\mathbf{5} = \mathbf{123} = \mathbf{45} = -\mathbf{1234} = -\mathbf{1235} = \mathbf{12345}$$

which gives the confounding pattern

12345	2345	1345	345
−1235	−235	−135	−35
−1234	−234	−134	−34
45	145	245	1245
123	23	13	125
−5	−15	−25	−124
−4	−14	−24	3
I	**1**	**2**	**12**

If this analysis is made for the remaining blocks, it possible to compute all main effects and all two-variable interaction effects from the aliases obtained for each block.

The complete design partitioned into four blocks will thus be

Reactor I
4 = −1

THF
5 = −1

1	2	3	*Exp no*
−1	−1	1	5
1	−1	−1	2
−1	1	−1	3
1	1	1	8

Reactor II
4 = +1

THF
5 = −1

1	2	3	*Exp no*
−1	−1	−1	9
1	−1	1	14
−1	1	1	15
1	1	−1	12

Dioxane
5 = +1

1	2	3	*Exp no*
−1	−1	−1	*1*
1	−1	1	*6*
−1	1	1	*7*
1	−1	−1	*4*

Dioxane
5 = +1

1	2	3	*Exp no*
−1	1	1	*13*
1	−1	−1	*10*
−1	1	−1	*11*
1	−1	1	*16*

9. All runs in a fractional factorial design are useful

9.1. Projections down to a space spanned by the significant variables

After a fractional factorial design has been run with a view to screen the experimental variables some variable may turn out to be insignificant. In such cases, the experiments which have been run can be used to obtain more detailed information on the remaining variables if the insignificant variable is left out. It is possible to consider the experiments already performed as a larger fraction of design in the remaining variables.

We can make a projection of the original design down to the space spanned by the remaining variables. The principles are illustrated in Fig. 6.12, in which a 2^{3-1} fractional factorial design is projected down to the planes spanned by the remaining variables.

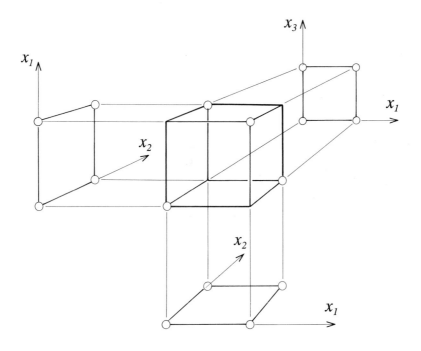

Fig.6.14: A fractional factorial design can be projected down to a lower-dimensional variable space if it should turn out that some variables are insignificant.

Fig. 6.14 shows, that if x_3 does not influence the response at all, i.e. if the spread of the points in the x_3 direction is without influence, the whole design can be regarded as a full factorial design in x_1 and x_2. It is therefore also possible to estimate the interaction effect β_{12} from the experiments. This effect was confounded with β_3 in the original design.

The consequences for the confounding patterns, if some variable turns out to be not significant, can be determined from the confounding pattern of the fractional design as all effects associated with the insignificant variables can be removed from the aliases obtained from each column. This technique is not entirely fool-proof and should not be routinely applied. It is sometimes a useful technique for obtaining new ideas. These ideas must then, of course, be checked through experiments.

9.2. Example: Two insignificant variables in a 2^{6-3} fractional factorial design

Assume that you have run a fractional factorial design 2^{6-3} (**4** = **12**, **5** = **13**, **6** = **23**), showing that the variables **1** and **5** are probably not significant, the following interpretation of the estimated effects can be attempted.

The set of generators for the 2^{6-3} design is

$$I = 124 = 135 = 236 = 2345 = 1346 = 1256 = 456$$

which gives the confounding pattern of the estimates as follows:

1	= **24** = **35** = **1236** = **12345** = **346** = **256** = **1456**
2	= **14** = **1235** = **36** = **345** = **12346** = **156** = **2456**
3	= **1234** = **15** = **26** = **245** = **146** = **12356** = **3456**
12	= **4** = **235** = **136** = **1345** = **2346** = **56** = **12456**
13	= **5** = **234** = **126** = **1245** = **46** = **2356** = **13456**
23	= **6** = **134** = **125** = **45** = **1246** = **1356** = **23456**
123	= **34** = **25** = **16** = **145** = **246** = **356** = **123456**

If we remove all "words" which contain the insignificant variables **1** and **5** from the relations above, we obtain

I	= **236**
24	= **346**
2	= **36**
3	= **26**
4	= **2346**
46	= **234**
23	= **6**
34	= **246**

If we assume that interaction effects between three or more variables are negligible, a reinterpretation of the experiments already performed, might be that the estimates are the following aliased effects:

176

From column		
I	β_0	
1	β_{24}	
2	$\beta_2 + \beta_{36}$	
3	$\beta_3 + \beta_{26}$	
12	β_4	
13	β_{46}	
23	$\beta_6 + \beta_{23}$	
123	β_{34}	

A complementary set of experiments should therefore be designed so that the coefficients which still are confounded can be separated.

References

1. G.E.P. Box and J.S. Hunter
   Technometrics 3 (1961) 311, 449

2. C. Vallejos
   Diss. IPSOI, Marseille 1976.

3. N.R. Draper and H. Smith
   Applied Regression Analysis, 2nd Ed.
   Wiley, New York 1981, Chapter 6.

4. G.E.P. Box, W.G. Hunter and J.S. Hunter
   Statistics for Experimenters
   Wiley, New York 1978.

5. D. Brunel, J. Itier, A. Commeyras, R. Phan-Tan-Luu and D. Mathieu
   Bull. Soc. Chim. Fr. (1979) 249.

6. C. Daniel
   Technometrics 1 (1959) 311.

7. R. Carlson, T. Lundstedt and R. Shabana
   Acta Chem. Scand. B 40 (1986) 534.

8. *(a)* C. Willgerodt
   Ber. dtsch. Chem. Ges. 20 (1887) 2467.

   (b) K. Kindler
   Liebigs Ann. Chem. 431 (1923) 193, 222.

9. N.R. Draper and H. Smith
   Applied Regression Analysis, 2nd Ed.
   Wiley, New York 1981, pp 141–192.

10. C. Daniel
    Application of Statistics to Industrial Experimentation
    Wiley, New York 1976.

11. G.E.P. Box and N.R. Draper
     Empirical Model-Building and Response Surfaces
     Wiley, New York 1987, pp. 131–134.

12. Å. Nordahl and R. Carlson
     Acta Chem. Scand. B 42 (1988) 28.

13. *(a)* G.E.P. Box, W.G. Hunter and J.S. Hunter
     Statistics for Experimenters
     Wiley, New York 1978;

     (b) R.A. Fisher
     The Design of Experiments, 8th Ed.
     Oliver and Boyd, Edinburgh 1966;

     (c) W.G. Cochran and G.M. Cox
     Experimental Design
     Wiley, New York 1957.

Suggestions to further reading

G.E.P. Box, W.G. Hunter and J.S. Hunter
Statistics for Experimenters
Wiley, New York 1978.

G.E.P. Box and N.R. Draper
Empirical Model-Building and Response Surfaces
Wiley, New York 1987.

Chapter 7

Other designs for screening experiments

Redundancy can be expensive

In the preceding chapter it was discussed how designs for screening experiments can be obtained from two-level factorial designs. In such designs the number of runs will be a power of two, 4, 8, 16, 32 .... 2^k. There are, however, situations where this may be inconvenient. An example of this would be if we wish to run a screening experiment with 17 variables. The smallest fractional factorial design which can accomodate that many variables would be a 2^{17-12} design with 32 runs. In a screening experiment with that many variables, it would be sufficient to have estimates of the average result and the main effects, i.e. to approximate the response surface by a plane. For this, 18 experiments is the minimum. A fractional design with 32 runs would therefore contain a number of unnecessary runs, which are not used for the purpose of the experiment. This would be wasteful, especially if the individual runs are time-consuming or expensive.

In this chapter, two other principles for the construction of screening designs are discussed: design by a Hadamard matrix and D-optimal design. Through such designs the inconveniences mentioned above can be overcome. However, other inconveniences which will emerge instead.

1. Screening design by Hadamard matrices: Plackett-Burman designs

To assess the influence of many variables it would be too demanding to account also for the interaction effects. A screening design which can estimate the main effects will therefore be the best choice. We have seen that an orthogonal design such as two-level fractional factorial designs allow for *independent* estimates of the main effects. An orthogonal design fulfils the criterion that $\mathbf{X'X}$ is a diagonal matrix. For the factorial and fractional factorial designs the diagonal elements are equal to the number of experiments, n, and $\mathbf{X' X} = n \cdot \mathbf{I_n}$. The dispersion matrix is given by $(\mathbf{X'X})^{-1} = 1/n \cdot \mathbf{I_n}$.

A *Hadamard matrix*, $\mathbf{H_n}$, is a mathematical toy with the following properties:

$\mathbf{H_n}$ is a (n x n) matrix with the elements (-1) or ($+1$), and which obeys the relation

$$\mathbf{H'H} = n \cdot \mathbf{I_n}$$

This is equivalent to saying that the columns in a Hadamard matrix are orthogonal.

We see that this applies to the model matrices from factorial designs and fractional factorial designs: these matrices are Hadamard matrices. It is possible to construct Hadamard matrices by other principles, and it was shown by Plackett and Burman[1] how such matrices can be obtained for n = 4, 8, 12, 16, 20, 24, 28, 32,..., i.e. when n is a multiple of four.

Plackett and Burman have shown the construction of all matrices for n up to 100, with one exception, (92 x 92) matrix, which they failed to construct. This matrix has been reported later.[2] The Plackett-Burman paper[1] discusses the use of Hadamard matrices to define screening experiments, in which ($n - 1$) variables can be tested in n runs.

The construction of a design by a Hadamard matrix is simple. Plackett and Burman have determined how the first row in the design matrix should be constructed so that the remaining rows can be obtained by cyclic permutations of the first row. An example to show this is given below. For a given n there can be many different n x n matrices which are Hadamard matrices. The Plackett-Burman matrices are by no means the unique solutions.

The first row of the design matrices are given below as a *sequence of signs*. The signs correspond to the levels of the variables. The Hadamard matrix is obtained from the design matrix by identifying the ($-$) with (-1) and the ($+$) sign with ($+1$), and by adding a column $\mathbf{I}$ of ones (used to compute the average).

Sign sequences for experimental designs with n runs for n = 8, 12, 16, 20, 24, 28 are given here. It is rare that more than 27 variables are involved in synthetic procedures. Readers who are interested in constructing larger designs, should consult the paper by Plackett and Burman.

First row in the Placett- Burman designs

$n = 8$	+ + + − + − −
$n = 12$	+ + − + + + − − − + −
$n = 16$	+ + + + − + − + + − − + − − −
$n = 20$	+ + − − + + + + − + − + − − − − + + −
$n = 24$	+ + + + + − + − + + − − + + − − + − + − − − −
$n = 28$	**A B C**
	B C A
	C A B

The sub-matrices use to define the (28 x 28) matrix are

```
A                          B                          C
+ − + + + + − − −          − + − − − + − − +          + + − + − + + − +
+ + − + + + − − −          − − + + − − + − −          − + + + + − + + −
− + + + + + − − −          + − − − + − − + −          + − + − + + − + +
− − − + − + + + +          − − + − + − − − +          + − + + + − + − +
− − − + + − + + +          + − − − − + + − −          + + − − + + + + −
− − − − + + + + +          − + − + − − − + −          − + + + − + − + +
+ + + − − − + − +          − − + − − + − + −          + − + + − + + + −
+ + + − − − + + −          + − − + − − − − +          + + − + + − − + +
+ + + − − − − + +          − + − − + − + − −          − + + − + + + − +
```

To show how a design matrix is constructed from the sequences of signs above, the construction of an eight-run design from the sequence of seven signs is given.

Exp no	*Variables*						
	1	**2**	**3**	**4**	**5**	**6**	**7**
1	+	+	+	−	+	−	−
2	+	+	+	−	+	−	−	+

$\downarrow$ ————————————————— $\uparrow$

The second row is obtained from the first row by leaving out the first sign, writing the sequence starting from the second sign, and adding the left-out sign in the last column. The procedure is then continued until the seventh row has been written. The design matrix is completed by adding a last row of only (−) signs.[1]

---

[1] It is also possible to make the cyclic permutation column-wise, and add a last row of (−) signs.

The following design matrix is obtained

Exp no	1	2	3	4	5	6	7
1	+	+	+	−	+	−	−
2	+	+	−	+	−	−	+
3	+	−	+	−	−	+	+
4	−	+	−	−	+	+	+
5	+	−	−	+	+	+	−
6	−	−	+	+	+	−	+
7	−	+	+	+	−	+	−
8	−	−	−	−	−	−	−

The Hadamard matrix is obtained by appending a column of ones, identifying the signs by −1 or +1. In this case the Hadamard matrix will be

$$
\mathbf{H} = \begin{bmatrix}
1 & 1 & 1 & 1 & -1 & 1 & -1 & -1 \\
1 & 1 & 1 & -1 & 1 & -1 & -1 & 1 \\
1 & 1 & -1 & 1 & -1 & -1 & 1 & 1 \\
1 & -1 & 1 & -1 & -1 & 1 & 1 & 1 \\
1 & 1 & -1 & -1 & 1 & 1 & 1 & -1 \\
1 & -1 & -1 & 1 & 1 & 1 & -1 & 1 \\
1 & -1 & 1 & 1 & 1 & -1 & 1 & -1 \\
1 & -1 & -1 & -1 & -1 & -1 & -1 & -1
\end{bmatrix}
$$

This matrix can be used to compute the coefficients of a linear model in the seven variables.

$$y = \beta_0 + \Sigma \, \beta_i x_i + e$$

$$\mathbf{b} = (\mathbf{X'X})^{-1}\mathbf{X'y} = 1/8 \cdot \mathbf{I_8 X'y} = 1/8 \cdot \mathbf{X'y}$$

The computations to obtain the estimates can be carried out by hand directly from the sign table as was shown for the factorial designs.

These designs will be Resolution III designs, and the estimated main effects will be confounded with two-variable interactions. However, the confounding patterns are not easy to elucidate directly from the design matrices. We do not have the generator of the fractional factorial designs. Confoundings must be analyzed by other means. In Appendix 7A is shown how the confoundings in the above design can be determined.

2. Screening by D-optimal designs

In Chapter 5 was discussed how the properties of the model matrix $\mathbf{X}$ is transmitted to different criteria of the quality of the estimated model parameters. One such criterion is that the joint confidence region should be as small as possible. This is obtained if the determinant of the dispersion matrix, $|(\mathbf{X'X})^{-1}|$, is as small as possible.

Since

$$|(\mathbf{X'X})^{-1}| = 1 / |\mathbf{X'X}|$$

this is equivalent to saying that the determinant $|\mathbf{X'X}|$ should be as large as possible.

The model matrix $\mathbf{X}$ is defined from the postulated model and the design matrix. Provided that a reasonable model can be suggested, it will thus be possible to select a series of experiments in such a way that $|\mathbf{X'X}|$ has a maximum value. These experiments will then allow the parameters of the suggested model to be estimated with a maximum precision. Such a design is said to be *D-optimal* (D stands for "determinant"). The D-optimality criterion can be used to select experiments for screening.

2.1 When is a D-optimal design appropriate?

There are situations when a D-optimal design is the preferred choice:

* *The experimental domain is constrained*: There may be certain combinations of the experimental variables which cannot be used, e.g. due to safety specifications of the equipment to be used, see Fig. 7.1. The available domain is truncated by constraints. A two-level design of any kind would necessitate the shrinking of the explored domain to accomodate the constraints. This would leave interesting parts of the domain uncovered by the design. A D-optimal design would select experiments to cover the *whole* domain of interest.

* *A minimum number of experiment is mandatory*: There are situations where one cannot afford the luxury of redundancy, and it is necessary to select a minimum set of experiments to answer the

184

questions posed on the influence of the experimental variables. The
D-optimality criterion allows a minimum set of experiments to be
selected so that the estimated effects have the maximum possible
precision.

* *Selection of complementary experiments to improve a model*: When a
series of experiments has been run to establish an initial model it is
sometimes found that the model shows a lack of fit. If the
experimenter can give a reasonable explanation for this, and he/she
wishes to improve the model by adding some corrective terms, a D-
optimal design can be used to select those complementary
experiments which should be run to obtain maximum precision
estimates of the parameters of the corrected model.

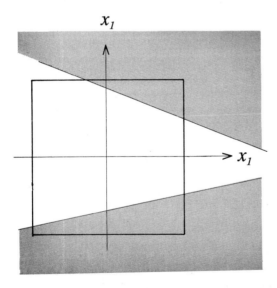

Fig. 7.1: The experimental domain is restricted due to constraints.

It should be remembered that factorial, fractional factorial and Plackett-Burman
designs are D-optimal designs. These designs can easily be constructed by hand. In

the general case, however, D-optimal designs cannot be constructed by hand. It will be necessary to use computers. There are several commercially available programs which have modules for the construction of D-optimal designs.[3] Several algorithms to construct these designs have been described.[4] Two such algorithms are briefly discussed in Appendix 7B.

A review on D-optimal regression designs has been given by St John and Draper.[5]

The use of a D-optimal design for screening is illustrated by an example on enamine synthesis.

2.2 Example: D-optimal design for screening of variables in enamine synthesis

A procedure was studied by which enamines are produced by the condensation of the parent ketones and a secondary amine in the presence of molecular sieves. The molecular sieves are used to trap the water which is formed.[6] The enamine formation is acid catalyzed; both Brönsted acids and Lewis acids can be used as catalysts.[7]

The condensation of methyl isobutyl ketone and morpholine in the presence of two commercially available zeolite-type molecular sieves (Type 3A and 5A) was used as a model reaction. The procedure was known with other systems, although the conclusions on the amounts of amine to be used were contradictory.[8] These ambiguities needed to be clarified.

To determine suitable experimental conditions for preparative runs, the important experimental variables should be known. To determine this, a D-optimal design was used. The variables which were studied are summarized in Table 7.1.

Some interactions between the variables could be expected:

* The amount of morpholine and the amount of solvent can be expected to influence the rate of enamine formation, and hence the evolution of yield over time. To account for this, a term $b_{12}x_1x_2$ should be added to the model.

* The amount of molecular sieves and the type of sieve can be

expected to show compensating effects: A small amount of an efficient sieve could produce the same result as a large amount of a less efficient one. To account for such effects, a term $b_{34}\,x_3x_4$ shold be included in the model.

*   The same arguments as above apply to the amounts and the type of acid catalyst, and a term $b_{56}\,x_5x_6$ should be added to the model.

Table 7.1: Variables and experimental domain in enamine synthesis over molecular sieves

Variables	Domain	
	−1	1
x_1: The amount of morpholine / ketone (mol/mol)	1.0	3.0
x_2: The amount of solvent[a] /ketone (ml/mol)	200	800
x_3: The amount of molecular sives/ketone (g/mol)	200	600
x_4: Type of molecular sieve	3A	5A
x_5: Type of acid catalyst[b]	SA	TFA
x_6: Amount of acid catalyst/ketone		
SA (g/mol)	50	200
TFA (ml/mol)	0.5	12.5

[a] Cyclohexane was used. [b] Two different types of acid catalysts were tested: *SA* was a solid Silica-Alumina, Lewis acid-type of catalyst. It is used in petroleum cracking; *TFA* was trifluoroactic acid, a proton acid.

The attempted model will thus be

$$y = \beta_0 + \Sigma\,\beta_i\,x_i + \beta_{12}\,x_1x_2 + \beta_{34}\,x_3x_4 + \beta_{56}\,x_5x_6 + e$$

There are ten parameters in the model. To estimate these, a design with twelve experimental runs was assumed to be sufficient. Each variable is varied on two levels, and there are $2^6 = 64$ different possible runs. The number of ways by which twelve experiments can be selected is:

$$64! \; / \; (52! \cdot 12!) \approx 3.28 \cdot 10^{12}$$

It is evident, that a random selection of twelve experiments runs the risk of being a very poor selection. To obtain a sub-set of twelve runs which can give good estimates of the model parameters the D-optimality criterion was used.

To establish a design, the model and the desired number of experimental runs must be specified. An arbitrary initial set of twelve runs (which may be drawn at random) is fed into the computer. By an iterative procedure, the initial set of experiments is then modified by adding and deleting experiments so that the determinant $|\mathbf{X'X}|$ is maximized. In this case, the algorithm by Mitchell was used, see Appendix 7B. The design obtained through this procedure is given in Table 7.2. The model matrix obtained by this design gave $|\mathbf{X'X}| = 2.038 \cdot 10^{10}$. The corresponding dispersion matrix is given in Table 7.3.

Table 7.2: D-optimal design in enamine synthesis, and yields, y_i obtained[a].

Exp no	x_1	x_2	x_3	x_4	x_5	x_6	y_{28}	y_{52}	y_{76}	y_{98}
1	−1	−1	1	−1	1	1	58	66	77	76
2	−1	−1	−1	−1	1	−1	8	13	18	23
3	−1	−1	1	1	−1	1	60	61	65	67
4	−1	1	1	−1	−1	−1	40	60	69	74
5	1	−1	1	−1	1	−1	17	30	40	48
6	−1	1	1	1	1	−1	80	89	98	99
7	1	−1	−1	1	−1	−1	32	46	54	64
8	1	1	1	1	1	1	47	57	80	82
9	1	1	−1	−1	−1	−1	12	15	21	24
10	−1	1	−1	1	1	1	19	26	40	48
11	1	−1	−1	−1	−1	1	29	33	45	54
12	−1	1	−1	1	−1	1	46	40	57	60

[a] The indicies of the response variables, y_i, is the reaction time in hours.

One experiment, *Exp no 6*, gave almost quantitative yield, and the conditions of this experiment must be close to the optimum conditions.

Table 7.3: Dispersion matrix $(\mathbf{X'X})^{-1}$ from the D-optimal design

I	x_1	x_2	x_3	x_4	x_5	x_6	x_1x_2	x_3x_4	x_5x_6
0.81	0	0	0	0	0	0	0	0	0
	0.10	0	0.025	0	0.025	0	0	0	0
		0.11	0	−0.04	0	0.01	0	−0.01	−0.04
			0.10	0	−0.025	0	0	0	0
				0.11	0	−0.04	0.01	0	0.01
					0.10	0	0	0	0
Symmetric						0.11	−0.04	0.01	0
							0.11	−0.04	−0.01
								0.11	0.04
									0.11

The dispersion matrix is not a diagonal matrix. There are correlations between the model parameters. Hence, they are not independently estimated. Through a D-optimal design the parameters are estimated as independently as possible. This is the sacrifice which must be made when the number of experiments does not permit an orthogonal design. The covariances of the model parameters are rather small and the correlations are weak and will, hopefully, not lead to erroneous conclusions as to the influence of the variables. The estimated model parameters are summarized in Table 7.4.

The results can be interpreted as follows: There are two estimated effects which are significant beyond doubt:

b_3, which is positive. This implies that x_3, the amount of molecular sieves, should be at high level.

b_4, which is also positive and implies that x_4 should be at high level, (i.e. a 5A type of molecular sieve shall be used).

The remaining variables do not influence the final yield (98 h). For short reaction time it is indicated that x_1 should be at its low level, $b_1^{28\,h} = -5.1$. This corresponds to a stoichiometric amount of morpholine. This effect can be interpreted as an effect of adsorption of amine on the surface of the molecular sieves. An excess of amine can therefore block the acid catalytic sites, as well as the pores of the molecular sieve, which results in a lowered rate of enamine formation. These effects probably contribute to explain the slight increase of the interaction effect b_{12} over time. A more dilute solution will lower the amine sorption and favour the reaction. Due to this, a stoichiometric amount of amine is indicated.

Table 7.4: Estimated model parameters in enamine synthesis

Parameter	Reaction time (h)			
	28	56	72	98
b_0	35.6	45.2	54.8	59.9
b_1	−5.1	−3.7	−0.8	−0.3
b_2	−1.5	1.1	2.9	1.4
b_3	14.4	15.7	16.3	15.1
b_4	7.9	8.3	8.7	8.3
b_5	−3.9	−4.8	−2.7	−2.4
b_6	2.5	1.1	3.4	3.8
b_{12}	−3.2	−2.6	−4.4	−6.1
b_{34}	5.3	0.2	1.3	0.7
b_{56}	1.3	−0.4	0.4	1.2

The following conditions are therefore suggested for the preparative applications:

x_1:   should be on (−1) level; a stoichiometric amount of morpholine.

x_2:   is not too important. For economic reasons, use a small amount.

x_3:   should be on (+1) level. A large amount of molecular sieves should be used. The amount should probably be increased to achieve a more rapid reaction.

x_4:   Molecular sieves of 5A type should be used.

x_5:   The type of catalyst is not critical.

x_6:   The amount of catalyst is sufficient at the low level.

For the final optimization of the procedure, *Silica-Alumina* was chosen as catalyst. It is a solid catalyst which is regenerated by the same procedure used to regenerate the molecular sieves. Thus, an optimum mixture of the catalyst and the molecular sieves was to last several runs after regeneration. By using 750 g of molecular sieves and 250 g of solid catalyst per mol of ketone, afforded a quantitative yield after 70 h. The pure enamine was obtained after filtering the mixture and evaporation of the solvent.

2.3   A word of caution on the use of D-optimal designs

To use the D-optimality criterion to select a design it is *required* that the experimenter can specify a plausible model. To do so, a good portion of chemical intuition is required. As intuition is a function of experience, this can be difficult with

190

totally *new* reactions where there is no prior experience. If a D-optimal design with a minimum number of experiments is constructed, there will be no, or very few, degrees of freedom left for checking the adequacy of the model. Thereby, the experimenter runs the risk of fooling himself/herself if the suggested model is bad.

The moral of this is, that whenever there are any doubts as to the form of the model, it is always better to use a fractional factorial design for screening experiments. With fractional factorial designs, analysis of the confounding pattern may give clues to how the model should be refined.

References

1. R.L. Plackett and J.P Burman
   Biometrika 33 (1946) 305, 328.

2. l. Baumert, S.W. Golumb and M. Hall
   Am. Math. Soc. Bull. 68 (1962) 237.

3. *(a) ECHIP*
   Expert in a Chip
   RD1 Box 384Q
   Hockessin, DE 19707, USA;

   (b) NEMROD
   LPRAI, *Att.* Prof R. Phan-Tan-Luu
   Université d'Aix-Marseille III
   Av Escadrille Normandie Niemen
   F-13397 Marseille Cedex 13, France

   (c) RS1/Discover
   BBN Software Products Corp.
   10 Fawcett Street
   Cambridge, MA 02238, USA.

4. R.D. Cook and C.J. Nachtsheim
   Technometrics 22 (1980) 315.

5. R.C. StJohn and N.R. Draper
   Technometrics 17 (1975) 15.

6. R. Carlson, R. Phan-Tan-Luu, D. Mathieu, F.S. Ahouande, A. Babadjamian and J. Metzger
   Acta Chem. Scand. B 32 (1978) 335.

7. G.A. Cook (Ed.)
   Enamines: Synthesis, Structures and Reactions
   Marcel Dekker, New York 1969.

8. *(a)* K. Taguchi and F.H. Westheimer
   J. Org. Chem. 36 (1971) 1570;

   (b) D.P. Roelofsen and H. van Bekkum
   Recl. Trav. Chim. Pays-Bas 91 (1972) 605.

Suggestions to further reading

A textbook largely dealing with designs based on Hadamard matrices is

W.J. Diamond
Practical Experimental Designs for Engineers and Scientists
Van Nostrand Reinhold Co, New York 1981.

Appendix 7A: Confounding pattern in Plackett-Burman designs

To analyze the confounding pattern in Plackett-Burman designs, a general method must be used. This method is to determine the *alias matrix*.

The model used to assess the influence of the variables by a Plackett-Burman design is a linear model:

$$y = \beta_0 + \Sigma \beta_i x_i + e$$

If this model is correct, a least squares fit of the model will give an unbiassed estimate, **b**, of the vector of the "true" model parameters **b**. However, if the model should contain also interaction terms, the estimated model parameters would be biassed by confounding with interaction effects. This bias occurs since the Taylor expansion is truncated after the linear terms. To analyze *which* second order interaction effects contaminate the true linear effects, we shall write down the augmented model with the second order interaction terms included.

$$y = \beta_0 + \Sigma \beta_i x_i + \Sigma\Sigma \beta_{ij} x_i x_j + e$$

Let X_1 be the model matrix which corresponds to the linear model, and let X_2 be a complementary matrix which is used to append the interaction terms to the linear model. The columns in X_2 are the vectors of variable cross-products. Let b_1 be the vector of "true" parameters of the linear model, and let b_2 be the vector of "true" cross-product coefficients.

$$
b_1 = \begin{bmatrix} \beta_0 \\ \beta_1 \\ \cdot \\ \cdot \\ \cdot \\ \beta_k \end{bmatrix}
\qquad
b_2 = \begin{bmatrix} \beta_{12} \\ \beta_{13} \\ \cdot \\ \cdot \\ \cdot \\ \beta_{ij} \end{bmatrix}
$$

If the linear model is correct, the expectation values will be

$$E[y] = X_1 b_1 \quad \text{and}$$
$$E[b] = b_1$$

But if the interaction model is correct, these expectations will be

$$E[y] = X_1 b_1 + X_2 b_2 \quad \text{and}$$
$$E[b] = b_1 + (X_1'X_1)^{-1}X_1'X_2 b_2$$

These relations can be used to determine the confoundings in Plackett-Burman designs. As an example to illustrate the principles, the design with seven variables in eight runs described on p. 182 is used.

To show the computation in detail, the matrices will be printed with small characters.

The model matrix, X_1, of the linear model is

```
I  1   2   3   4   5   6   7
1  1   1   1  -1   1  -1  -1
1  1   1  -1   1  -1  -1   1
1  1  -1   1  -1  -1   1   1
1 -1   1  -1  -1   1   1   1
1  1  -1  -1   1   1   1  -1
1 -1  -1   1   1   1  -1   1
1 -1   1   1   1  -1   1  -1
1 -1  -1  -1  -1  -1  -1  -1
```

The matrix, X_2, of the omitted second order interaction terms is

```
12 13  14  15  16  17  23  24  25  26  27  34  35  36  37  45  46  47  56  57  67

 1   1  -1   1  -1  -1   1  -1   1  -1  -1  -1   1  -1  -1  -1   1   1  -1  -1   1
 1  -1   1  -1  -1   1  -1   1  -1  -1   1  -1   1   1  -1  -1  -1   1   1  -1  -1
-1   1  -1  -1   1   1  -1   1   1  -1  -1  -1  -1   1   1   1  -1  -1  -1  -1   1
-1   1   1  -1  -1  -1  -1  -1   1   1   1   1  -1  -1  -1  -1   1  -1   1   1   1
-1  -1   1   1   1  -1   1  -1  -1  -1   1  -1  -1  -1   1   1   1  -1   1  -1  -1
 1  -1  -1  -1   1  -1  -1  -1  -1   1  -1   1   1  -1   1   1  -1   1  -1   1  -1
-1  -1  -1   1  -1   1   1   1   1  -1   1  -1   1  -1  -1   1  -1  -1   1  -1
 1   1   1   1   1   1   1   1   1   1   1   1   1   1   1   1   1   1   1   1   1
```

$$E[b] = b_1 + (X_1'X_1)^{-1}X_1'X_2b_2$$

as the Plackett-Burman design is orthogonal this can be simplified to

$$E[b] = b_1 + 1/8 \cdot I_8 X_1'X_2 b_2$$
$$E[b] = b_1 + 1/8 \cdot X_1'X_2 b_2$$

The last term in the expression above, $1/8 \cdot X_1'X_2 b_2$, is called the alias matrix. To compute this matrix, the multiplication $X_1'X_2$ is done first. This gives the following matrix $X_1'X_2$:

	12	13	14	15	16	17	23	24	25	26	27	34	35	36	37	45	46	47	56	57	67
I	0	0	0	0	0	0	0	0	0	0	0	0	0	0	0	0	0	0	0	0	0
1	0	0	0	0	0	0	0	0	0	-8	0	-8	0	0	0	0	0	0	0	-8	0
2	0	0	0	0	-8	0	0	0	0	0	0	0	0	0	-8	-8	0	0	0	0	0
3	0	0	-8	0	0	0	0	0	0	0	-8	0	0	0	0	0	0	0	-8	0	0
4	0	-8	0	0	0	0	0	0	-8	0	0	0	0	0	0	0	0	0	0	0	-8
5	0	0	0	0	0	-8	0	-8	0	0	0	0	0	-8	0	0	0	0	0	0	0
6	-8	0	0	0	0	0	0	0	0	0	0	-8	0	0	0	0	0	-8	0	0	0
7	0	0	0	-8	0	0	-8	0	0	0	0	0	0	0	0	0	0	-8	0	0	0

The alias matrix is then obtained by first multiplying the above matrix by $1/8$ and then multiplying the result by the vector of cross-product coefficient, b_2. This gives:

$$1/8 \cdot X_1'X_2 b = \begin{bmatrix} 0 \\ -\beta_{26} - \beta_{34} - \beta_{57} \\ -\beta_{16} - \beta_{37} - \beta_{45} \\ -\beta_{14} - \beta_{27} - \beta_{56} \\ -\beta_{13} - \beta_{25} - \beta_{67} \\ -\beta_{17} - \beta_{24} - \beta_{36} \\ -\beta_{12} - \beta_{35} - \beta_{47} \\ -\beta_{15} - \beta_{23} - \beta_{46} \end{bmatrix}$$

From this follows, that the coefficients, $b = [b_0, b_1...b_7]'$, estimated for the linear model are aliases in which the true linear coefficients are confounded with the cross-product coefficients as follows

$$b_0 = \beta_0$$
$$b_1 = \beta_1 - \beta_{26} - \beta_{34} - \beta_{57}$$
$$b_2 = \beta_2 - \beta_{16} - \beta_{37} - \beta_{45}$$
$$b_3 = \beta_3 - \beta_{14} - \beta_{27} - \beta_{56}$$
$$b_4 = \beta_4 - \beta_{13} - \beta_{25} - \beta_{67}$$
$$b_5 = \beta_5 - \beta_{17} - \beta_{24} - \beta_{36}$$
$$b_6 = \beta_6 - \beta_{12} - \beta_{35} - \beta_{47}$$
$$b_7 = \beta_7 - \beta_{15} - \beta_{23} - \beta_{46}$$

The above method must be used to elucidate the confounding pattern of Plackett-Burman designs. It is seen, that a fold-over design would separate the main effects from confounding with the two-variable interactions. A fold-over design would switch the signs of the X_1 matrix and hence switch the signs of the alias matrix.

Appendix 7B: Algorithms for the construction of D-optimal designs

The volume of the joint confidence region is proportional to the square root of the determinant of the dispersion matrix

$$Volume \propto SQR \; |(X'X)^{-1}|$$

Since $(X'X)^{-1}$ is a symmetric matrix the determinant is positive. It is seen that the minimum value of the volume of the confidence region is obtained when the matrix $|(X'X)^{-1}|$ has its minimum value. Since $|(X'X)^{-1}| = 1 \;/\; |X'X|$, this is equivalent to saying that $|X'X|$ should have its maximum value. Two different algorithms by which it is possible to select experiment by an iterative procedure to maximize the determinant $|X'X|$, are briefly sketched below.

To understand the algorithms it is neccesary to define a new concept, the *variance function*:

When a model is fitted to experimental data, the experimental error is transmitted to the coefficients of the model. These parameters will therefore have a probability distribution. When the model is then used to predict the response for a given combination of the experimental variables, $x_i = (x_{1i}, x_{2i},...,x_{ki})'$, the prediction is also afflicted by the probability distribution of the model parameters. We will have a certain degree of uncertainty in the predictions. There will be a variance of the predicted response, $V(y^{Pred})$. The variance of prediction is described by the following relation

$$V(y^{Pred}) = z_i'(X'X)^{-1}z_i \cdot \sigma^2 = d_i \cdot \sigma^2$$

The vector z_i is the vector of all variables in the model. This vector is obained from x_i by adding the elements which correspond to the constant term, cross-product terms, square terms ... etc.

Algorithm of exchange by Mitchell [1]

The algorithm uses the *variance function*, $d_i = z_i'(X'X)^{-1}z_i$ described above.

Let $D_{N(0)}$ be an initial design matrix with N runs, which gives the *information matrix* $X_{N(0)}'X_{N(0)}$. If one more experiment, i, is added to D_N to give a design with $N + 1$ runs, we obtain the information matrix $X_{N+1}'X_{N+1}$. It can be shown that these matrices are related by the following:

$$X_{N+1}'X_{N+1} = X_{N(0)}'X_{N(0)} + z_iz_i'$$

The corresponding determinants are also related:

$$|X_{N+1}'X_{N+1}| = |X_{N(0)}'X_{N(0)}| \cdot (1 + d_i)$$

If an experiment, j, is removed from the design with $N + 1$ runs to give another design $D_{N(1)}$ with N runs, the following relations apply

$$X_{N(1)}'X_{N(1)} = X_{N+1}'X_{N+1} - z_jz_j'$$

and

$$|\mathbf{X}_{N81)}\!'\mathbf{X}_{N(1)}| = |\mathbf{X}_{N+1}\!'\mathbf{X}_{N+1}| \cdot (1 - d_j)$$

From the above is seen, that an increased value of $|\mathbf{X}'\mathbf{X}|$ can be obtained by introducing experiments i with large values of d_i, followed by a removal of experiments j with a feeble value of d_j. This defines the algorithm of exchange given by T. Mitchell. The exchanges are made, first by a single exchange

$$N \rightarrow N+1 \rightarrow N \quad \text{etc.}$$

then by double exchange

$$N \rightarrow N+1 \rightarrow N+2 \rightarrow N+1 \rightarrow N \quad \text{etc.}$$

then by triple exchange, and so on, until the determinant has reached a sufficiently large value. There is no criterion of convergence for this algorithm, and it can therefore not be concluded that the *maximum* value of the determinant has been attained. Hopefully, the resulting design will be good enough for its purposes.

Algorithm of exchange by Fedorov [2]

This algorithm allows the simultaneous *interchange* of two experiments i and j in an experimental design. Let $\mathbf{D}_{N(0)}$ be the initial design with N runs, and let $\mathbf{X}_{N(0)}$ be the corresponding model matrix. Let z_i be the vector of the model variables of an experiment i, which belongs to the designs $\mathbf{D}_{N(0)}$, and with the settings of the experimental variables x_i. Let z_j be the corresponding for a candidate experiment j which could be included in the design to replace i. The design matrix which is obtained by this exchange is $\mathbf{D}_{N(1)}$, and the corresponding model matrix is $\mathbf{X}_{N(1)}$.

The following relations apply between these designs

$$\mathbf{X}_{N(1)}\!'\mathbf{X}_{N(1)} = \mathbf{X}_{N(0)}\!'\mathbf{X}_{N(0)} - z_i z_i' + z_j z_j'$$

and

$$|\mathbf{X}_{N(1)}\!'\mathbf{X}_{N(1)}| = |\mathbf{X}_{N(0)}\!'\mathbf{X}_{N(0)}| \cdot (1 + \delta_{ij})$$

The increment function, δ_{ij} is defined as

$$\delta_{ij} = [d_j - (d_i d_j - d_{ij}^2) - d_i]$$

The variance functions, d_i and d_j, are obtained from the dispersion matrix of the first design

$$d_i = z_i'(\mathbf{X}_{N(0)}\!'\mathbf{X}_{N(0)})^{-1} z_i$$

$$d_j = z_j'(\mathbf{X}_{N(0)}\!'\mathbf{X}_{N(0)})^{-1} z_j$$

The covariance function, d_{ij} is defined as

$$d_{ij} = z_i'(X_{N(0)}'X_{N(0)})^{-1}z_j = z_j(X_{N(0)}'X_{N(0)})^{-1}z_i$$

It is seen that the value of the determinant $|X'X|$ will have a maximum increase if interchanges which give the maximum value of the increment function δ_{ij} are chosen. In principle, this algorithm will stop when the possibilities to make interchanges which give positive values of δ_{ij} have been exhausted.

Computers must be used to construct designs by these algorithms

These algorithms cannot be used for hand calculations. A computer must be used. The time of computations will be very different for the two algorithms sketched above.

A comparison of different algorithms has been made by Cook and Hachtsheim.[3]

For a simple exchange, the algorithm by Mitchell, it is necessary to compute the variance function for $N + N_c$ experimental points (N_c is the number of candidate points).

With the algorithm by Fedorov, for each interchange, it is necessary to compute $N \cdot (N_c - 1)$ values of the increment function δ_{ij}, and this necessitates the computations of $N_c \cdot (N + 1)$ variance and covariance functions. This implies a rather extensive amount of computation. The increase in the value of the determinant $|X'X|$ is, however, more rapid by this algorithm, than by the algorithm of Mitchell.

It has also been found, that the algorithm by Fedorov, in general, leads to larger values of $|X'X|$ than those obtained by the algorithm by Mitchell.[4]

References

1.   T.J. Mitchell
      Technometrics 16 (1974) 203.

2   V.V. Fedorov
      Theory of Optimal Experiments
      Academic Press, New York 1972.

3.   R.D. Cook and C.J. Nachtsheim
      Technometrics 22 (1980) 315.

4.   D. Mathieu
      Diss. Université d'Aix-Marseille III, Marseille 1981.

Appendix 7C: Some comments on the "optimality" of a design

There are a number of criteria which have been suggested to select the "best" design. The problem is, however, to define what is meant by "best". What is usually meant is that the performance or the properties of the model established by the design have been optimized in certain respects. These criteria are all associated with the corresponding dispersion matrix, $(X'X)^{-1}$.

Some of the criteria are related to the shape and extension of the joint confidence region, discussed in Chapter 5.

Other criteria are related to the quality of the predictions from the model.

Criteria related to the shape of the confidence region

These criteria are all related to the eigenvalues, $[\lambda_1, \lambda_2....,\lambda_k]$, of the dispersion matrix. The eigenvalues are the roots of the secular equation:

$$|(X'X)^{-1} - \lambda \cdot I| = 0$$

It can be shown that the determinant of the dispersion matrix is the product of its eigenvalues.

$$|(X'X)^{-1}| = \Pi\lambda_i = \lambda_1 \cdot \lambda_2 \cdot ... \cdot \lambda_k$$

D-optimality: The D-optimality criterion will therefore specify that the product of the dispersion matrix eigenvalues should be as small as possible.

A-optimality: This criterion refers to the average variance of the estimated model parameters, (A stands for Average variance). An experimental design is A-optimal if the sum of the dispersion matrix eigenvalues

$$A = \Sigma\lambda_i$$

is as small as possible. Since the eigenvalues are related to the length of the main axes of the confidence ellipsoid, the A-criterions will attempt to render the confidence ellipsoid as spherical as possible.

E-optimality: This criterion is fulfilled when the largest eigenvalue of the dispersion matrix is as small as possible. This minimizes the largest variance of the estimated model parameters.

However, in order to use these criteria in the selection of an experimental design from a set of possible candidate experiments, certain assumptions must be fullfilled: *(1)* The mathematical form of the model to be fitted is perfectly known. *(2)* The region in which the experiments can be run, i.e. the experimental domain is known, and it is excluded that experiments *can* be run outside this region.

It is seen, that such assumptions are rarely valid when the problem is to establish suitable experimental conditions in synthetic chemistry. Neither the feature of the response function nor the optimum experimental domain is known beforehand.

For an illumination discussion on "Aspects of Alphabetc Optimal Design Theory", see Box and Draper.[1]

Criteria related to the performance of the model

Some of these criteria use the *variance function, d_i*, defined in Appendix 7B.

G-optimality: A design is G-optimal for which the largest value of d_i is minimized for all possible settings of the experimental variables, x_i, in the experimental domain. This criterion is equivalent to D-optimality for discrete setting of the experimental variables.

Rotatability: This property of experimental designs has been introduced by Box and Hunter.[2] A design is said to be rotatable if d_i is constant for all settings of the experimental variables, x_i, at the same distance, r, from the center point of the experimental design. This means that the predictions from the model will have the same precision, for all points at the distance r from the center.

The criterion of rotatability is much used to establish designs for second order, quadratic response surface models. These aspects will be further discussed in Chapter 12.

Uniform precision: It is possible to establish experimental designs for determining second order quadratic response surface models, so that the prediction variance is fairly constant in the explored domain, i.e. $d_i \approx constant$. This criterion is linked to rotatability.

The J-Criterion: When a model is fitted to experimental data, there will always be an error term, e. This error will have two components. One of these is due to the approximate nature of the model, which gives a deviation, *bias*, between the true value of the response, η, and the value predicted from the model y^{Pred}. The other component of the error term is due to the consequences of the random experimental error, which will give a probability distribution of the model parameters. This is then transmitted to the predicted response, which therefore will have a prediction error variance. A criterion, the J-criterion, which take these two error components into account has been introduced by Box and Draper.[3]

$$J = B + V$$

In the above expression, B is the averaged contribution of the bias, and V is the averaged contribution of the prediction variance in the experimental domain covered by the design. A good design should minimize J. The derivation of the expression for J is rather complicated and beyond the scope of this book. A detailed account is given in the original paper by Box and Draper[3], (see also a thorough discussion in the book by the same authors).[4]

References

1. G.E.P. Box and N.R. Draper
    Empirical Model-Building and Response Surfaces
    Wiley, New York 1987, pp. 490–495.

2. G.E.P. Box and J.S. Hunter
    Ann. Math. Statist. 28 (1957) 195.

3. G.E.P. Box and N.R. Draper
    Technometrics 31 (1971) 731.

4. See Ref. [1], pp. 423–476.

Chapter 8

Summary of screening experiments

Objectives

The objectives of a screening experiment is to proceed from a stage where very little is known with certainty about the roles played by the experimental variables, up to a stage where the relative importance of the variables can be assessed. In synthetic chemistry this entails the identification of those variables which have a significant influence on, for example, yield and selectivity. To achieve this, an approximate response function is established. This model contains linear terms of the variables of interest, and possibly also cross-product terms to describe interaction effects.

Models for screening:

$$y = \beta_0 + \Sigma \, \beta_i \, x_i + e$$

$$y = \beta_0 + \Sigma \, \beta_i \, x_i + \Sigma\Sigma \, \beta_{ij} \, x_i x_j + e$$

To establish these models, a series of experiments is designed so that the values of the model parameters can be estimated by a least-squares fit to known experimental results. The number of experiments must be at least as many as there are unknown parameters in the model.

The significant variables will be identified as those for which the corresponding coefficient in the model is significantly larger than the experimental error. The significance can be assessed by statistical tests, such as t-tests, F-tests on ANOVA tables, or from cumulative normal probability plots.

The variables found to be significant by the above methods, are probably the most rewarding ones to proceed with.

A good rule of thumb is: Do not use more than 25 % of the available resources in the screening phase.[1] *The remaining resources should be available for optimization and for validation of the procedure.*

1. Steps to be taken in a screening experiment

The following suggestions should be regarded as advice and not as the rules of the game.

1.1. Select variables to be studied

Heuristics by which potentially important variables can be selected were discussed in Chapter 4. It is better to include variables assumed to have a minor influence in a first design and have these assumptions confirmed, than to introduce overlooked variables at a later stage.

1.2. Run a pilot experiment

The first thing to do is to run a pilot experiment and observe all that can be observed:

* *Are there colour changes?* A transient colour indicates the presence of an intermediate. This may indicate that the *rate of addition of the reagents* can be critical.

* *Is there evolution of heat?* It is strongly recommended that the pilot experiment is run with a thermometer inserted in the reaction mixture. Evolvement of heat may indicate that the stirring rate *and* the efficiency of the temperature control are critical.

* *Which by-products are formed?* Analyse the reaction mixture and *identify* the principal by-products. When the experimental conditions are varied by a design, rather large variations of the reaction conditions are explored. Sometimes, the amounts of by-products and their distributions will show a significant variation in the explored domain. If the secondary products are known beforehand, they can be evaluated as secondary responses in the design, and thereby give clues to a better understanding of the underlaying reaction mechanisms.

1.3. Make replicate runs of the pilot experiment

Sometimes it is impossible to reproduce an experimental result. The magnitude of variation is too large to be due to a random error. When this happens, the cause of the failure is often that some important experimental factor has been overlooked and is out of control. Under such circumstances, it is rather pointless to attempt to run an experimental design.

The ability to reproduce the pilot experiment is therefore a good check to see if a screening design will be meaningful. There is another benefit of a replicated pilot experiment: the observed variations in response give an estimate of the experimental error, which then can be used to evaluate the significance of estimated effects.

1.4. Determine a suitable range of variation of the experimental variables

Especially with new reactions, it can be difficult to assign the range of variation of the experimental variables which will produce an observable effect. The experimental domain is by no means clearly fixed. If there are any doubts in this respect, make sure that a sufficiently large domain has been chosen by running a pilot test using the two extreme experiments (one with all variables at their low level, and another with all variables at their high level). If the results in these experiments are significantly different, a screening design will probably reveal *which* variables are responsible for the observed variation.

1.5. Choose a design for the screening experiment

The choice of design will depend on how detailed the desired information is to be. If it is strongly suspected that certain variables will interact, it is recommended that terms for these interaction should be included in the model, and a design be selected which can estimate these parameters. Use a Resolution V design by which each two-variable interaction can be estimated, or use a Resolution IV design which will confound the interaction effect. These can then be isolated by complementary runs.

Full factorial designs: Such designs are the best choice when the number of variables is four, or less. A full four-variable factorial design gives estimates of all main effects and two-variable interaction effects, and also an estimate of the experimental error variance. This is obtained from the residual sum of squares after a least squares fit of a second-order interaction model, see (Example: Catalytic hydrogenation, p. 112). A full factorial design should be used if individual estimates of the interaction effects are desired. Otherwise, it is recommended first to run a half fraction 2^{4-1} ($I = 1234$), and then run the complementary fraction, if necessary, (see Example: Synthesis of a semicarbazide, p. 135).

Fractional factorial designs: The first choice when there are more than four variables, should always be to attempt a fractional factorial design. The confounding patterns are easily obtained from the generators. It is also easy to append complementary runs to resolve any ambiguities.

Resolution III designs can be constructed to accomodate three variables in four runs, up to seven variables in eight runs *etc.* Such designs can be augmented to Resolution IV designs by fold-over. Resolution IV designs are useful, since they make it possible to detect presence of strong interaction effects.

It is, of course, possible to construct Resolution IV designs directly, by using three-variable interaction column to define the "extra" variables. This is advantageous with eight variables, which by this technique can give a resolution IV design in 16 runs, 2^{8-4} (**I = 1235 = 1246 = 1347 = 2348**).

With more than eight variables it is probably more practical to use a Resolution III design first, and run a complementary fraction by fold-over.

With more than 15 variables, a Plackett-Burman design will reduce the number of experimental runs.

To establish a complete second order interaction model, a Resolution V (or higher) design must be used. Such designs can be constructed by using four-variable interaction columns to define the "extra" variables, e.g. 2^{5-1} (**I = 12345**), 2^{10-5} (**I = 12346 = 12357 = 12458 = = 13459 = 2345·10**)

This is, however, not the first choice with many variables involved. A more economical strategy is to establish such models when the significant main effects have been estimated through an initial, first-order model design.

Plackett-Burman designs: Do not consider a Plackett-Burman design as a first choice. These designs are Resolution III designs, but the confounding pattern in such designs is more difficult to analyze than that of fractional factorial designs.

A Plackett-Burman design may be the preferred choice if: *(a)* a linear model is assumed to be satisfactory; *(b)* the number of variables is too high to permit the number of runs of a fractional factorial design.

A twelve-run Plackett-Burman design can accomodate eleven variables. With 12 - 15 variables, a fractional factorial design is better (confounding pattern can be analyzed). With more than 15 variables a Plackett-Burman is the preferred choice: 19 variables in 20 runs, 23 variables in 24 runs. However, that many variables rarely need to be considered. The original paper by Plackett and Burman[2] gives details on how to construct even larger designs.

D-optimal designs: There are certain situations when a D-optimal design is a good choice. It should, however, be remembered that to construct such designs, the experimenter must *specify a reasonable model*, and that the design is optimized to estimate the parameters of this model and *nothing else!* D-optimal designs are very useful when only a few variables are likely to have interaction effects. A model with linear terms and the *specified* interaction terms would then be reasonable.

It is possible to construct D-optimal designs to obtain estimates of the model parameters from a minimum of experiments. Therefore, a D-optimal design should be used, only when the model is assumed to be known and when a minimum of experiments is necessary.

Another case where a D-optimal design is suitable is, when the experimental domain is restricted by constraints, which make it impossible to use two-level designs to cover the entire domain of interest.

A disadvantage of D-optimal designs is that they cannot be constructed by hand. It is necessary to use a computer.

1.6. Screening is just the beginning

Evaluation of the screening experiment will identify the significant variables. Some variables have probably turned out to have a negligible or only a minor influence. The insignificant variables and variables with only minor influence can be fixed at favourable or convenient settings when the reaction is further investigated.

Check also the consequences for the interpretation of the result if the insignificant variables are "removed" from the design by projecting the screening design down to the space spanned by the significant variables.

Try to fit a response model using the significant variables only. Make diagnostic plots of the residuals to analyze the fit. If an independent estimate of the experimental error is available, use analysis of variance of the reduced model to find out if there is a significant lack of fit. If the reduced model withstands the diagnostic tests, it is a good indication that the important variables have been found.

The best validation of a model is, of course, that predictions from the model can be verified by running the corresponding experiment

Often, it turns out that the experimental domain used in the first design was not well chosen, and that improvements can be expected outside the explored domain. In Chapters 10 and 11 is discussed how a better domain can be established from an intial screening experiment. This will actually be the beginning of an optimization of the procedure. The techniques for this are discussed in the chapters to follow.

References

1.   G.E.P. Box, W.G. Hunter and J.S. Hunter
     Statistics for Experimenters
     Wiley, New York 1978.

2.   R.L. Plackett and J.P. Burman
     Biometrika 33 (1946) 305.

Chapter 9

Introduction to optimization

1.   The problem

The problem of optimizing a synthetic reaction can be illustrated geometrically as in Fig. 9.1

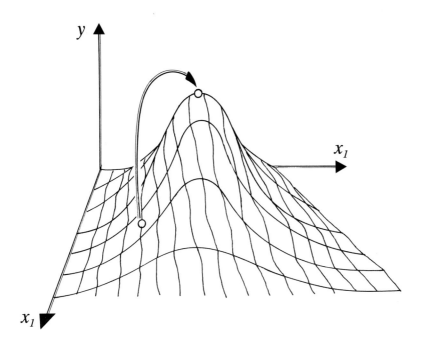

Fig.9.1: The optimization problem.

The investigation often starts under conditions which are far from the optimum conditions and the problem is to reach the optimum with a minimum of effort.

In addition to just *locating* the optimum conditions, it is also important to know *how sensitive* the optimum result is to variations in the experimental conditions around the optimum. Thus, we need to know if the response surface is very pointed, which implies that the conditions must be carefully controlled, or if it is rather flat and can tolerate variations without a serious drop in the performance of the reaction.

In the three chapters to follow, we shall see that the techniques for determining the optimum conditions are not at all complicated. It is therefore astonishing that the problem of optimization often is approached by means of inappropriate methods.[1]

* The experimental conditions have been examined by adjusting one variable at a time, a notoriously poor method when there are interactions among the variables.

* The problem is avoided by adding a foot-note "*yields are not optimized*", even if the paper claims to be presenting *a new method*.

* By presenting the best result ever obtained as the "optimum result", without any comment on how this conclusion as to the optimality has been reached.

* By a more elaborate, but equally misleading, one-variable-at-a time analysis, viz. to use an hypothesized reaction mechanism to *derive* the optimum conditions from physical chemical models. Such models are as a rule univariate models, and as such they cannot account for interaction effects.

2.   The methods

There are three straightforward methods to establish optimum conditions by experimental studies:[2]

---

[1] The examples of inappropriate methods given, were found as a result of personal experience: Seven years ago (1984) I surveyed the recent literature on new synthetic methods with the objectives to select good candidates for experimental testing in an advanced course on "Preparative organic chemistry" for graduate students. I scrutinized more than 2000 recent papers, some of them published in "prestige" journals. Of these papers, *only four! (4)* presented new methods which had been adequately optimized.

[2] There is a branch of mathematics which deals with "optimization theory". In this context the *object function* is assumed to be known. In our context, the object function is the response surface. An analytical expression for this is definitely *not known*.

The first two methods are useful for locating a near-optimum region, but they are not well suited for a more precise location of the specific optimum.

The method of steepest ascent determines the direction from an initial experimental domain which has the steepest slope upwards along the response surface, and hence point towards the optimum conditions. A series of experiments can then be run along this steepest ascent vector. This will lead to rapid improvements. The method was introduced by Box and Wilson[1] and was the first method for systematic multivariate optimization experiments in chemistry.

The sequential simplex search[2] does the same job as the method of steepest ascent, but with this method, the experiments are run one at a time. *Which* experiment should be performed next is determined from the previous experiments. Thus, an iterative progression along the path of the steepest ascent is accomplished.

Response surface modelling is used to locate the detailed optimum conditions. The principle is to establish a response surface model which maps the optimum region. The map can then be used for navigation in the optimum region. Close to an optimum the surface is curved and to describe the general features of the response surface quadratic models are used. The response surface methods have been and are being developed by Box and coworkers.[3]

3. The requisites

3.1. Define a criterion to optimize

To analyze the results it is necessary to have a criterion of optimality. Often, the most interesting response, e.g. the yield of the desired product, is used as the criterion. The objective is the to maximize the response, or to reach a level which is satisfactory.

The method of steepest ascent and the simplex search can handle only one criterion, while the response surface methods allow simultaneous mapping of several responses. Response surface modelling can therefore be used to optimize several responses simultaneously. The problem of multiple responses is elegantly handled by PLS modelling. This technique is discussed in Chapter 17.

3.2. Use the significant variables in the optimization experiments

An initial screening has revealed *which* variables have a significant influence. To achieve a rapid progression towards the optimum conditions, it is evident that manipulation of the significant variables will be the most rewarding procedure. Fortunately, in most cases there are rarely more than a handful of significant variables. This implies that the number of experiments which needed to reach the optimum can be limited.

3.3. Use continuous variables for optimization

To be able to adjust the experimental conditions toward an optimum performance, *all variables must be continuous*. This means, that when we have run a screening experiment in which discrete variations were also included, we should use the result from the screening to determine the settings of the discrete variables so that they define the most promising *experimental system*: Solvent, catalyst, type of reagent, equipment etc. The optimization experiments are then run with this system.

3.4. An optimum synthetic method

The fact that conditions for a quantitative *conversion* to the desired product have been established does not mean that an optimum *method* has been obtained. A quantitative conversion is, of course, an excellent result which may lead to an optimum method.

A method will define all intermediary steps from the starting material to the *isolated* product. Therefore, the work-up procedure must also be considered and eventually be adjusted before an optimum method has been established.

References

More detailed reference lists are given in the chapters which describe the methods.

1. G.E.P. Box and K.B. Wilson
   J. Roy. Statist. Soc. Ser. B 13 (1951) 1.

2. W. Spendley, G.R. Hext and F.R. Himsworth
   Technometrics 4 (1962) 441.

3. G.E.P Box
   Biometrics 10 (1951) 16.

Chapter 10

Steepest ascent

1.   Principles

The method of steepest ascent was the first method by which multivariate experiments could be designed with a view to achieving a systematic improvement of the result. It was described by Box and Wilson[1] as early as 1951, and has been much used over the years, especially in industrial experimentation. The underlying principles are simple as can be seen in the following geometric illustration.

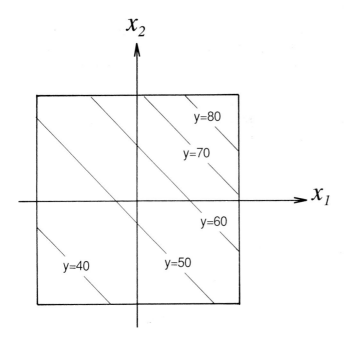

Fig.10.1:   A response surface, showing that the best result is obtained at the limit of the explored domain.

A linear response surface is shown in Fig. 10.1. It is seen that the best (highest) response within the domain is at the limit. It is also seen that an increased response can be expected outside the explored domain, in a direction which describes the steepest path upwards along the response surface. When this direction is known, it is possible to run a series of experiments in which the settings of the variables are adjusted to follow the path of the steepest ascent, see Fig 10.2.

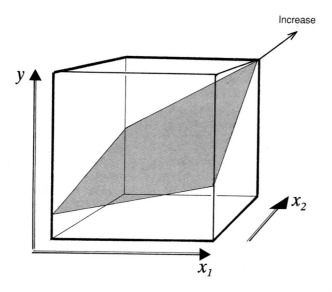

Fig 10.2: The method of the steepest ascent.

The experiments are carried out in this direction as long as they lead to improvements. This will cease to be the case when the experiments have reached a domain where the response surface is no longer monotonous, e.g. the path of the steepest ascent has passed over a rising ridge, Fig 10.3a, or has reached a domain where the surface is saddle-shaped, Fig. 10.3b.

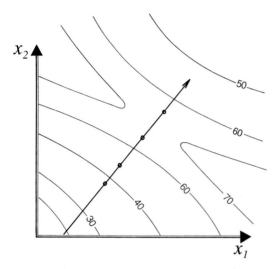

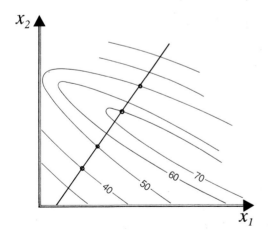

Fig.10.3: When the path of the steepest ascent has reached a ridge or a saddle point on the response surface, the experiments will not give any further improvements.

214

When this happens, there are four possibilities: *(1)* A satisfactory result has been obtained, and further investigations are unnecessary. *(2)* A new direction of the steepest ascent is determined and the investigation is continued in this direction, Fig. 10.4. *(3)* The turning point is probably close to the optimum conditions and to locate the optimum more precisely, a second order response surface model is established to map the optimum domain. *(4)* The improvements obtained are not significant enough, and the investigation is abandoned.

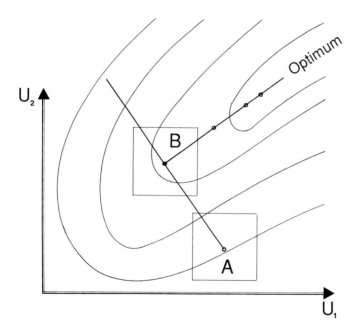

Fig.10.4:   When no further improvements are obtained along the path of the steepest ascent first determined, new direction can be determined from a series of experiments in the near-optimum domain.

The method will thus not permit a detailed localization of the optimum conditions. It is a method by which it is possible to rapidly arrive at a *near-optimum* experimental domain.

1.1 Direction of the steepest ascent

It is realized that the direction of the steepest ascent is related to the slope of the response surface. Thus, to determine this direction, it is necessary first to determine the slope of the response surface by means of a linear approximation of the response surface model.

$$y = \beta_0 + \Sigma \, \beta_i \, x_i + e$$

Often, an initial screening experiment can be used to determine the model.

How the experimental variables, x_i, should be changed, $x_i + \Delta x_i$, to give a maximum increase, Δy, of the response can be explained as follows.[1]

$$y + \Delta y = b_0 + \Sigma \, b_i \, (x_i + \Delta x_i) + e$$

which can be simplified to

$$\Delta y = \Sigma \, b_i \, \Delta x_i + e$$

In matrix notation, the above expression is written

$$\Delta y = [b_1 \; b_2 \; .... \; b_k] \begin{bmatrix} x_1 \\ x_2 \\ . \\ . \\ x_k \end{bmatrix} + e = b'\Delta x + e$$

---

[1] It is also possible to derive the procedure as a constrained maximization problem, by using the method of *Lagrange's* multipliers. This procedure is more elaborate, and can be applied also to second and higher order response surface models. This is in essence the method of *Ridge analysis*[2]. We do not go into this here.

This is a scalar product of the vectors b and Δx, which also can be written

$$b' \Delta x = \|b\| \cdot \|\Delta x\| \cdot \cos(\phi)$$

where $\|b\|$ and $\|\Delta x\|$ are the norms of the vectors b and Δx;

$$\|b\| = (b'b)^{1/2} = (\Sigma b_i^2)^{1/2} \geq 0$$

$$\|\Delta x\| = (\Delta x' \Delta x)^{1/2} = (\Sigma \Delta x_i^2)^{1/2} \geq 0$$

$\phi =$ is the angle between the vectors b and Δx

The scalar product has its maximum valued when $\cos(\phi) = 1$, i.e. the angle ϕ is zero, which is obtained when the vectors are parallel.

For a maximum increase in the response, each variable, x_i, should therefore be adjusted by a change Δx_i which is proportional to b_i.

It is therefore possible to design a series of experiments to explore the path of the steepest ascent, by giving increasing values to the contant k, see Fig. 10.5.

$$\Delta x = b \cdot k \quad (k \text{ is a constant} > 0)$$

$$
\begin{bmatrix} \Delta x_1 \\ \Delta x_2 \\ . \\ . \\ . \\ \Delta x_k \end{bmatrix}
=
\begin{bmatrix} b_1 \\ b_2 \\ . \\ . \\ . \\ b_k \end{bmatrix}
\cdot k
$$

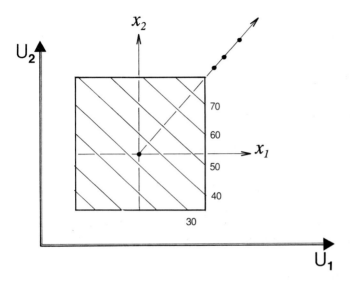

Fig.10.5:   A series of experiments is defined by a successive displacement along the direction of the steepest ascent.

The coordinates for points along the direction of the steepest ascent, i.e. the settings of the experimental variables are given by the following expression

$x = x_0 + \Delta x$

$$
\begin{bmatrix} x_1 \\ x_2 \\ . \\ . \\ . \\ x_k \end{bmatrix} = \begin{bmatrix} x_{10} \\ x_{20} \\ . \\ . \\ . \\ x_{k0} \end{bmatrix} + \begin{bmatrix} \Delta x_1 \\ \Delta x_2 \\ . \\ . \\ . \\ \Delta x_k \end{bmatrix}
$$

Where $x_0 = [x_{10}\ x_{20}...x_{k0}]'$ is starting point for the steepest ascent path. For convenience, this point can be the center of the experimental domain ($x_1 = x_2 = ... = x_k = 0$). This gives

$$x = b \cdot k$$

A starting value of k can be set so that the first experiment is done just ouside the explored domain. How to do this is illustrated by an example.

1.2 Example: Olefin rearrangement

To illustrate how a design should be laid out, we shall use the example of olefin rearrangement which was discussed in Chapter 5. Three variables were studied in a factorial design 2^3. Table 10.1 summarizes the variables and the explored domain.

Table 10.1: Variables and experimental domain in the olefin rearrangements study

Natural variable	Coded variable	Levels of coded variables		
		-1	0	1
u_1	x_1: Amount of catalyst/substrate (mol/mol)	0.01	0.03	0.05
u_2	x_2: Reaction temperature ($^\circ$ C)	40	50	60
u_3	x_3: Reaction time (h)	0.5	0.75	1.0

The scaling of the x-variables have been done by the following relation:

$$x_i = (u_i - u_i^\circ) / \delta_i$$

where u_i° is the average of the high and low levels of u_i, and δ_i is the range of variation from u_i° to the high level. The following applies to the variables given

$$x_1 = (u_1 - 0.03) / 0.02$$
$$x_2 = (u_2 - 50) / 10$$
$$x_3 = (u_3 - 0.75) / 0.25$$

It is also possible to go backwards from the scaled to the natural variables

$$u_i = u_i^\circ + \delta_i \cdot x_i$$

To apply the method of steepest ascent a linear response surface model is required

$$y = \beta_0 + \beta_1 x_1 + \beta_2 x_2 + \beta_3 x_3 + e$$

From the factorial experiment given in Chapter 5, the following model is obtained

$$y = 80.0 + 2.0 x_1 + 3.5 x_2 + 4.5 x_3 + e$$

The highest yield, 91 %, was obtained with all variables at their high level.

A series of experiments along the path of steepest ascent is obtained by

$$x = x_0 + b \cdot k$$

By setting $x_0 = o$ we obtain

$$x = b \cdot k$$

$$\begin{bmatrix} x_1 \\ x_2 \\ x_3 \end{bmatrix} = \begin{bmatrix} 2.0 \\ 3.5 \\ 4.5 \end{bmatrix} \cdot k$$

By assigning increasing values to k we can leave the explored domain by the path of the steepest ascent. A series of such experiments is given in Table 10.2.

From the relations which define the scaling of the variables, translate Table 10.2 into the natural variables, u_i. This is given in Table 10.3.

However, it is not to be taken for granted that the experiments in Table 10.3 really are located on the path of steepest ascent if the response surface model *should be expressed in the natural variables*. The slopes in different directions of the response surface *are not invariant* when the variables are transformed. When the coded variables, x_i, are translated back to the natural variables, u_i, the step of variation of the natural variables, δ_i, will intervene and this may change the direction of the steepest ascent path. The direction is invariant to a change of variables, only if the steps of variation are *equal* for all variables. If there are different units of measurement, e.g. ° C, h, equivalents of reagents, etc., they are likely to be different.

To see how the path of the steepest ascent is changed when the model is expressed in the natural variables, we shall see an example with two variables.

Table 10.2: Experiments along the path of steepest ascent

Exp no	k	x_1	x_2	x_3	y^{Pred}
1	0.24	0.48	0.84	1.08	88.70
2	0.30	0.60	1.05	1.35	90.95
3	0.36	0.72	1.26	1.62	93.14
4	0.42	0.84	1.47	1.89	95.33
5	0.48	0.96	1.68	2.16	97.52
6	0.54	1.08	1.89	2.43	99.71
7	0.60	1.20	2.10	2.70	101.90

1.3. Different directions of the steepest ascent path due to scaling of the variables

Assume that a linear response surface model in two variables has been obtained.

$$y = B_0 + B_1 x_1 + B_2 x_2 + e$$

The features of the response surface can be illustrated by a contour plot, which can be obtained as follows.

Table 10.3: Experiment in Table 10.2 translated into natural variables

Exp no	Catalyst u_1	Temp. u_2	Time u_3
1	0.0396	58.4	1.02
2	0.0420	60.5	1.09
3	0.0444	62.6	1.16
4	0.0468	64.7	1.22
5	0.0492	66.8	1.29
6	0.0516	68.9	1.36
7	0.0540	71.0	1.43

The model can be rewritten

$$b_2 x_2 = (y - b_0) - b_1 x_1$$

which is equal to

$$x_2 = (y - b_0) / b_2 - (b_1 / b_2) \cdot x_1$$

For any given value of y, the above expression defines a straight line which shows the relation between x_1 and x_2 for which the response is constant $= y$. By assigning increasing values to y, a score of isoresponse lines can be drawn to show the variation of the response over the plane spanned by x_1 and x_2, see Fig. 10.1.

It is seen in Fig. 10.5, that the path of steepest ascent will be in a direction which is perpendicular to the isoresponse lines. An equation for a line in the direction[2] to the isoresponse lines will thus be

$$x_2 = Constant - [- 1/(b_1 / b_2)] \cdot x_1$$

i.e.

$$x_2 = Constant + (b_2 / b_1) \cdot x_1$$

From this follows, that along the path of the steepest ascent, a change of x_1 by Δx_1 is related to a change of x_2 by Δx_2 according to

$$\Delta x_2 = (b_2 / b_1) \cdot \Delta x_1$$

If we choose to change Δx_1 in proportion to b_1, i.e. $\Delta x_1 = b_1 \cdot k$, it is seen that

$$\Delta x_2 = b_2 \cdot k$$

which is exactly what was expected.

Now, suppose that x_1 and x_2 have been defined from the natural variables u_1 and u_2 by the usual scaling so that x_i spans the interval $[-1$ to $+1]$; u_i^0 is the average of the high and low level of u_1, and δu_i is the step from u_i^0 to the high level.

---

[2] A straight line $y = a + bx$ will have a normal with the slope $= -(1 / b)$.

$$x_1 = (u_1 - u_1^0) / \delta u_1$$

$$x_2 = (u_2 - u_2^0) / \delta u_2$$

The response surface model

$$y = b_0 + b_1 x_1 + b_2 x_2 + e$$

can thus be written

$$y = b_0 + b_1 \cdot [(u_1 - u_1^0) / \delta u_1] \cdot u_1 + b_2 \cdot [(u_2 - u_2^0) / \delta u_2] \cdot u_2 + e$$

which can be simplified to

$$y = C + (b_1/\delta u_1) \cdot u_1 + (b_2/\delta u_2) \cdot u_2 + e$$

With the constant $C = b_0 - b_1 u_1^0/\delta u_1 - b_2 u_2^0/\delta u_2$. The model can be rewritten as

$$(b_2/\delta u_2) \cdot u_2 = (y - C) - (b_1/\delta u_1) \cdot u_1$$

i.e.

$$u_2 = [(y - C)/(b_2/\delta u_2)] - [(b_1/\delta u_1)/(b_2/\delta u_2)] \cdot u_1 + e$$

The direction of the steepest ascent in this case will be

$$-\{- 1/[(b_1/\delta u_1)/(b_2/\delta u_2)]\} = [(b_2/\delta u_2)]/[(b_1/\delta u_1)] = (b_2/b_1) \cdot (\delta u_1/\delta u_2)$$

An equation to describe the variation of u_1 and u_2 along the path of steepest ascent will therefore be

$$u_2 = Constant + [(b_2/b_1) \cdot (\delta u_1/\delta u_2] \cdot u_1$$

From this follows that the vector, Δu, describing the variation of u_1 and u_2 in the direction of the steepest ascent will be

$$\Delta u = \begin{bmatrix} \Delta u_1 \\ \Delta u_2 \end{bmatrix} = \begin{bmatrix} (b_1 / \delta u_1) \\ (b_2 / \delta u_2) \end{bmatrix}$$

This is *not* the same direction as was obtained for the scaled variables x_1 and x_2.

The increments of each natural variable, u_i, is determined by the response surface slopes, b_i, in the space spanned by the x_i variables (x-space), *and* the step of variation, δu_i, used to scale the natural variables to unit variation in the x-space. From this it is seen, that the directions of the steepest ascent in the x-space and the u-space are equal, (Δx parallel to Δu), only if the scaling factors, δu_i, are all equal.

2. Advantages and disadvantages of the method of steepest ascent

2.1. Advantages

The method makes it possible to rapidly move the investigation to an experimental domain which is closer to the optimum conditions than the domain first considered. This fits well into the general feature of all experimental studies as an interactive process, in which the questions become more detailed as more experience is gained. In retrospect, the initial assumptions often seem naive when more detailed knowledge has been acquired.

The method can be used directly after an initial screening. The effects of the variables as determined from the screening experiment can be used to define the direction of the steepest ascent. This allows of a stepwise approach.

To determine the direction of the steepest ascent, *all* estimated slopes along the variable axes are used. This means that even variables with a minor influence are used for maximum profit. We cannot be *sure* that variables associated with small linear coefficients are *without* any influence. If they have a small influence, this is picked up by the model parameter, and the method of steepest ascent will pay account for even such small influences.

2.1. Disadvantages

The method is sensitive to the range of variation explored in the initial design. The direction of the steepest ascent will depend on the scaling of the variables.

The linear model, used to define the direction of the steepest ascent, is a *local* model, which is valid only in the explored domain. When experiments are run along the direction of the steepest ascent, this is actually an extrapolation from the known domain. It is evident that such extrapolations (like all extrapolations) become more and more unreliable the longer we elongate from what is really known.

References

1. G.E.P. Box and K.B. Wilson
   J. Roy. Statist. Soc. B 13 (1951) 1.

2. *(a)* A.E. Hoerl
   Chem. Eng. Progr. 55 (1959) 69;

   (b) N.R. Draper
   Technometrics 5 (1963) 469.

Chapter 11

Simplex methods

1.   A sequential technique

In the preceding chapter was discussed how a near-optimum experimental domain can be established through the method of steepest ascent. A disadvantage of this method is that it requires that the slopes of a linear approximation of the response surface model is established first. When it is reasonable to assume that the most influencing experimental variables are known, it is possible to locate a near-optimum domain through alternative procedures, viz. the simplex methods. In these methods, a minimum initial set of experiments is used to span a variation in the experimental domain. From these experiments, it is then possible to determine in which direction an improved response is to be expected. A new experiment is run in this direction. If the new experiment afforded an improvement, it is then used to determine a new direction in which a further improvement is to be expected. This is continued until a satisfactory result has been obtained. The simplex methods are defined by a set of heuristic rules which describe how the conditions of the next experiment are determined from the conditions used in the previous experiments. After the initial set of experiments used to determine a first move towards an improved domain, all subsequent experiments are run one at a time. The simplex methods are interactive and sequential. They are sometimes referred to as *sequential simplex methods*.

1.1.   What is a simplex?

The word *simplex* is a geometrical term which denotes a polytope with $(k + 1)$ vertices ("corners") in a k-dimensional space:

In two dimensions, $k = 2$, a simplex is a triangle; in three dimensions, $k = 3$, it is a tetrahedron etc. see Fig. 11.1. Although we cannot imagine the shape of simplexes in higher dimensions than three, they exist as mathematical constructions.

226

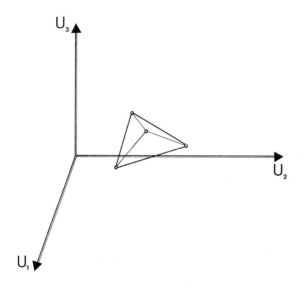

Fig. 11.1: A simplex in three dimensions.

The important issue is: We cannot draw a straight line which fits to all vertices of a triangle. There will always be one vertex which does not fit. We cannot fit a plane to all vertices of a tetrahedron.[1]

Assume that there are k experimental variables. Let each variable define a coordinate axis. If a simplex is placed in this k-dimensional space, the coordinates of the vertices will define different settings of the experimental variables, see Fig. 11.2.

As the vertices of a simplex are not linearly related to each other, an experimental design defined by a simplex forces the experimenter to change each variable in relation to each other in an uncorrelated manner over the set of $(k + 1)$ experiments. A consequence of this is that the experimental result will carry

---

[1] Mathematically this means that the vectors defined by the coordinates of the vertices of a simplex are linearly independent.

information on the influence of the individual variables. This information can then be used to adjust the variables to improve the result.

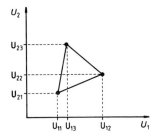

Fig.11.2: Settings of the experimental variables defined by a simplex.

2. How to use a simplex for optimization

2.1. Principles

The basic principles are *illustrated* with two variables; the simplex is a triangle. It will be evident how the principles can be extended to any number of variables.

A contour plot of a response surface is shown in Fig. 11.3. The aim is to locate the optimum conditions. Neither the shape of the response surface, nor the location of the optimum domain is known when we begin the experimental studies.

If three experiments are laid out so that the variable settings define the coordinates of the vertices of a simplex, one of the experiments is likely to give a poorer result than the other two. This experiment has been run under conditions which are more elongated from the optimum conditions than the conditions used for the two "better" experiments. We shall therefore move away from the poor conditions. The next step will be to run a new experiment so that the variables settings of the new experiment and the settings of the remaining two "better" experiments form a *new* simplex. The new simplex will be oriented away from the poorest conditions of the first simplex, see Fig. 11.4a.

The results obtained in the vertices of the new simplex are compared with each other and the poorest vertex is discarded and replaced by another vertex oriented in such a way that it forms a third simplex with the remaining two "better" vertices of the second simplex, etc. After a series of experiments the simplex has approached the optimum domain by a systematic moving away from the poor experimental conditions, see Fig. 11.4b.

228

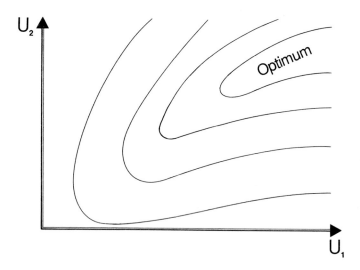

Fig.11.3: At the outset of an experimental study the location of the optimum is not known.

In practice, this means that the series of simplex experiments will describe a zig-zag path along the direction of the steepest ascent. The decision in which direction to move is made on the basis of the last simplex run. Hence, the method does not involve any drastic extrapolations. The movement upwards will be adjusted to follow the direction of the steepest ascent, even if the slopes of the response surface is changed when we move away from the initial experimental domain.

When the simplex method is used, it is not common to run replicated experiments. If a measured response should be erroneous, and thus result in a move in the wrong direction, this will be detected after a few experiments, and the movements will be corrected to follow the right path again. It could also be that a response erroneously was assumed to be a very good result. The rules of the simplex method say that if the same vertex is retained after k moves, the corresponding experiment should be repeated. This makes it possible to detect erroneously "good" responses. The method is therefore *robust*.

Close to an optimum, further attempts to improve the result will fail. The simplexes will encircle the optimum conditions. To locate the optimum conditions

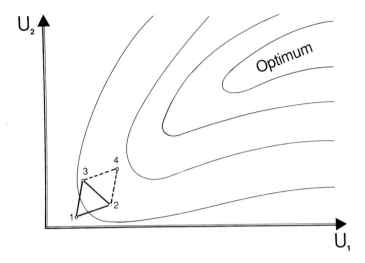

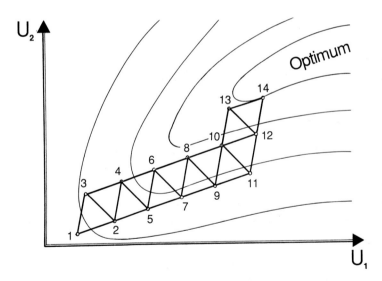

Fig. 11.4: Optimization by a simplex search: *(a)* Starting simplex and the first move from the poor conditions, *(b)* Progression of the simplex towards the optimum conditions.

more precisely, it is possible either to shrink the size of the simplex, or to run a set of complementary experiments to obtain a second order response surface model which can map the optimum domain.

The method described above is, in essence, the *Basic simplex method* which was described 1962 by Spendley, Hext and Himsworth [1] as a procedure for improvement of industrial processes. In this method, the experiments are laid out by a *regular simplex* (all vertices are at the same euclidian distance to each other). Any simplex can be made to be a regular simplex by a proper scaling of the variables axes. The starting simplex can therefore always be regarded as a regular simplex. In the Basic simplex method, each move has a fixed step-lenghts and defines a new regular simplex. This may result in a slow progress towards an optimum, especially if the the initial simplex was small.

To permit a more rapid convergence towards an optimum, several modified simplex methods have been suggested.[2] By these modified methods, the step-length of the next move is adjusted depending on the degree of improvement.[2] If a new vertex should give a considerable improvement, it is rather natural to try to move further in that direction. One such modified simplex method is given in detail below, after a presentation of the basic simplex method.

2.2. The number of variables to use in a simplex search

One should always keep in mind *why* the experiments are being run. The simplex method is used when we wish to find a *better* experimental domain than was initially considered. For this, it will be sufficient to adjust the most influencing variables. These can be determined by an initial screening experiment. In principle, it is possible to investigate any number of variables by means of the simplex method. The method is efficient when the number of variables, k, is not too high. With 2 – 4 variables, the method is excellent; with $k \geq 8$ it becomes rather clumsy to use.

2.3. A design matrix of a regular simplex

There are innumerable ways to *orient* a simplex in a k-dimensional space. The detailed orientation of a starting simplex does not have any significant influence on how rapidly the method converges towards an optimum. A study of this was given in

---

[2] Several modifications have been developed and analyzed by simulation using mathematically defined response functions. For instance the *Rosenbrook banana function*, $y = C(x_1 - x_2)^2 + (1 - x_1)^2$, has been very popular in this context. Often several hundreds of iterations have been used to show that the particular modification presented has an increased rate of convergence compared to its competitors. Such exercises have little relevance for practical experimentation.

the original paper by Spendley, Hext and Himsworth [1], and the same conclusions have been reached by others.[2]

In this section, a useful design is given to define a regular starting simplex. The design matrix is given in scaled design variables, x_i. To obtain an experimental design expressed in natural variables, u_i, the following procedure should be applied.

Define a starting value, u_i^0, of the natural variable, and a step-length for its variation, δu_i. The step length will define how large a part of the experimental domain is covered by the design, i.e. the "size" of the simplex.

The setting of the natural variable, in experiment j is obtained from the scaled design variables by the following relation:

$$u_{ij} = u_i^0 + x_{ij} \cdot \delta u_i$$

The simplex is oriented so that one vertex corresponds to the best conditions hitherto known. The settings of the variables in this experiment is used as starting values for the variation. The other vertices are placed so that the simplex is oriented symmetrically from the origin into the first quadrant ($k = 2$), octant ($k = 3$), n-tant ($n = 2^k$) of the space spanned by the *design variables*, see Fig. 11.5. The orientation of the simplex in the space of the *natural variables* can be in any direction. The orientation depends on the *sign* of the step-length. When the starting simplex is defined, we can express a desire as to in which direction the simplex should try to move. For instance, it would be a good result if the total reaction time could be shortened. To explore this, a negative δu_{Time} is used. It would also be a good result if the amount of a precious metal catalyst could be reduced. For this a negative $\delta u_{Catalyst}$ is used. If our intentions can be met by the experimental observations, this will result in a rather rapid progress. If the experiments show that our intentions cannot be met, the simplex will move in other directions.

A general design matrix is given in Table 11.1. The values of the coordinates p and q which depend on the number of variables are summarized in Table 11.2.

When the *design variable*, x_i is transformed into the *natural variable*, u_i, the value of the setting can be rounded to handy figures, e.g. if the temperature of an oil-bath is suggested to be 73.86 °C, this would be rather difficult to realize in practice. A set-value of 74 °C shall of course be used in such cases.[3]

---

[3] I have heard of one experiment in an industrial laboratory where the experimenter had tried in vain to adjust the pH to be exactly as specified by the fourth decimal place in a computer-generated design. This was a rather senseless attempt to achieve unnecessary perfection.

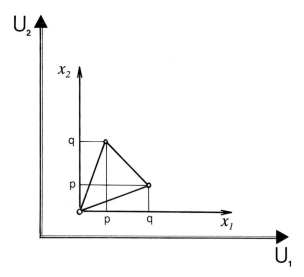

Fig. 11.5: A starting simplex in two dimensions.

To define a *regular* simplex, the values of p and q should be chosen so that each vertex has the same euclidian distance to all other vertices. If this distance, for convenience, is set to $d = 1$, the values of p and q will be as described in Table 11.2. below. How these values have been determined is described below.

The distance from vertex 1 to a vertex j is

$$[(p - 0)^2 + (k - 1)(q - 0)^2]^{1/2} = 1$$

After taking the squares of both sides, we obtain

$$p^2 + (k - 1)q^2 = 1$$

Table 11.1: Design matrix

Exp. no	x_1	x_2	x_3	x_4	x_i	x_k
1	0	0	0	0	0	0
2	p	q	q	q	q	q
3	q	p	q	q	q	q
4	q	q	p	q	q	q
5	q	q	q	p	q	q
.						
.						
.						
j	q	q	q	q	p	q
.						
.						
.						
k + 1	q	q	q	q	q	p

The distance between two vertices i and j is

$$[(p - q)^2 + (q - p)^2 + (k - 1)(q - q)^2]^{1/2} = 1$$

After taking the squares of both sides, we obtain

$$2(p - q)^2 = 1$$

These relations give the following system of equations

$$p^2 + (k - 1)q^2 = 1$$

$$(p - q)^2 = 0.5$$

with the roots

$$p = 1/k\sqrt{2} \cdot [k - 1 + \sqrt(k + 1)]$$

$$q = 1/k\sqrt{2} \cdot [\sqrt(k + 1) - 1]$$

From this gives the values of p and q summarized in Table 11.2. were computed for different values of k

Table 11.2: Values of the coordinates p and q of the vertices of the simplex design

Number of variables, k	Coordinates p	q
2	0.9658	0.2580
3	0.9478	0.2357
4	0.9256	0.2185
5	0.9121	0.2050
6	0.9011	0.1940
7	0.8918	0.1847
8	0.8839	0.1768

3. The Basic simplex method

3.1. Rules

In the Basic simplex method, each vertex is at equal distance to other vertices, the simplex is *regular*. In two dimensions it is an equilateral triangle; in three dimensions it is a regular tetrahedron. Each move will therefore have a fixed step-length due to this constraint. The method is defined by a set of rules:

Start

With k variables, run $(k + 1)$ experiments so that the variables settings define coordinates of the vertices of a simplex in the space spanned by the k variable axes.

Rule 1

Discard the vertex which has given the poorest result, and run a new experiment such that the settings of the variables of the new experiment define a vertex of a new simplex defined by the k remaining better experiments and the new experiment. The coordinates of the new experiments are obtained by reflecting (mirroring) the worst experiments in the centroid of the (hyper)plane spanned by the k better experiments. With two variables, this is equivalent to going from the worst vertex, **W**, through the midpoint, **P**, of the opposing side of the triangle and placing the new vertex, **R**, so that the distance **WP** is equal to **PR**. With three variables, this entails going from **W** through the centroid, **P**, of the opposing side of the tetrahedron, and placing **R** so that the distance **WP** and **PR** are equal, see Fig. 11.6.

Continue this process as long as the reflected vertex **R** gives an improved result.

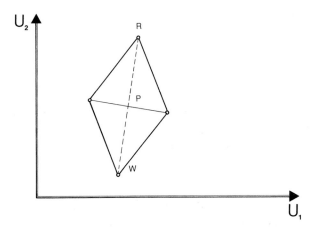

Fig. 11.6: Reflection of the worst vertex.

Rule 2

If the response in a reflected vertex **R** is the worst in the the new simplex, run a new experiment under conditions which are defined by reflecting the *second worst* vertex, either in the preceeding simplex or in the new simplex, then go back to *Rule 1*.

This situation will occur if the simplex moves over a rising ridge. *Rule 2* has been introduced to prevent the simplex from oscillating between two poor conditions. By reflecting the second worst vertex, the simplex can move in another direction.

Rule 3

If a vertex remains after $k + 1$ steps, repeat this experiment and use the newly determined response to evaluate the result.

Such a vertex may be close to the optimum conditions. In that case, a repeated experiment will still give the best result in this vertex. If a satisfactory result is obtained, then stop and use these conditions as the preferred conditions. It is, of course, also possible to continue and more precisely locate the optimum conditions by reducing the size of simplex, or by determining a response surface model and using it as a map to locate the optimum conditions.

The old results may also have been erroneously assumed to be the best result. A repeated run will then give a correction and allow the simplex to proceed. If this is the case, go back to *Rule 1*.

Replacing the old value by the new one gives a protection against time-dependent systematic errors.

Rule 4

If a reflected point should be outside the *possible* experimental domain, assign a very poor result to this vertex, e.g. the yield = 0%.
Go back to *Rule 1*.

Examples of such events are: The suggested temperature is above the boiling point of the solvent. The suggested concentration of a reactant is *negative*.

This rule makes it possible to cope with various constraints imposed on the experimental conditions and forces the simplex to remain in the possible experimental domain.

3.2. Reflection of the worst vertex: Calculation of the variable settings of a new experiment

The calculations will be shown in the natural variables, u_i; the design variables are only used to lay out the starting simplex. After the experiments of the starting simplex have been carried out, the vertex, **W**, which has given the worst (poorest) response should be replaced by a reflected vertex, **R**. The settings of the variables in **R** obtained by mirroring **W** in the (hyper)plane spanned by the remaining vertices, are given by

$$u_{iR} = 2/k \cdot [\Sigma\, u_j] - u_{iW} \quad \text{(where the summation is taken over all } j \text{ except } \mathbf{W}\text{)}$$

These calculation can easily be carried out by hand or by using a pocket calculator.

To show how the above formula has been obtained, we shall use vector notation for the general case. The principles are *illustrated* by a simplex in two variables, Fig. 11.7.

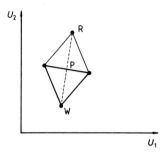

Fig. 11.7: Mirroring the worst vertex in the centroid.

Let **P** be the centroid of the (hyper)plane spanned by the k better vertices, and let **O** be the origin of the variable space. The coordinates of the centroid are given by the vector **OP**, and these coordinates will be the average of the variable settings in the k better experiments.

$$\textbf{OP} = \begin{bmatrix} u_{1P} \\ u_{2P} \\ u_{3P} \\ . \\ . \\ u_{kP} \end{bmatrix} \quad \text{with } u_{iP} = 1/k \cdot \Sigma \, u_{ij} \text{ (where the summation is taken over all } j \text{ except } \textbf{W})$$

The coordinates of the worst vertex are given by the vector **OW**. It is seen in Fig. 11.8 that

OW = OP + PW

238

from which follows

$$-PW = OP - OW$$

A reflection implies that

$$PR = -PW$$

Hence, we can compute

$$OR = OP + PR$$

This can be written

$$OR = OP + (OP - OW) = 2OP - OW$$

which corresponds to the formula given above.

The vector notations are introduced since the computations involved in the modified simplex method are more easily expressed in such notations.

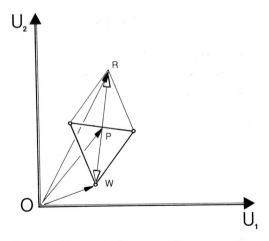

Fig.11.8: Vector additions are used to compute the coordinates of the reflected vertex.

$$
\text{OR} \;=\; 2 \;\cdot\;
\begin{bmatrix} u_{1P} \\ u_{2P} \\ u_{3P} \\ \cdot \\ \cdot \\ u_{kP} \end{bmatrix}
\;-\;
\begin{bmatrix} u_{1S} \\ u_{2S} \\ u_{3S} \\ \cdot \\ \cdot \\ u_{kS} \end{bmatrix}
$$

3.3. Example: Optimization of a Friedel-Crafts alkylation by the Basic simplex method

The alkylation of chlorobenzene with *tert*-butylchloride was studied.[4] The new feature was that ferric chloride ($FeCl_3$) was used as a catalyst. The reaction was conducted at room temperature.

The amount of chlorobenzene was 0.6 mol in all runs and the variables were adjusted according to this constraint. The variables considered are summarized in Table 11.3. It is seen that negative step-lengths were used. The objective was to develop a cheap and convenient procedure, and it was believed that both the reaction time and the relative amounts of reagents could be lowered.

Table 11.3: Variables in the alkylation of chlorobenzene

Variables, u_i	Starting level, u_i^0	Step-length, δu_i
u_1: Time of addition of t-BuCl (min)	60	−30
u_2: Total reaction time (min)	180	−60
u_3: The ratio PhCl / t-BuCl (mol/mol)	3.0	−1.0
u_4: The amount of calatalyst (g)	0.400	−0.100

The simplex design for four variables shown in Table 11.4 was used to define the starting simplex. From Table 11.2 is seen that with four variables, the corresponding values of p and q are:

$$p = 0.9256$$
$$q = 0.2185$$

Table 11.4: Design matrix for the starting simplex in the alkylation of chlorobenzene

Exp no	x_1	x_2	x_3	x_4
1	0	0	0	0
2	p	q	q	q
3	q	p	q	q
4	q	q	p	q
5	q	q	q	p

The design variables, x_i, are transformed into the natural variables by

$$u_{ij} = u_i^0 + x_{ij} \cdot \delta u_i$$

which give the experimental settings shown in Table 11.5.

To demonstrate how the variable settings of the reflected poorest vertex are calculated, the setting of u_1 in *Experiment 6* is shown. The worst vertex is *Experiment 4*.

The relation

$$u_{iR} = 2/k \cdot [\Sigma \, u_j] - u_{iW} \text{ (where the summation is taken over all } j \text{ except } \mathbf{W})$$

corresponds to

$$u_{16} = 2/4 \cdot [60 + 32 + 54 + 54] - 54 = 46$$

The other settings are computed analogously.

Table 11.5: Simplex experiments in the optimization of the alkylation of chlorobenzene

Starting simplex	Exp no	u_1	u_2	u_3	u_4	Yield (%)
	1	60	180	3.00	0.400	92
	2	32	167	2.78	0.378	93
	3	54	125	2.78	0.378	80
	4	54	167	2.07	0.378	65
	5	54	167	2.78	0.300	96
New simplex (1, 2, 3, 5)	+6	46	152	3.60	0.350	99.5
New simplex (1, 2, 3, 5, 6)	+7	42	208	3.30	0.336	99.7
Replicates of 6	8	46	152	3.60	0.350	99.4
	9					99.7

In the last simplex *(Exp no 1, 2, 5, 6, 7)* all vertices afforded a yield > 92 %. The settings of the variables show a rather large variation. This gives important information on the sensitivity of the experimental procedure to variation in yield. The optimum domain is obviously rather flat.

For new synthetic methods, it would be most helpful for future users, if the inventors include a sensitivity test of the suggested experimental conditions, e.g. by a simplex design, when the method is published.

4. Modified simplex methods

4.1. Some initial remarks

Several modifications of the Basic simplex method have been suggested over the years.[2] The common feature of all these modified methods is that the step-length of the next move depends on the result of the last reflection. Some tricks have been suggested by which the simplex can be forced to converge more rapidly. Examples of suggested modifications are: After a failure to improve the result by reflections, the simplex is subjected to a *massive contraction* towards the best vertex by replacing the remaining vertices with the centroids of the opposing (hyper)planes.[2f] Another modification, *Super-modified simplex*[2c,d], determines the location of the next vertex by determining a maximum of the response along the direction from the worst vertex through the centroid of the opposing side. This maximum is obtained by a polynomial regression. In the *Centroid-weighted modified simplex* method, the direction in which one should move is determined after consideration of the response values of all the remaining vertices as well.[2b]

The modified simplex methods have gained considerable popularity in analytical chemistry, especially for the optimization of instrumental methods. Applications in organic synthesis are, however, remarkably few. There are several reasons for this difference:

Instrumental analysis: At the outset of the study, the experimental domain is not firmly settled. The variable settings can be fairly rapidly adjusted (currents in shim-coils to improve the resolution in NMR; potentiometer setting to focus the mass spectrometers; flow-rate in FIA analysis; carrier gas flow, column temperature, temperature program in gas chromatography, etc). The response can be recorded directly as an output signal. Each experiment is rapidly run, and it is not very important if 25 or 40 iterations are necessary to establish the optimum conditions.
Under these circumstances, the modified simplex methods are very convenient.

Organic synthesis: The experimental domain of interest is usually fairly well known. Each experiment is, however, time-consuming. The response must be evaluated by a separate analytical procedure. The number of experiments to achieve an "optimum" result is very important.

4.2 Modified simplex method

A modified simplex method is described below. The procedure is close to the modifications suggested by Nelder and Mead.[2a] Other modifications are described in the works given in the reference list.

The method is described by a set of rules (as for the Basic simplex method). The principles are *illustrated* by a simplex in two variables. The formulae for computations are gereral and can be used for any number of variables. Vector notations are used to describe the method. It is assumed that a *maximum* response is desired. If a minimum is to be found, all relations *greater than* ($>$) and *less than* ($<$) should be reversed in the following rules.

Start

With k variables, run $(k + 1)$ experiments defined by a simplex design. Let **O** be the origin of the variables space, and let **P** be the centriod of the (hyper)plane opposing **W**.

Range three of the vertices: **W** (worst), **N** (next to worst), and **B** (best). The coordinates of these vertices are given by the vectors **OW**, **ON**, and **OB**.

Rule 1

Discard the worst vertex, **W**, and run an experiment at the reflected vertex, **R**, such that

$$\mathbf{PR} = -\mathbf{PW}$$

This gives

$$\mathbf{OR} = \mathbf{OP} + \mathbf{PR} = 2\mathbf{OP} - \mathbf{OW}$$

which is the same as for the Basic simplex method, i.e.

$$u_{iR} = 2/k \cdot [\Sigma u_j] - u_{iW} \quad \text{(where the summation is taken over all } j \text{ except } \mathbf{W})$$

The next step depends on the response obtained in **R**. (The notation used to signify the vertex will also be used to denote the corresponding response).

If **R** > **B**, go to *Rule 2*
If **R** < **W**, go to *Rule 3*
If **W** < **R** < **N**, go to *Rule 4*
If **N** < **R** < **B**, keep **R** as the new vertex and continue according to *Rule 1*
If **R** is outside the possible experimental domain, go to *Rule 3*.

Rule 2

Run an experiment in **E** (extended reflection, see Fig. 11.9), such that

$$\mathbf{PE} = 2\mathbf{PR} = -2\mathbf{PW}$$

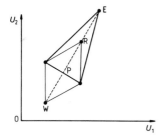

Fig. 11.9: Extended reflection of the worst vertex.

It is seen that

OE = OP + 2PR = 3OP − OW   i.e.

$u_{iE} = 3/k \cdot [\Sigma u_j] - u_{iW}$ (where the summation is taken over all j except **W**)

(a) If **E > R**, keep **E** as the new vertex;
(b) If **E ≤ R**, keep **R** as the new vertex.
Go back to *Rule 1*

Rule 3

Run an experiment in **C$_W$** (contracted towards **W**, see Fig. 11.10), such that

PC$_W$ = 0.5·PW

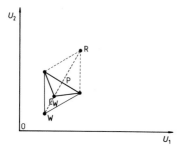

Fig. 11.10: Contraction of the simplex towards the worst vertex.

It is seen that

$$OC_W = OP + 0.5 \cdot PW \quad \text{i.e.}$$

$u_{iCW} = 1/2k \cdot [\Sigma u_j] - 0.5 \cdot u_{iW}$ (where the summation is taken over all j except **W**)

(a) If $C_W \leq W$, run an experiment in the vertex defined by reflecting the next to worst vertex, **N**;

(b) If $C_W > W$, keep C_W as the new vertex.
Go back to *Rule 1*

Rule 4

Run an experiment in C_R (contracted reflection, see Fig. 11.11), such that

$$PC_R = 0.5 \cdot PR$$

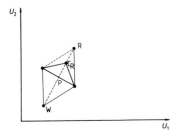

Fig. 11.11: Contracted reflection of worst vertex.

It is seen that

$$OC_R = OP + 0.5 \cdot PR \quad \text{i.e.}$$

$u_{iCR} = 3/2k \cdot [\Sigma u_j] - 0.5 u_{iW}$ (where the summation is taken over all j except **W**)

(a) If $C_R < W$, keep **R** as the new vertex.

(b) If $R < C_R < N$, keep C_R as the new vertex.

(c) If $W < C_R < R$, keep **R** as the new vertex.

(d) If $N < C_R$, keep C_R as the new vertex.

Go back to *Rule 1*.

Rule 5

If a vertex is retained after $(k + 1)$ moves, it may be close to the optimum conditions. Run this experiment again and use the newly determined response for evaluation.

When to stop

The search is terminated when a satisfactory result has ben obtained, or when either of the following happens: *(a)* The reflections do not give any further improvement. Use the best vertex to specify the preferred conditions. *(b)* The responses of the vertices of the last simplex are all similar and assumed to be within the noise level of the experimental error. Use the average of these settings to specify the preferred conditions.

It is, of course, also possible to map the near-optimum domain by a response surface model, and then use this model to locate the detailed optimum conditions.

5. A few comments on the choice of simplex method

If the objective is to *improve* the yield and to locate a near-optimum experimental domain, any of the simplex methods will do the job.

If the objective is to establish the *maximum possible yield*, none of the methods will achieve this without an excessive number of experiments. Response surface modelling is often more efficient for this purpose.

5.1. Design

If a screening experiment has been run and the principal variables have been identified, as well as their main effects on the response, the method of steepest ascent can be used. The direction of the path of the steepest ascent is, however, sensitive to the scaling of the variables used in the screening experiment. The ambiguities associated with this can be overcome if a simplex method is used, and in this case the simplex design given above is appropriate. The signs of the step-length when the design variables are translated into the natural variables should then be the same as the signs of the corresponding main effects from the screening experiment. This will orient the simplex *towards* the direction of the steepest ascent path, and this may afford a rapid progress.

If no such details as to the effects of the experimental variables are known, it is a matter of taste which design should be used to define a starting simplex; personally I prefer the simplex design given above.

5.2. Choice of simplex method

Be flexible in the choice of simplex method. It can be difficult to state *a priori* which method will be the preferred in a given situation. If a reflected vertex should give a considerable improvement, it would be natural to try an experiment in an extended vertex.

The Basic simplex method is good when the experimental domain is not too large. If prior knowledge of the influence of the variables is available, the size of the simplex can be chosen to give a good progression along the path of the steepest ascent. Close to an optimum, the simplex will encircle the optimum conditions. As these experimental points are evenly distributed, they can be incorporated in a equiradial design (see Chapter 12) which can be used to fit a response surface model in the optimum domain.

The modified simplex method is good when prior knowledge on the possible experimental domain is scarce. The size of the simplex will be corrected automatically to ensure a reasonable progress: A too large initial simplex will be contracted, a too small simplex will expand. The modified simplex will, in general, give a rapid initial improvement by quickly moving away from a poor first domain. Close to an optimum it will be less efficient. As the experimental points in the optimum domain will be unevenly distributed, they are not well suited for response surface modelling.

References

1. W. Spendley, G.R. Hext and F.R. Himsworth
    Technometrics 4 (1962) 441.

2. *(a)* J.A. Nelder and R. Mead
    Computer J. 7 (1964) 308;

    (b) B. Ryan, R.L. Barr and H.D. Todd
    Anal. Chem. 52 (1980) 1460;

    (c) P.B. Routh, P.A. Swartz and M.B. Denton
    Anal. Chem. 49 (1977) 1422;

    (d) P.F.A. van der Wiel
    Anal. Chim. Acta 122 (1980) 421;

    (e) E.R. Åberg and A.G.T. Gustavsson
    Anal. Chim. Acta 144 (1982) 39;

    (f) R.R. Ernst
    Rev. Sci. Instr. 39 (1968) 988.

3. R. Carlson, L. Hansson and T. Lundstedt
    Acta Chem. Scand b 40 (1986) 444.

248

4.   N. Effa
      Diplome d'Etude Approfondie, IPSOI, Marseille 1978.

Suggestions for further reading

Tutorial articles:

K.W.C. Burton and G. Nickless
Optimization via Simplex. Part I. Background, Definitions and Simple Application
Intell. Lab. Syst. 1 (1987) 135.

E. Morgan, K.W. Burton and G. Nickless

Optimization Using the Modified Simplex Method
Intell. Lab. Syst. 7 (1990) 209.

E. Morgan, K.W. Burton and G. Nickless
Optimization Using the Super-Modified Simplex Method
Intell. Lab. Syst. 8 (1990) 97.

Chapter 12

Response surface methods

1. Preliminaries

1.1. Why use response surface models?

The important technique of response surface modelling for the study of how response variables are related to variations in the experimental conditions was introduced by George E. P. Box and coworkers early in the fifties. [1] The technique has been extensively used in many areas of experimental work. It is especially well suited to the study of synthetic chemical procedures, and this was actually one of the first areas to which the method was applied.[2]

In the preceding chapters it was discussed how a near-optimum experimental domain can be found by the method of steepest ascent or by a simplex search. However, these methods cannot be used to efficiently *locate* the optimum conditions. For this, response surface modelling is a far more efficient technique.

By means of a response surface model it is not only possible to locate the optimum conditions, but also to analyze how sensitive the optimum conditions are to variations in the settings of the experimental variables. Another advantage of response surface models is that it is possible to make different projections which provide graphic illustrations of the shape of the surfaces, thus allowing a visual interpretation of the functional relations between the response and the experimental variables. This may give an intuitive "feeling" of what is going on in the system studied. Interdependencies among the experimental variables will influence the actual shape of the response surface. Such relations can be examined by *canonical analysis* of the response surface model. This may give valuable clues to mechanistic details of the reaction. Canonical analysis is discussed later in this chapter.

The principles behind response surface modelling are simple. If the experimental domain is not too large, an unknown "theoretical" response function, f

$$\eta = f(x_1, x_2, ... x_k)$$

can be approximated by a *Taylor expansion* (see Chapter 3). The variable settings corresponding to the optimum conditions will give the best result, all other settings will give an inferior result. This means that the response surface is curved around the optimum. It will therefore be necessary also to incorporate squared terms, $\beta_{ii}x_i^2$, in the Taylor expansion to arrive at an adequate description of the response surface in the optimum domain.

$$y = \beta_0 + \Sigma \beta_i x_i + \Sigma\Sigma \beta_{ij} x_i x_j + \Sigma\Sigma \beta_{ii} x_i^2 + e$$

From the model it is then possible to analyze the roles played by the experimental variables. The model will be a map of the explored domain.

The technique to establish the response surface models is also simple: Experiments are laid out by a design which spread the settings of the experimental variables over the experimental domain of interest. The model parameters are then estimated by a least squares fitting of the model to the experimental results obtained in the design points. We will use b_0, b_i, b_{ij} and b_{ii} to denote the *estimated* values of the "true" Taylor expansion coefficients β_0, β_i, β_{ij} and β_{ii}. Whenever the *attempted* model is presented, the "true" β coefficients are used. When the parameters of this model have been estimated, they will be denoted by b coefficients.

In the previous chapters it was seen that the linear coefficients, β_i, and the rectangular (interaction) coefficients, β_{ij}, can be efficiently estimated by a two-level factorial or fractional factorial design. To determine also the square coefficients, β_{ii}, it will, however, be necessary to explore the variations of the experimental variables on more than two levels. One possibility would be to use a multi-level factorial design to define a grid of experimental points in the domain. However, with r levels and k variables, the number of experiments, r^k, increases rapidly and becomes prohibitively large when the number of levels and the number of variables increase.

With two experimental variables, a three-level factorial design, 3^k, is convenient. One example is shown below. With more than two variables, it is much more convenient to use a composite design. Such designs are discussed after the example below.

1.2 Example: Formation of an ether by acid catalyzed dimerisation of an alcohol. Response surface model from a 3^2 factorial design. [3]

This example is given to show the geometric illustration of a response surface model.

The formations of an ether by acid catalyzed dimerisation of an alcohol[1] was studied.

$$2\ R_3CCH_2OH \xrightarrow{\text{Nafion}^{\circledR}} R_3CCH_2OCH_2CR_3 + H_2O$$

The objective was to determine how the yield of the ether depends on the *reaction temperature* and the *amount of acid catalyst* (Nafion®). The catalyst is a strongly acidic perfluorinated sulfonic acid ion-exchange resin. The levels of the variables are given in Table 12.1, and the experimental design and yields obtained are given in Table 12.2. The yields obtained after 10 h are given.

A least squares fit of a quadratic response surface model to the data in Table 12.2 gave the following model

$$y = 10.30 - 0.47\ x_1 - 0.45\ x_2 - 2.45\ x_1x_2 + 0.26\ x_1^2 - 4.59\ x_2^2 + e$$

A plot of this response surface is shown in Fig. 12.1.

Table 12.1: Variables and experimental domain in the alcohol dimerisation study

Variables	Levels		
	−1	0	1
x_1: Amount of Nafion®/substrate (mol/mol)	6	9	12
x_2: Reaction temperature (°C)	150	160	170

The three extra experiments at the center point ($x_1 = x_2 = 0$) were included to allow an estimate of the lack of fit of the model. We will not perform any statistical analysis of this particular model. The lack-of-fit tests will be discussed in the context of the composite designs.

---

[1] Due to secrecy agreements with a chemical company, the detailed structure of the alcohol cannot be given.

Table 12.2: 3^2 Factorial design and yields obtained in the alcohol dimerisation study

Exp. no	x_1	x_2	Yield, y, (%)
1	−1	−1	3.3
2	0	−1	7.6
3	1	−1	8.1
4	−1	0	11.9
5	0	0	10.3
6	1	0	9.3
7	−1	1	8.7
8	0	1	3.9
9	1	1	3.7
10	0	0	10.6
11	0	0	10.3
12	0	0	9.9

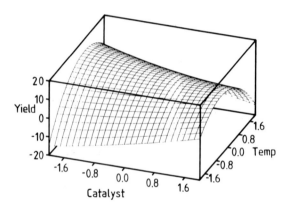

Fig. 12.1: Response surface obtained in the dimerisation of an alcohol.

It is seen in Fig. 12.1 that the reaction temperature has a strong non-linear influence, and that the variation in yield shows a maximum along the temperature axis. The curvature in the other direction is feeble and presumably not significant. This feeble curvature has an opposite sign compared to the temperature curvature, and the shape of the surface is as an elongated saddle. In fact, the surface is very close to a maximum ridge which passes diagonally through the experimental domain. The direction of the ridge is determined largely by the interaction term. Such ridges are common in chemical systems. A more detailed account will be given in the section on *Canonical analysis* below.

2. Step-wise strategy by composite designs

At the outset of an experimental study, the shape of the response surface is not known. A quadratic model will be necessary only if the response surface is curved. It was discussed in Chapters 5 and 6 how linear and second-order interaction models can be established from factorial and fractional factorial designs, and how such models might be useful in screening experiments. However, these models cannot describe the curvatures of the surface, and should there be indications of curvature, it would be convenient if a complementary set of experiments could be run by which an interaction model could be augmented with squared terms.

In fact, such a strategy is available through the composite designs. These designs were introduced by Box and coworkers[1,4] and have been very much used. The composite designs are linked to the criterion of rotatability, mentioned in Appendix 7C.

2.1. Rotatability

A response surface model which has been determined by regression to experimental data can be used to predict the response for any given settings of the experimental variables. The presence of a random experimental error is, however, transmitted into the model and gives a probability distribution of the model parameters. Hence, the precision of the predictions by the model will depend on the precision of the parameters of the response surface model. The error variance of a predicted response, $V(y_i)$, for a given setting, $x_i = [x_{1i}, x_{2i}, \dots x_{ki}]'$ of the experimental variables is determined by the *variance function*, d_i, introduced in Chapter 7.

$$V(y_i) = z_i'(\mathbf{X'X})^{-1}z_i' \cdot \sigma^2 = d_i \cdot \sigma^2$$

The matrix $(\mathbf{X'X})^{-1}\sigma^2$ is the variance-covariance matrix of the experimental design which has been used to establish the response surface model. The vector z_i is the vector of variables of the model, $z_i = [1, x_{1i}, x_{2i}, \dots, x_{ki}, x_{1i}x_{2i}, \dots, x_{ri}x_{ji}, x_1^2, \dots, x_k^2]'$.

An experimental design is *rotatable* if the variance function, d_i, is constant for all experimental settings, $x_i = [x_{1i}, \dots, x_{ki}]'$, at the same euclidian distance r from the center point, $x_0 = [0, \dots, 0]'$, of the design, see Fig. 12.2.

$$\|x_i - x_0\| = r \implies d_i = \text{constant}$$

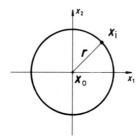

Fig. 12.2: Rotatability: The variance function is constant for all settings x at the same distance from the center point.

To achieve rotatability, certain requirements are imposed on the experimental design. To derive the general concept of rotatability would require a rather elaborate treatment, which is beyond the scope of this book. For a thorough treatment, see Box and Hunter [4]; see also Myers [5].

2.2. Central composite rotatable designs

These designs consist of three parts:

(1) A factorial or fractional factorial design, which is used to estimate the coefficients of the linear and the interaction terms.

(2) At least one experiment at the center of the experimental domain $(x_1 = x_2 = ... = x_k = 0)$. This experiment has two purposes: *(a)* If it is replicated it affords an estimate of the experimental error variance, and *(b)* it gives an opportunity to assess the presence of curvature.

(3) Experiments symmetrically spaced at $\pm \alpha$ along the variable axes. These experiments are used to estimate the coefficients of the square terms. The value of α depends on the number of experiments, N_F, of the factorial part of the design. For obtaining rotatability, α should be equal to the fourth root of N_F.

$$\alpha = (N_F)^{1/4}$$

Examples of designs matrices are given below and the corresponding geometric illustrations of the distribution of the experimental points are given in Fig. 12.3.

Central composite rotatable design for two variables

	x_1	x_2
Factorial points	-1	-1
	1	-1
	-1	1
	1	1
Center points	0	0
	0	0
	$\cdot$	$\cdot$
	0	0
Axial points	-1.414	0
$\alpha = (4)^{1/4} \approx 1.414$	1.414	0
	0	-1.414
	0	1.414

2.3. How many experiments should be run at the center point?

To fit a quadratic model, it is necessary to run at least one experiment at the center point, otherwise the $X'X$ matrix will be singular and cannot be inverted to give a least squares fit. Although the property of rotatability is obtained with only one experiment at the center, there is a lot to be gained if more than one such experiment is run:

(1) Estimate of the experimental error: Replication of the center point experiment gives an independent estimate, s^2, of the experimental error variance, σ^2, which can be used to asses the significance of the model. It can be used to evaluate the lack of fit by comparison with the residual mean square, as well as to assess the significance of the individual terms in the model. *(2) Check of curvature*: If a linear or a second-order interaction model is adequate, the constant term b_0 will correspond to the expected response, $y(0)$, at the center point. If the difference $y(0) - b_0$ should be significantly greater than the standard deviation of the experimental error as determined by the t-statistic

$$|y(0) - b_0| / s > t_{\alpha/2}^{\text{Crit.}}$$

there is a clear indication that the response surface is curved.

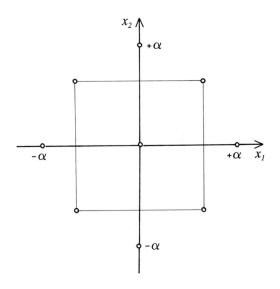

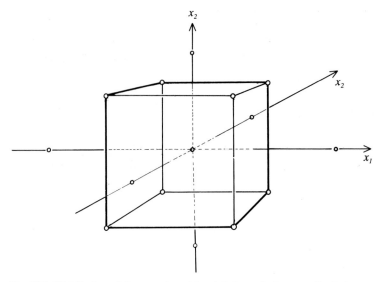

Fig. 12.3: Distribution of the experimental points in central composite designs.

Central composite design for three variables

	x_1	x_2	x_3
Factorial points	−1	−1	−1
	1	−1	−1
	−1	1	−1
	1	1	−1
	−1	−1	1
	1	−1	1
	−1	1	1
	1	1	1
Center points	0	0	0
	0	0	0
	.	.	.
	.	.	.
	0	0	0
Axial points	−1.682	0	0
$\alpha = 8^{1/4} \approx 1.682$	1.682	0	0
	0	−1.682	0
	0	1.682	0
	0	0	−1.682
	0	0	1.682

When linear and second order interaction models are determined from a two-level design, the estimate of the constant term will be an *alias* in which the "true" intercept β_0 is confounded with all square coefficients.

$$b_0 = \beta_0 + \beta_{11} + \beta_{22} + \ldots + \beta_{kk}$$

Experiments at the center point will give an estimate, b_0, of the "true" intercept β_0. The observed difference $b_0 - y(\mathbf{0})$ will estimate the sum of the square coefficients

$$\Sigma\, b_{ii} = b_0 - y(\mathbf{0})$$

To separate the aliased square coefficients, it will be necessary to run experiments on more than two levels so that the curvatures in different directions can be discerned. This is accomplished by the experiments in the axial points.

From the above it is evident how a step-wise strategy can be used:

(1) Run experiments by a factorial or a fractional factorial design and fit a second order interaction model.

(2) Run replicates at the center point and check the curvature.[2]

(4) If a significant curvature is found, run the axial experiments and fit a quadratic model. Use the independent estimate of the experimental error from the replicated center point experiments to check the fit of model.

The number of experiments at the center point will influence the properties of the fitted quadratic model.

2.4. Uniform precision

Rotatability implies that the value of the variance function d_i in a point x_i is only dependent on its distance r to the center point. As d_i depends on the whole design used to fit the model, it will also depend on *the number of experiments at the center.* It is possible to select the number of center point experiments so that d_i is fairly *constant* in the whole experimental domain. This will give almost the same precision of the *predicted* response, $y^{Pred.}$ for all possible settings of the experimental variables in the explored domain. Such a design is called a *uniform precision design.*[4] It is evident that this is a desirable property if the model is to be used for simulations. The number of center points that should be included to obtain a uniform precision design is given in Table 12.3.

2.5. Orthogonal designs

An orthogonal design is a design for which the dispersion matrix $(\mathbf{X'X})^{-1}$ is a diagonal matrix. An orthogonal design ensures independent estimates of the model parameters. For a central composite rotatable design, the model matrix $\mathbf{X}$ will not have such properties that $(\mathbf{X'X})^{-1}$ directly will be diagonal matrix. The column

---

[2] One important conclusion from this is, that it is *always* useful to include at least one experiment at the center point when a factorial or fractional factorial design is run. This will not cause any difficulties if the design is evaluated by hand-calculation: A least squares fit of the linear coefficients, b_i, and the interaction coefficients, b_{ij}, are determined as usual from the factorial points. The intercept, b_0, is computed as the average of *all* experiments. The standard error of the estimated coefficient will, however, be slightly different. Let N be the total number of experiments, and let N_F be the number of factorial points. If σ^2 is the experimental error variance, the standard error of the intercept b_0 will be $\sigma/\sqrt{N}$ and for the linear coefficients and interaction coefficients the standard error will be $\sigma/\sqrt{N_F}$.

Table 12.3: Summary of Central Composite Designs[*]

Number of variables:	2	3	4	5	5	6	6
Factor points:							
Design	2^2	2^3	2^4	2^{5-1}	2^5	2^{6-1}	2^6
Number of factor points, N_F	4	8	16	16	32	32 .	64
Center points:							
Number of center points, N_0							
Uniform precision	5	6	7	6	10	9	15
Orthogonality	8	9	12	10	17	15	24
Axial points:							
Value of α	1.414	1.682	2.000	2.000	2.378	2.378	2.828
Number of axial points, N_A	4	6	8	10	10	12	12
Total number of experiments, N:							
Uniform precision	13	20	31	32	52	53	91
Orthogonality	16	23	36	36	59	59	100

[*] Data are taken from Box and Hunter[4]

vectors of **X** corresponding to the squared variables be neither orthogonal to each other, nor orthogonal to the constant column **I**. It is, however, possible to make a slight transformation of the square columns by subtracting the average of the squared variable from each element of these columns to make all columns of x orthogonal to each other. The model is then fitted with the term $\beta_{ii}(x_i^2 - \overline{x}_i^2)$ to replace the square terms. This transformation will only alter the estimate of the intercept term, b_0. An example of this technique is shown in Appendix 12A. By an orthogonal design it is possible to fit the model by hand-calculations.

The number of center point experiments which should be included to obtain a design with orthogonal properties is also given in Table 12.3. It is seen, that more experiments are needed to achieve an orthogonal design than for uniform precision. A model established by an orthogonal central composite design will therefore yield predictions with smaller variance for variable settings near the center than for settings towards the periphery of the explored domain.

3. Validation of the model

3.1. Lack-of-fit

From the replicated experiments at the center point, we can obtain an estimate, s_1^2 with $(N_0 - 1)$ degrees of freedom, of the experimental error variance, σ^2.

$$s_1^2 = (y_i - \bar{y}_i)^2 / (N_0 - 1)$$

If the fitted model (containing p parameters) is adequate, the residuals,

$$e = y^{\text{Pred.}} - y^{\text{Obs}}$$

will depend only on the experimental error. Therefore, the residual sum of squares, $\Sigma\ e^2$, should be nothing but a summation of the squared errors. This sum can be partitioned into two parts: One due to the lack-of-fit, and the other due to pure error. The lack of fit sum of squares is obtained as the difference between the residual sum of squares and the sum of squared deviations in the center point experiments.

The lack-of-fit sum of squares would thus give another estimate, s_2^2, of the error variance, with $(N - p - N_0 - 1)$ degrees of freedom, *provided that the model is good*.

$$s_2^2 = [\Sigma\ e^2 - \Sigma(\ y_{i0} - \bar{y}_0)] / (N - p - N_0 - 1)$$

The estimates of s_1^2 and s_2^2 are obtained with different degrees of freedom and may thus appear to be different. To check wether or not they are *significantly* different, the F statistic can be used. The ratio $F = s_2^2 / s_1^2$ should not exceed the critical F ratio for the significance level α, and $(N - p - N_0 - 1)$ and $(N_0 - 1)$ degrees of freedom. Usually the significance level $\alpha = 5\ \%$ is used.

If $F = s_2^2/s_1^2 > F^{\text{Crit.}}$ the residuals are too large to be attributed to by the experimental error and the conclusion must be that the model gives a poor description.[3]

A statistical test of the model fit *should always be made* before any conclusions are drawn from the model, e.g. on the location of the optimum conditions. However, an insignificant lack of fit *is not* a sufficient condition for safe conclusions. The only truly

---

[3] Sometimes it is seen that the residual mean square $\Sigma e_i^2 / (N - p)$ is compared to an estimate of the pure error from the replicated center point experiments. This is not quite correct, but will reveal a highly significant lack-of-fit.

reliable way to validate predictions from a model is to compare these predictions with the observed experimental results under the same conditions. If a model suggests a quantitative yield for a given set of experimental conditions, and if this can then be confirmed by experiments, the predictions were correct and the model is reliable. Therefore, any predictions as to favourable experimental conditions *must be confirmed* by experimental runs.

3.2. Analyze residuals

The response surface model is used as a map of the explored domain and will, hopefully, give a reliable description. The model should therefore withstand all possible diagnostic tests. An analysis of the residuals, as was discussed in Chapter 6, may often be informative and it is advisable always to make the following plots as a safe-guard against potential inadequacies of the model:

(a) A normal probability plot of residuals: If the model is adequate, it is assumed that the residuals depend only on a normally distributed random experimental error. A cumulative probability distribution plot of the residuals on normal probability paper should therefore give a straight line.

(b) A plot of the residuals against the predicted response: This plot should not show abnormal behaviour, see section 6.5.4. Sometimes it is found that such plots display a funnel-like scattering of the residual points. This indicates that the experimental error is dependent on the magnitude of the response. In such cases, a transformation of the response is indictated. A short discussion on how this can be done in given in Appendix 12B.

3.3. Example: Optimization of a synthetic procedure by response surface modelling from a central composite design. Enamine synthesis by a modified $TiCl_4$-method

A standard method for enamines synthesis from carbonyl compounds is to heat the parent aldehyde or ketone and a secondary amine in benzene or toluene and to remove the eliminated water by azeotropic distillation. However, this method fails with methyl ketone substrates which are prone to self-condensation under these conditions. These difficulties could be overcome by a procedure using anhydrous titanium tetrachloride as water scavenger.[6] In the original procedure, titanium tetrachloride was added dropwise to a cold solution of the ketone and the amine, followed by prolonged stirring at room temperature. It was later found that the reaction time could be considerably shortened by a modified procedure, in which the carbonyl compound is added to a pre-formed complex between the amine and titanium tetrachloride.[7]

For the modified procedure, it was found that the relative amounts of $TiCl_4$/ketone and amine/ketone which should be used to achieve a rapid conversion were highly dependent on the structure of the ketone.[7,8] In the example below, the formation of the morpholine enamine from methyl isobutyl ketone was studied.

The experimental domain is given in Table 12.4. A uniform precision central composite rotatable design (5 center point experiments) was used. The design and the yields of enamine obtained after 15 minutes are given in Table 12.5.

Table 12.4: Experimental domain in enamine synthesis by a modified $TiCl_4$-method

Variables	Domain				
	−1.414	−1	0	1	1.414
x_1: Amount of $TiCl_4$/ketone (mol/mol)	0.50	0.57	0.75	0.93	1.00
x_2: Amount of morpholine/ketone (mol/mol)	3.0	3.7	5.5	7.3	8.0

A least-squares fit of a quadratic response surface model to the results in Table 12.5 gives

$$y = 93.46 + 2.22\, x_1 + 9.90\, x_2 + 3.42\, x_1 x_2 - 4.37\, x_1^2 - 10.32\, x_2^2 + e$$

A rather large experimental error is evident in the center point experiments. The analysis of variance given in Table 12.6 does not show a significant lack of fit.

A normal probability plot of the residuals is shown in Fig. 12.4. There are evidently some abnormally high errors associated with the replicated center point experiments. Maybe they were run on a different occasion. The remaining residuals

Table 12.5: *Experimental design and yields of enamine by the modified TiCl₄-method*

Exp no	Variables		Yield (%)
	x_1	x_2	y
1	−1	−1	73.4
2	1	−1	69.7
3	−1	1	88.7
4	1	1	98.7
5	−1.414	0	76.8
6	1.414	0	84.9
7	0	−1.414	56.6
8	0	1.414	81.3
9	0	0	96.8
10	0	0	96.4
11	0	0	87.5
12	0	0	96.1
13	0	0	90.5

show a very good fit to normal distribution. The model is only slightly changed if the suspected aberrant results are excluded from the regression analysis. It is therefore concluded that the model fitted to all the experiments gives an acceptable description of the yield variation. Plots of the response surface are shown in Fig. 12.5. It is seen that there is a maximum within the explored domain. This maximum is therefore assumed to represent the optimum conditions. This is further discussed in the next section.

Table 12.6: *ANOVA of the response surface modelling in the enamine synthesis*

Source	SS	df	MS	F
Total		94519.4	13	
Regression				
due to b_0	92637.4	1	92637.4	
due to variable terms	1677.2	5	335.44	
Residuals	204.7	7	29.24	
Partition of RSS				
Lack of fit	133.65	3	44.55	2.51[*]
Pure error	71.05	4	17.76	

[*] $2.51 < F^{Crit.} = 6.59$ (with 3 and 4 degrees of freedom and $\alpha = 0.05$)

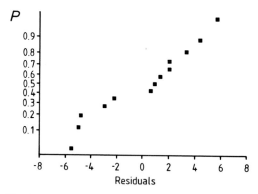

Fig.12.4: Normal probability plot of residuals in the enamine study.

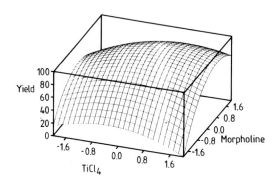

Fig.12.5: Response surface in the enamine study.

4.   Optimum conditions

4.1. A maximum or a minimum within the explored domain

If there is a true optimum *within the explored domain* there will be a turning point on the response surface which will either be a maximum or a minimum. A tangential plane to the surface at this point will have a zero slope in all directions. This is equivalent to saying that the partial derivatives of the response surface model with regard to all experimental variables will be *zero* at this point.

$$\partial y / \partial x_i = 0 \quad \text{for all } x_i$$

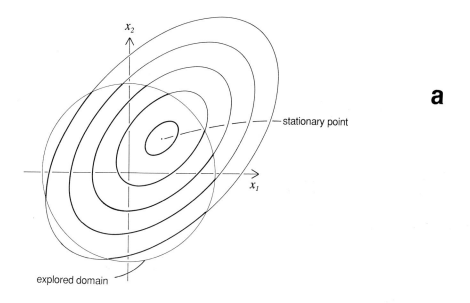

a

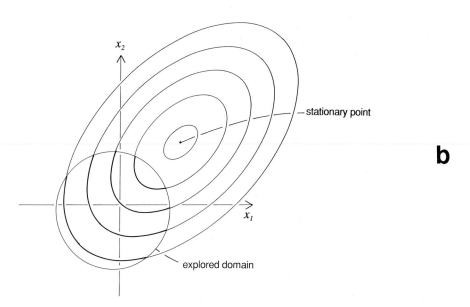

b

Fig.12.6: Stationary points on response surfaces: *(a)* A maximum or minimum within the explored domain; *(b)* A maximum or minium outside the explored domain.

The above relation defines a system of equations and its solution gives the coordinates of the maximum (or minimum) point, see Fig. 12.6a. A point on the response surface in which all partial first derivatives are zero will hereafter be called *a stationary point*.

A stationary point does not always correspond to the optimum conditions. It can be a saddle-point (minimax) at which the surface passes through a maximum in certain directions, and through a minimum in other directions. Saddle points are rather common. To improve (increase) the response, the directions in which the response surface increases should be explored.

The nature of a stationary point is conveniently determined by the canonical analysis.

As an example of how the optimum conditions are determined from the response surface model, the enamine synthesis described above is used. It was seen in Fig 12.5 that there was a maximum in the explored domain. The coordinates for the maximum point are computed as follows:

The response surface model is

$$y = 93.46 + 2.22\,x_1 + 9.90\,x_2 + 3.42\,x_1x_2 - 4.37\,x_1^2 - 10.32\,x_2^2 + e$$

Setting the partial derivatives with respect to x_1 and x_2 equal to zero gives the following system of equations

$$2.19 + 3.42\,x_2 - 8.74\,x_1 = 0$$

$$9.90 + 3.42\,x_1 - 20.64\,x_2 = 0$$

Solution of this equation system gives the roots

$$x_1 = 0.4721 \approx 0.47$$

$$x_2 = 0.5580 \approx 0.56$$

This corresponds to the following settings of the natural variables, u_1 (amount of $TiCl_4$/ketone) and u_2 (amount of morpholine/ketone)

$$u_1 = 0.75 + 0.47 \cdot 0.18 \approx 0.83$$

$$u_2 = 5.5 + 0.56 \cdot 1.8 \approx 6.51$$

Running the reaction under these conditions afforded 98 ± 2 % yield after 15 min. This confirmed the optimum conditions.

4.2. A stationary point outside the explored domain

The most common situation encountered is, however, that the response surface does not have a stationary point within the explored domain. In such cases, the surface is monotonous in certain directions along which it is described by a rising or a falling ridge. It can also be described by a constant ridge. In other directions, the surface may pass over a minimum or a maximum. The best experimental conditions in such cases will be found on the ridge. With rising or falling ridges these conditions will be at the limit of the explored domain, see Fig 12.6b. When constant ridges occur, the optimum conditions will not be in a single point. Any point along the ridge will be satisfactory. Ridge-shaped surfaces are very common when chemical reactions are modelled, and to explore such systems, the most efficient method is to make a canonical analysis of the response surface model.

5. Canonical analysis

5.1. What is canonical analysis, and why is it useful?

A second-order model with square terms can describe a variety of differently shaped response surfaces. It is rather difficult to comprehend how the surface is shaped by mere inspection of the algebraic expression of the model. A canonical analysis constitutes a mathematical transformation of the original model into a form which only contains quadratic terms:

$$y = y_S + \lambda_1 z_1^2 + \lambda_2 z_2^2 + \ldots + \lambda_k z_k^2 + e$$

This transformation is made by replacing the original coordinate system $\{x_1 \ldots x_k\}$ of the experimental space defined by the experimental variables with a new orthogonal coordinate system $\{z_1 \ldots z_k\}$ in which the origin is located at the stationary point of the surface, and in which the z_i axes are rotated so that they are parallel to the main axes of the quadratic surface, see Fig. 12.7. The constant y_S is the calculated response value at the stationary point.

A description of how the rotation of the coordinate system is effected is given below. For the moment it is sufficient to realize that the algebraic expression of the model is highly simplified in the $\{z_1 \ldots z_k\}$ system. As the squared variables z_i^2 in the model cannot be negative we shall see that the shape of the surface is determined by the *sign* and the *magnitude* of the coefficients $\lambda_1 \ldots \lambda_k$:

If all coefficients are negative, the surface will have *maximum* in the stationary point and y_S will be the maximum response. By analogy, it is seen that if all coefficients are positive, y_S will be the minium value.

If the coefficients have different signs, the surface is saddle-shaped and the stationary point is a saddle-point. The surface will pass over a maximum in those z_i directions associated with negative coefficients λ_i, while there will be a minimum in those directions corresponding to positive coefficients. To improve the result in such cases, it would be interesting to explore those z_i directions which have $\lambda_i > 0$, while adjusting the experimental variables so that their settings correspond to $z_j = 0$ for all $\lambda_j < 0$.

It may also be that one or more coefficients are small and have values close to or equal to zero (within the limits of the experimental error). This implies that there are z_i directions in which the response is almost constant and insensitive to variations of the experimental settings, i.e. the surface describes a ridge. When ridges are found, they usually indicate that there are some kind of functional dependence between experimental variables. This, in turn, can give clues to an understanding of the underlying chemical mechanisms.

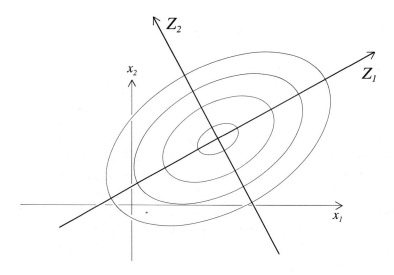

Fig. 12.7: Transformation of the experimental variables into canonical variables.

Rotation of the coordinate system removes the crossproduct terms from the model. A change of origin to the stationary point removes the linear terms. It is obvious that any conclusions as to the nature of the stationary point are reasonable only if the stationary point is within or in the close vicinity of the explored domain. However, it is often found that the stationary point is remote from the design center and that the constant y_S in the canonical model corresponds to a totally unrealistic response value, e.g. a yield > 100 %. It may also occur that the experimental conditions at the stationary point are impossible to attain, e.g. they may involve negative concentrations of the reactants. Under such circumstances, the response surface around the stationary point does not represent any real phenomenon. It should be borne in mind that a polynomial response surface model is a Taylor expansion of an underlying, but unknown, "theoretical" response function, $\eta = f(x_1..x_k)$, and that the polynomial approximation is a *local* model which is valid only in the explored domain. When the stationary point is found to be remote from the explored domain, it is often helpful to use another form of the canonical model, which is obtained by rotation of the coordinate system to remove the crossproduct terms but keeping the original origin at the center point. We shall use capital letters Z_i to denote the the axes of such rotated systems. This will give a canonical model of the form

$$y = \beta_0 + \Theta_1 Z_1 + ... + \Theta_k Z_k + \lambda_1 Z_1^2 + ... + \lambda_k Z_k^2 + e$$

The model contains linear terms in the canonical variables. At first sight, this may not look like a simplification. The following discussion will explain why this form of the model is useful for exploring ridge systems. Ridges occur when the curvature of the surface is feeble in certain directions. These will correspond to Z_i (and z_i) directions associated with small quadratic coefficients λ_i. The corresponding linear coefficients, Θ_i, will describe the *slope* of the ridge *in the explored domain*. From the values of linear coefficients it is therefore possible to determine whether the surface is described by a rising ridge, by a falling ridge, or by a constant ridge. When a synthetic procedure is studied and it is found that the corresponding response surface is described by a ridge in the experimental domain, the conditions which will give the maximum yield will be found at the limit of the domain and in the direction of a rising, maximum ridge, see Fig. 12.6b.

5.2. Calculations

In this section, matrix calculation will be used. For readers who are unfamiliar with matrix calculus, a short summary is given in Appendix: Matrix calculus at the end of this book.

First, the general principles are described. Then, a worked-out example is given to illustrate the calculations.

General principles

There are four steps to be taken to make the transformation to the canonical model:

Assume that the response surface model is:

$$y = B_0 + \Sigma\, B_i\, x_i + \Sigma\Sigma\, B_{ij}\, x_i x_j + \Sigma\Sigma\, B_{ii}\, x_i^2 + e$$

Step (1):

Determine the coordinates of the stationary point, x_S. This is equivalent to determining the roots of the systems of equations obtained by setting all partial derivatives $\partial y / \partial x_i = 0$.

Let y_S be the calculated response at x_S.

Step (2):

Define a new coordinate system $\{w_1 ... w_k\}$ by the transformation

$$\mathbf{w} = \mathbf{x} - \mathbf{x}_S$$

which gives

$$w_1 = x_1 - x_{1S}$$
$$w_2 = x_2 - x_{2S}$$

.

.

.

$$w_k = x_k - x_{kS}$$

This variable transformation moves the origin of the $\{w_1 ... w_k\}$ system to the stationary point, see Fig. 12.8a.

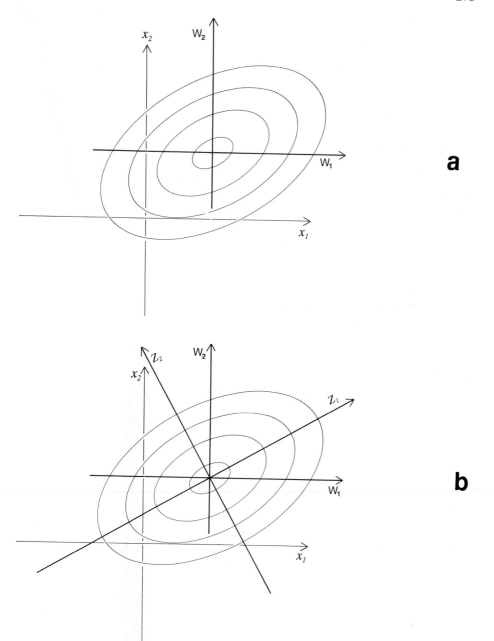

Fig.12.8: Variable transformation in the canonical analysis.

When the response surface model is expressed in the transformed variables, $x_i = w_i + x_{iS}$, the linear terms vanish and the model will be

$$y = y_S + \Sigma\Sigma\, b_{ij}\, w_i w_j + \Sigma\Sigma\, b_{ii}\, w_i^2 + e$$

In matrix notation this can be written

$$y = y_S + \mathbf{w'Bw} + e$$

The coefficient matrix $\mathbf{B}$ is symmetric; the diagonal elements are the square coefficients, the off-diagonal elements are the cross-product coefficients divided by two.

$$
y = y_S + [w_1 \cdots w_k]
\begin{bmatrix}
b_{11} & \tfrac{1}{2}b_{12} & \tfrac{1}{2}b_{13}\cdots & \tfrac{1}{2}b_{1j}\cdots & \tfrac{1}{2}b_{1k} \\
\tfrac{1}{2}b_{12} & b_{22} & \tfrac{1}{2}b_{23}\cdots & \tfrac{1}{2}b_{2j}\cdots & \tfrac{1}{2}b_{2k} \\
\tfrac{1}{2}b_{13} & b_{23} & b_{33}\cdots & \tfrac{1}{2}b_{3j}\cdots & \tfrac{1}{2}b_{3k} \\
\cdot & \cdot & \cdots & \cdots & \cdot\cdot \\
\cdot & \cdot & \cdots & \cdots & \cdot\cdot \\
b_{1j} & \tfrac{1}{2}b_{2j} & & b_{jj} & \cdot\cdot \\
\cdot & \cdot & \cdots & \cdots & \cdot\cdot \\
\cdot & \cdot & \cdots & \cdots & \cdot\cdot \\
\tfrac{1}{2}b_{1k} & \cdot & \cdots & \cdots & b_{kk}
\end{bmatrix}
\begin{bmatrix}
w_1 \\ w_2 \\ w_3 \\ \cdot \\ \cdot \\ w_j \\ \cdot \\ \cdot \\ w_k
\end{bmatrix}
+ e
$$

Step (3):

Determine the eigenvalues and the corresponding eigenvectors to the coefficient matrix $\mathbf{B}$.

An eigenvector to the matrix $\mathbf{B}$ is any vector $v \neq \mathbf{0}$ which fulfils the following relation

$$\mathbf{B}v = \lambda v$$

This means that the vector which is obtained by multiplying v by $\mathbf{B}$ is parallel to v. The number λ is a constant of proportionality which describes the change of metric in this transformation. This number λ is called the eigenvalue of the corresponding eigenvector.

To determine the eigenvalues and the eigenvectors, the following relations apply:

$$\mathbf{Bv} = \lambda v$$

$$\mathbf{Bv} = \lambda \mathbf{I} \ v \quad \text{where } \mathbf{I} \text{ is the identity matrix of the same dimensions as } \mathbf{B}.$$

$$(\mathbf{B} - \lambda \mathbf{I}) \ v = \mathbf{0}$$

This equation will have a non-trivial solution ($v \neq \mathbf{0}$) only if the matrix

$$(\mathbf{B} - \lambda \mathbf{I})$$

is singular, which means that its determinant is zero. The values of λ for which this is fulfilled are the eigenvalues of the matrix $\mathbf{B}$. To determine these eigenvalues, the secular equation is solved.

$$\det(\mathbf{B} - \lambda \mathbf{I}) = 0$$

$$\begin{vmatrix} b_{11} - \lambda & \tfrac{1}{2}b_{12} & \tfrac{1}{2}b_{13} \cdots & \tfrac{1}{2}b_{1j} \cdots & \tfrac{1}{2}b_{1k} \\ \tfrac{1}{2}b_{12} & b_{22} - \lambda & \tfrac{1}{2}b_{23} \cdots & \tfrac{1}{2}b_{2j} \cdots & b_{2k} \\ \cdot & & b_{33} - \lambda & \text{etc} & \cdot \\ \cdot & \text{Symmetric} & b_{jj} - \lambda & & \cdot \\ \cdot & \cdot & \cdot & \cdot & b_{kk} - \lambda \end{vmatrix} = 0$$

This defines an equation of degree k and its roots are the eigenvalues of $\mathbf{B}$. Remember that the eigenvalues will all be real if the matrix is symmetric. Remember also that eigenvectors corresponding to different eigenvalues are orthogonal to each other. The eigenvalues to $\mathbf{B}$ are coefficients of the canonical model. The corresponding eigenvectors are used to define the variable tranformation. The eigenvectors to each eigenvalue are determined from the relation:

$$(\mathbf{B} - \lambda_i \mathbf{I}) \ v_i = \mathbf{0}$$

However, this will not give a unique solution, since any vector αv ($\alpha \neq 0$) can be a solution. Let $v_i = \alpha \cdot [v_{i1}, v_{i2}....v_{ik}]'$ be an eigenvector to λ_i. The norm of $v_i = \|v\|$ is computed as

$$\|v\| = [(\alpha \cdot v_{i1})^2 + (\alpha \cdot v_{i2})^2 + ... + (\alpha \cdot v_{ik})^2]^{1/2} = \alpha \cdot [\Sigma (v_{ij})^2]^{1/2}$$

The summation is taken for $j = 1$ to k.
A normalized eigenvector m_i is obtained by

$$m_i = (1/\|v_i\|) \cdot v_i$$

Let $\mathbf{M} = [m_1\ m_2\ ...\ m_k]$ be the the the matrix in which the column vectors, m_i, are the normalized eigenvectors for $i = 1$ to k.

$$\mathbf{M} \quad = [m_1 \qquad m_2 .... \qquad m_k\ ]$$

$$\mathbf{M} \quad = \begin{bmatrix} m_{11} & m_{12} ...... & m_{1k} \\ m_{21} & m_{22} ...... & m_{2k} \\ . & .... & . \\ . & .... & . \\ m_{k1} & m_{k2} ...... & m_{kk} \end{bmatrix}$$

$\mathbf{M}$ is an orthogonal matrix, which means that $\mathbf{M'} = \mathbf{M}^{-1}$.

The following important relation between $\mathbf{M}$ and $\mathbf{B}$ applies

$$\mathbf{M'B} \quad = \begin{bmatrix} \lambda_1 & 0 & 0 & 0... & 0... & 0 \\ 0 & \lambda_2 & 0 & 0... & 0... & 0 \\ 0 & 0 & \lambda_3 & 0... & 0...0 \\ . & . & . & .... & 0...0 \\ . & . & . & .... & \lambda_j... & 0 \\ . & . & . & .... & .... & 0 \\ 0 & 0 & 0 & .... & .... & \lambda_k \end{bmatrix} = \mathbf{L}$$

L is a diagonal matrix in which the diagonal elements are the eigenvalues. As **M** is an orthogonal matrix it is seen that the following relation applies

B = MLM'

This shows that any real symmetric matrix can be factorized into orthogonal eigenvector matrices and a diagonal matrix of the corresponding eigenvalues.[4]

Step (4):

Make the rotation of the coordinate system $\{w_1...w_k\}$. Let $\{z_1...z_k\}$ be the axes of the rotated coordinate system and define

$$z = \begin{bmatrix} z_1 \\ z_2 \\ . \\ . \\ . \\ z_k \end{bmatrix}$$

by the relation

z = M'w

This will describe a rotation of the coordinate system so that the z_i axes will be oriented along the symmetry axes of the response surface, (see Fig.12.8b). The rotation does not change the metrics of the original coordinate system.

The model expressed in $w_1... w_k$

$$y = y_s + w'Bw + e$$

can therefore now be written

---

[4] This important relation is used in principal components analysis which will be discussed in Chapter 15.

$$y = y_S + (Mz)'B(Mz) + e$$

This can be simplified into

$$y = y_S + z'(M'BM)z + e$$

$$y = y_S + z'Lz + e$$

The last equation corresponds to

$$y = y_S + [z_1 \ z_2 \dots z_k] \begin{bmatrix} \lambda_1 & 0 & 0 & 0 \dots & 0 \\ 0 & \lambda_2 & 0 \dots & 0 \dots & 0 \\ 0 & 0 & \lambda_3 & \dots & 0 \\ \cdot & \cdot & \cdot & \dots & 0 \\ \cdot & \cdot & \cdot & \dots & 0 \\ 0 & \cdot & \dots & \lambda_i \dots & 0 \\ \cdot & \cdot & \dots & \dots & 0 \\ 0 & 0 & 0 \dots & 0 \dots & \lambda_k \end{bmatrix} \begin{bmatrix} z_1 \\ z_2 \\ z_3 \\ \cdot \\ z_i \\ \cdot \\ \cdot \\ z_k \end{bmatrix} + e$$

This is the matrix notation of the canonical model

$$y = y_S + \Sigma \, \lambda_i \, z_i^2 + e$$

5.3. Example: Canonical analysis of the response surface model of the enamine synthesis

The model fitted[5] to the experiments in Table 12.5 is

$$y = 93.460 + 2.219 \, x_1 + 9.904 \, x_2 + 3.425 \, x_1 x_2 - 4.374 \, x_1^2 - 10.324 \, x_2^2 + e$$

The stationary point, x_S was determined to be

---

[5] To avoid round-off errors in the computation, we shall use three decimal precision in the model parameters for the calculations.

$$x_S = \begin{bmatrix} 0.4721 \\ 0.5580 \end{bmatrix}$$

Translation of the coordinate system by the transformation

$$w = x - x_S$$

gives

$$w = \begin{bmatrix} w_1 \\ w_2 \end{bmatrix} = \begin{bmatrix} x_1 - 0.4721 \\ x_2 - 0.5580 \end{bmatrix}$$

When the model is expressed in $x_1 = w_1 + 0.4721$, and $x_2 = x_2 + 0.5580$ we obtain

$$y = 93.46 + 2.219\,(w_1 + 0.4721) + 9.904\,(w_2 + 0.5580) + \\ + 3.425\,(w_1 + 0.4721)(w_2 + 0.5580) - \\ - 4.374\,(w_1 + 0.4721)^2 - 10.324\,(w_2 + 0.5580)^2 + e$$

After developing this it is seen that the linear terms vanish and the model is

$$y = 96.75 + 3.425\,w_1 w_2 - 4.374\,w_1^2 - 10.324\,w_2^2 + e$$

In matrix notation this is

$$y = 96.75 + [w_1 \; w_2] \begin{bmatrix} -4.374 & 1.7127 \\ 1.7125 & -10.324 \end{bmatrix} \begin{bmatrix} w_1 \\ w_2 \end{bmatrix}$$

The eigenvectors to the coefficient matrix is obtained by solving the secular equation

$$\begin{vmatrix} -4.374 - \lambda & 1.7125 \\ 1.7125 & -10.324 - \lambda \end{vmatrix} = 0$$

$$(-4.374 - \lambda)(-10.324 - \lambda) - 1.7125 \cdot 1.7125 = 0$$

$$\lambda^2 + 14.698\,\lambda + 42.2245 = 0$$

The roots are

$$\lambda_1 \approx -3.9163$$
$$\lambda_2 \approx -10.7817$$

The eigenvectors m_1 and m_2 are computed:

Let $v_1 = [v_{11} \ v_{12}]'$ be an eigenvector to $\lambda_1 = -3.9161$. For v_1 the following relation applies:

$$\begin{bmatrix} -4.374 - (-3.9161) & 1.7125 \\ 1.7125 & -10.324 - (-3.9161) \end{bmatrix} \begin{bmatrix} v_1 \\ v_2 \end{bmatrix} = \begin{bmatrix} 0 \\ 0 \end{bmatrix}$$

which gives

$$\begin{bmatrix} -0.4577 & 1.7125 \\ 1.7125 & -6.4977 \end{bmatrix} \begin{bmatrix} v_{11} \\ v_{12} \end{bmatrix} = \begin{bmatrix} 0 \\ 0 \end{bmatrix}$$

This gives the identical equations

$$-0.4577v_{11} + 1.7125v_{12} = 0$$

$$1.7125v_{11} - 6.4077v_{12} = 0$$

i.e. $v_{11} = 3.7415v_{12}$

It is seen that any vector

$$v_1 = \alpha \cdot \begin{bmatrix} 3.7415 \\ 1 \end{bmatrix}$$

is an eigenvector ($\alpha \neq 0$)

The norm $\|v_1\|$ of v_1 is computed

$$\|v_1\| = [(\alpha \cdot 3.7415)^2 + (\alpha \cdot 1)^2]^{1/2} = \alpha \cdot 3.8728$$

The normalized eigenvector $m_1 = (1/\|v_1\|) \cdot v_1$ is

$$m_1 = \begin{bmatrix} 0.9661 \\ -0.2582 \end{bmatrix}$$

The eigenvector m_2 corresponding to $\lambda_2 = -10.7817$ is computed analogously

$$m_2 = \begin{bmatrix} -0.2582 \\ 0.9661 \end{bmatrix}$$

The rotation matrix $M = [m_1 \ m_2]$ is therefore

$$M = \begin{bmatrix} 0.9661 & -0.2582 \\ 0.2582 & 0.9661 \end{bmatrix}$$

The transformations of w to z and *vice versa* are given by

$$w = Mz \quad \text{and} \quad z = M'w$$

After transformation to z the model

$$y = y_S + w'Bw + e$$

will be

$$y = y_S + z'(M'Bm)z + e$$

i.e. $y = y_S + z'Lz + e$

In the present case, this corresponds to

$$y = 96.75 + [z_1 \quad z_2] \begin{bmatrix} -3.9164 & 0 \\ 0 & -10.7819 \end{bmatrix} \begin{bmatrix} z_1 \\ z_2 \end{bmatrix} + e$$

This gives (after rounding) the canonical model

$$y = 96.75 - 3.92\, z_1^2 - 10.78\, z_2^2 + e$$

As λ_1 and λ_2 are both negative, it is seen that the stationary point is a maximum. The response surface is an elliptical hill, which is attenuated in the z_1 direction. The isocontour plot in which the canonical axes have been marked is shown in Fig. 12.9.

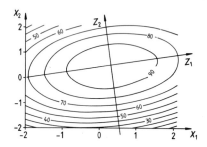

Fig.12.9: Response surface in the enamine study. The curvature of the surface is attenuated along the z_1 axis, cf. Fig.12.5a.

5.4. Canonical analysis and ridges

When it is found that a stationary point is remote from the design center, the response surface in the experimental domain will be described by some kind of ridge system. To determine the optimum conditions in such cases, it will be necessary to explore the ridges. This is most conveniently done by a canonical analysis. In a limiting case with a constant ridge, the corresponding eigenvalue, λ_i, will be exactly zero and the stationary point will be at infinity on the z_i axis. Such limiting cases are rare and eigenvalues which are *exactly* zero are not found in practice. Values *close* to zero are quite common, however. It is also common that certain eigenvalues are significantly smaller than others (in an absolute sense), and that the curvature of the response surface is attenuated in the corresponding z_i directions. These conditions will also give response surfaces which are approximated by ridge systems in the experimental domain. To analyze the response surface model with a view to exploring ridges, a canonical model obtained by rotation, but not involving the shift of origin, is the most convenient one. In this case too, the rotation of the coordinate system for the variable tranformation is given by the eigenvector matrix $\mathbf{M}$. Capital letters Z_i will be used for these variables, to distinguish them from the canonical variables discussed above.

$$x = \mathbf{M}Z \quad \text{and} \quad Z = \mathbf{M}'x$$

Derivation of the model

The estimated response surface model

$$y = b_0 + \Sigma \, b_i \, x_i + \Sigma\Sigma \, b_{ij} \, x_i x_j + \Sigma\Sigma \, b_{ii} \, x_i^2 + e$$

can be written in matrix notation as

$$y = b_0 + \mathbf{x}'\mathbf{b} + \mathbf{x}'\mathbf{B}\mathbf{x} + e$$

in which $\mathbf{b} = [b_1 \ b_2 \ ... \ b_k]'$ is the vector of linear coefficients and $\mathbf{B}$ is the matrix of the second-order coefficients.

In the rotated coordinate system $\{Z_1...Z_k\}$ the model will be

$$y = b_0 + (\mathbf{M}'Z)'\mathbf{b} + (\mathbf{M}Z)'\mathbf{B}(\mathbf{M}Z) + e$$

which gives

$$y = b_0 + Z'(\mathbf{M}'\mathbf{b}) + Z'\mathbf{L}Z + e$$

$$y = b_0 + Z'\Theta + Z'\mathbf{L}Z + e$$

The vector of the linear canonical coefficients $\Theta = [\Theta_1 \ \Theta_2 \ ... \ \Theta_k]'$ is obtained by multiplying $\mathbf{b}$ by the transposed eigenvector matrix, $\mathbf{M}'$. The canonical model will therefore be

$$y = b_0 + \Sigma \, \Theta_i \, Z_i + \Sigma \, \lambda_i \, Z_i^2 + e$$

The standard error of the coefficients in the canonical model will be approximately the same as for the linear and quadratic coefficients in the original model when a rotatable design has been used.

Analysis of the response surface

To increase the response, the model above suggests that the following measures should be taken:

The experimental conditions should be adjusted so that Z_i is zero for all terms with $\lambda_i < 0$. For those Z_i variables for which the eigenvalues are small ($\lambda_i \approx 0$), the experimental conditions should be adjusted along these Z_i axes so that the corresponding $\Theta_i Z_i$ terms give positive increments to y. For eigenvalues $\lambda_i > 0$, the experimental conditions should be adjusted along the corresponding Z_i axes so that both the linear terms and the square terms give an increase in y. If there are several possible such directions, the one which gives the maximum response should be used to define the optimum conditions.

This type of reasoning can be applied regardless of the number of experimental variables. It is therefore not *necessary* to resort to graphical illustration as the sole means of achieving a "feeling" for the phenomenon. Ridges occur when there are some kind of functional relations between the *experimental variables*. An analysis of the ridges by canonical analysis gives a description of the relations between the variables along the ridges. This in turn, may give clues to an understanding of the underlying chemical mechanisms.

5.5. Another method to obtain canonical models with linear terms

Canonical analysis is rather cumbersome to carry out by hand and several computer programs for response surface modelling are available.[9] However, some of these programs have routines for canonical analysis which only give the form with purely quadratic terms. In this section it is shown that it is easy to transform such models into a form with linear terms which is better suited for exploring ridge systems. The technique is to translate the origin of the canonical system back to the experimental domain, either to the design center or to another point within the domain by relocating the intersection of certain axes.

The transformation of the experimental variables into the canonical variables by translation *and* rotation is given by

$$
z = \begin{bmatrix} z_1 \\ z_2 \\ . \\ . \\ z_k \end{bmatrix} = M \begin{bmatrix} x_1 - x_{1S} \\ x_2 - x_{2S} \\ . \\ . \\ x_2 - x_{2S} \end{bmatrix}
$$

Let the origin in the $\{x_1 ... x_k\}$ system have the coordinates $z_0 = [z_{10} \ ... \ z_{k0}]'$ in the $\{z_1 ... z_k\}$ system. It is seen that

$$
z_0 = \begin{bmatrix} z_{10} \\ z_{20} \\ . \\ . \\ z_{k0} \end{bmatrix} = - Mx_0
$$

Then, define a coordinate system $\{Z_1 ... Z_k\}$ which is obtained by translating the $\{z_1 ... z_k\}$ to the center of the experimental domain. This tranformation is given by

$$
Z = \begin{bmatrix} Z_1 \\ Z_2 \\ . \\ . \\ Z_k \end{bmatrix} = \begin{bmatrix} z_1 + z_{10} \\ z_2 + z_{20} \\ . \\ . \\ z_k + z_{k0} \end{bmatrix}
$$

$z = z + z_0$ which gives

$z = z - z_0$

The canonical model can therefore be written

$$
y = y_S + \Sigma \ \lambda_i \ (Z_i - z_{i0})^2 + e
$$

After developing the quadratic parenthesis, we obtain

$$y = b_0 + \Sigma \, \Theta_i \, Z_i + \Sigma \, \lambda_i \, Z_i^2 + e$$

with $\Theta_i = -2 \, \lambda_i \, z_{i0}$

The linear coefficients of the canonical model is therefore easily computed from the corresponding eigenvalues and the coordinates of the stationary point.

By this technique it is possible to make the translation only for those z_i axes along which the coordinates of the design center are remote. To explore the ridges it is not necessary to transform those z_i axes which pass through the experimental domain. An example is shown in the next section.

5.6. Example: Synthesis of 2-trimethylsilyloxy-1,3-butadiene. Transformation to a canonical model for the exploration of a ridge system

A cheap and efficient method for synthesis of 2-trimethylsilyloxy-1,3-butadiene from methyl vinyl ketone was desired. The method had to be practical for molar scale preparations. Although a number of methods for the synthesis of silyl enol ethers from carbonyl compounds are known, none of them were applicable for this conversion. They were either too expensive for use on larger scale, or afforded poor yields on attempted synthesis.

A screening experiment[6] with different types of electrophilic additives showed that a promising procedure for future development was to treat methyl vinyl ketone (MVK) with chlorotrimethylsilane (TMSCl) in the presence of triethylamine (TEA) and lithium bromide in tetrahydrofuran.

To determine the amounts of reagent to be used for obtaining a maximum yield, a response surface study was undertaken. The experimental domain is given in Table 12.7 and the experimental design and yields obtained are summarized in Table 12.8.

---

[6] An account of this experiment is given in Chapter 16.

Table 12.7: Variables and experimental domain in the study of trimethylsilyloxy-1,3-butadiene synthesis.

Variables	Levels of coded variables				
	−1.75	−1	0	1	1.75
x_1: Amount of TEA/MVK (mol/mol)	1.56	1.75	2.00	2.25	2.44
x_2: Amount of TMSCl/MVK (mol/mol)	0.97	1.20	1.50	1.80	2.03
x_3: Amount of LiBr/MVK (mol/mol)	0.70	1.00	1.40	1.80	2.10

The response surface model determined from Table 12.8 was

$$y = 83.57 - 3.76\,x_1 - 0.96\,x_2 + 0.60\,x_3 + 1.71\,x_1x_2 - 0.41\,x_1x_3 - 0.46\,x_2x_3 - \\ - 0.63\,x_1^2 - 2.20\,x_2^2 - 0.88\,x_3^2 + e$$

Analysis of variance of the regression did not indicate any significant lack of fit, $F^{\text{Lack-of-fit}} = 4.74 < F^{\text{Crit}} = 9.01$ ($\alpha = 5\,\%$). The residual plots in Fig. 12.10 do not show abnormal behaviour. It was therefore assumed that the variation in yield was adequately described by the model. The isoresponse contour projections are shown in Fig.12.11

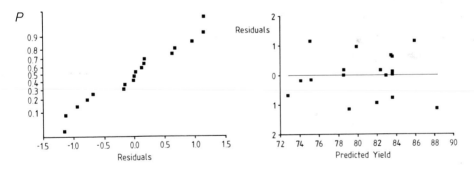

Fig.12.10: Residual plots in the silyloxydiene study: *(a)* Normal probability plot; *(b)* Plot of residuals against predicted response.

286

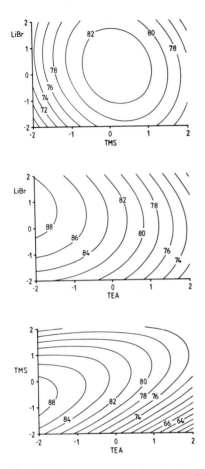

Fig. 12.11: Isocontour projections in the synthesis of silyoxidiene.

Canonical analysis of the fitted model

The stationary point x_S is clearly outside the explored domain

$$x_S = \begin{bmatrix} -8,6733 \\ -3.4906 \\ 3.2737 \end{bmatrix}$$

The value for x_1 at the stationary point corresponds to a negative amount of triethylamine. It is therefore obvious that the response surface does not describe any real chemical phenomenon at the stationary point.

Table 12.8: Experimental design and yields obtained in the study of 2-trimethylsilyloxy-1,3-butadiene

Exp no[*]	Experimental variables			Yield (%)
	x_1	x_2	x_3	y
1	−1	−1	−1	82.9
2	1	−1	−1	72.1
3	−1	1	−1	82.5
4	1	1	−1	77.9
5	−1	−1	1	87.0
6	1	−1	1	73.9
7	−1	1	1	84.1
8	1	1	1	78.5
9	0	0	0	83.8
10	−1.75	0	0	87.1
11	0	−1.75	0	75.0
12	0	0	−1.75	80.8
13	1.75	0	0	76.2
14	0	1.75	0	78.7
15	0	0	1.75	81.0
16	0	0	0	83.6
17	0	0	0	84.2
18	0	0	0	82.8

[*] The experiments were run in random order.

The canonical model

$$y = 99.2273 - 1.1528\, z_1^2 - 0.9842\, z_2^2 - 2.5937\, z_3^2 + e$$

is obtained by the transformation

$$x = \mathbf{M}z + x_S$$

$$\begin{bmatrix} x_1 \\ x_2 \\ x_3 \end{bmatrix} = \begin{bmatrix} 0.8451 & 0.3622 & -0.3934 \\ 0.3931 & 0.0774 & 0.9162 \\ -0.3624 & 0.9284 & 0.0765 \end{bmatrix} \begin{bmatrix} z_1 \\ z_2 \\ z_3 \end{bmatrix} + \begin{bmatrix} -8.6733 \\ -3.4400 \\ 3.2737 \end{bmatrix}$$

If we set $x = 0$, we can compute z_0 (the z_i coordinates of the design center) by the inverse transformation

$$z = M'(x - x_S)$$

This gives

$$\begin{bmatrix} z_{10} \\ z_{20} \\ z_{30} \end{bmatrix} = \begin{bmatrix} 0.8451 & 0.3931 & -0.3624 \\ 0.3622 & 0.0774 & 0.9284 \\ -0.3934 & 0.9162 & 0.0765 \end{bmatrix} \begin{bmatrix} -3.2737 \\ 8.6723 \\ 3.4400 \end{bmatrix}$$

$$\begin{bmatrix} z_{10} \\ z_{20} \\ z_{30} \end{bmatrix} = \begin{bmatrix} 9.8881 \\ 0.3723 \\ -0.4640 \end{bmatrix}$$

These coordinates show that the z_1-axis passes the design center at a distance which is far outside the explored range of variation of the experimental variables.

The origin of the canonical coordinate system is moved into the experimental domain by the transformation

$$Z_1 = z_1 - 9.8881$$

By this transformations the Z_1 axis passes through the design center. If we substitute z_1 by Z_1 in the canonical model we obtain

$$y = 99.2273 - 0.1528 (Z_1 + 9.8881)^2 - 0.9842 z_2^2 - 2.5937 z_3^2 + e$$

Developing this and rounding to handy figures gives

$$y = 84.29 - 3.02 Z_1 - 0.15 Z_1^2 - 0.98 z_2^2 - 2.59 z_3^2 + e$$

From this model it is seen that the response surface is described by a maximum ridge. The ridge is rising in the negative direction of the Z_1 axis. To avoid a decrease in yield, the experimental variables should be adjusted so that z_2 and z_3 both have zero values. If these requirements are fulfilled, an increase in yield is to be expected along a path described by the negative direction of the Z_1 axis. With small variations of Z_1 along this path, it is seen that the influence of the second degree term is negligible. If we now set $z_2 = z_3 = 0$ in the transformation below, we can compute the settings of $x_1 - x_3$ corresponding to different settings of Z_1.

$$
\begin{bmatrix} x_1 \\ x_2 \\ x_3 \end{bmatrix}
=
\begin{bmatrix} 0.8451 & 0.3622 & -0.3934 \\ 0.3931 & 0.0774 & 0.9162 \\ -0.3624 & 0.9284 & 0.0765 \end{bmatrix}
\begin{bmatrix} Z_1 \\ z_2 - 0.3723 \\ z_3 + 0.4640 \end{bmatrix}
$$

This gives

$$
\begin{aligned}
x_1 &= 0.8451Z_1 - 0.317 \\
x_2 &= 0.3931Z_1 + 0.3961 \\
x_3 &= -0.3624Z_1 - 0.3101
\end{aligned}
$$

These relations can be used to determine the variation of the natural variables along the negative direction of the Z_1 axis. To do so, we must rescale the x_i variable to the natural variable u_i. The following transformation is obtained from the scaling of the variables given in Table 12.7.

$$
\begin{aligned}
u_1 &= 0.25x_1 + 2.00 \\
u_2 &= 0.30x_2 + 1.50 \\
u_3 &= 0.40x_3 + 1.40
\end{aligned}
$$

Inserting the values of x_i as functions of Z_1 gives

$$
\begin{aligned}
u_1 &= 0.2113Z_1 + 1.9207 \\
u_2 &= 0.1179Z_1 + 1.6188 \\
u_3 &= -0.1496Z_1 + 1.2760
\end{aligned}
$$

The setting of the natural variables for decreasing values of Z_1 are shown in Table 12.9.

These data show that the yield increases when the relation between the amounts of TEA : TMSCl : LiBr approaches 1 : 1 : 2. For $Z_1 = -5.2$ a quantitaive yield is predicted. Although this is a drastic extrapolation outside the explored domain it makes chemical sense as these conditions correspond to equimolar amounts of TEA and TMSCl relative to methyl vinyl ketone and the double amount of lithium bromide. The canonical analysis thus revealed the stoichiometry of the reaction, which was not previously known.

The next step was to analyze whether or not an excess of the reagents in the relative proportions TEA: TMSCl: LiBr = 1: 1: 2 would improve the yield. By using 1.5 equivalent of TEA and TMSCl respectively and 3 equivalents of LiBr, 96.7 and 98.2 % yields were obtained in a duplicate run. This was considered to be close to the optimum conditions and these settings were then used for preparative applications of the method.[10]

Table 12.9: Settings of the natural variables along the rising ridge in the synthesis of 2-trimethylsilyloxy-1,3-butadiene

Z_1	TEA/MVK u_1	TMSCl/MVK u_2	LiBr/MVK u_3	Predicted yield $y^{Pred.}$
0	1.92	1.62	1.28	84.3
−0.5	1.82	1.56	1.35	85.2
−1.0	1.71	1.50	1.43	87.3
−1.5	1.60	1.44	1.50	88.8
−2.0	1.50	1.38	1.58	90.3
−2.5	1.39	1.32	1.65	91.8
−3.0	1.29	1.27	1.72	93.4
.				
.				
−5.0	0.86	1.03	2.02	99.4

6.   Visualization by projections

An appealing aspect of the response surface technique is that the relations between the response and the experimental variables can be illustrated graphically. We have seen how three-dimensional projections can be drawn to show the shape of the response surface over the plane(s) spanned by the experimental variables taken two at a time. Such projections look nice and can be very convincing[7], but actually they do not provide any information beyond that which can be obtained from the corresponding isocontour plots. In the isocontour plots for variables taken two at a time, the topography of the response surface is illustrated by isoresponse contour *lines*. It is also possible to make drawings of three-dimensional isocontours for variables taken three at a time. In such plots the isocontours will not be defined by lines, but by isoresponse *surfaces*. The shape of such three-dimensional isocontours can be understood by combining two-dimensional contour plots with the result of the canonical analysis. Some examples with three variables will illustrate this.

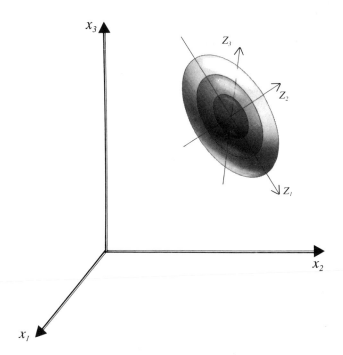

Fig.12.12:   Contour plots of a response surface when all eigenvalues have the same sign. The three-dimensional isocontours describe a score of concentric ellipsoids.

---

[7] The pedagogical and psychological aspect of nice-looking figures should not be neglected, especially not when results obtained by response surface technique are to be presented to persons who have aversions to statistical methods.

(a) If all eigenvalues are negative, and the stationary point is in the explored domain, the maximum response is obtained at the stationary point. The isocontour will define an elliptic shell with the center of the ellipsoid at the stationary point. Different isoresponse levels will thus define a score of concentric ellipsoidic shells. The ellipsoids will be elongated in the z_i direction corresponding to the numerically smallest eigenvalue. The situation may be as illustrated in Fig. 12.12. The intersections of the elliptic shells with the planes spanned by the experimental variables will give two-dimensional projections with elliptic contour lines. (A similar picture is obtained if the eigenvalues is positive, but in that case the stationary point is a minimum.)

It is, of course, also possible to make the projections in e.g. the x_1,x_2-plane for different settings of x_3 and thus obtain a tomographic representation of the three-dimensional structure.

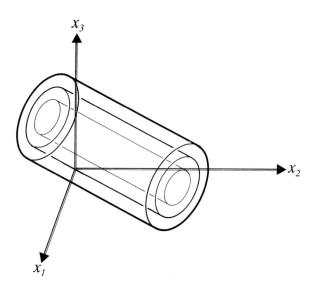

Fig.12.13: Contour plots of a response surface when two eigenvalues have the same sign, and one eigenvalue is zero. The three-dimensional isocontours describe a score of concentric cylinders.

(b) If all eigenvalues are negative, but the stationary point is remote from the design center, a ridge system is obtained. The ridge is oriented along the z_i axis corresponding to the smallest eigenvalue. If the ridge has a significant slope, it will be a maximum rising or falling ridge. In this case too, the isoresponse contours will be described by elongated ellipsoids, but as their center is remote, only the ends of the ellipsoids will dip into the experimental domain. Some projections to the experimental variable planes show elliptic contours, while there will be at least one projection which will describe a score of isocontour lines which are approximately parabolic, see Fig. 12.11.

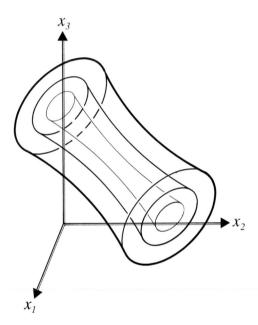

Fig.12.14:   Contour plots of a response surface when the eigenvalues have different sign. The three-dimensional isocontours describe a score of concentric hyperbolic cylinders.

(c) A limiting case of *(b)* is when one eigenvalue, λ_i, is exactly zero, and the stationary point is at infinity on the z_i axis. (This is never met in practice, but eigenvalues close to zero are common.) In such a case the response surface describe a constant ridge along the z_i axis. The three-dimensional isocountours will describe a

score of concentric cylinders. Intersections of these with the planes spanned by the experimental variables may well show elliptic contours in all projections, depending on the orientation of the canonical axes. However, a projection to the planes of the canonical axes will show a maximum ridge. The isocontours will be parallel lines, see Fig. 12.13.

(d) If the eigenvalues have different signs, and the stationary point is in the experimental domain, the isocontours describe a score of concentric hyperbolic cylinders, see Fig. 12.14. Their intersections with the planes spanned by the experimental variables will give at least one projection which describes a saddle point, while other projection(s) will show elliptic contours.

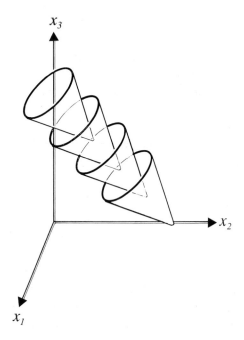

Fig.12.15:   Contour plots of a response surface when two eigenvalues have different sign, and one eigenvalue is zero. The three-dimensional isocontours describe a score of concentric hyperboloid.

(e) If the eigenvalues have different signs and the stationary point is elongated from the design center, a ridge system is obtained. The isoresponse surfaces will be described by a score of hyperboloids, see Fig. 12.15. At least one of the projections in the experimental variable planes will give isocontour lines which describe a score of hyperbolas.

7. Other designs for quadratic models

The composite designs discussed up to now are convenient in exploring synthetic methods since a step-wise strategy can be used. It is, of course, not *necessary* to use a step-wise approach. The designs can be directly used to fit quadratic models. The composite designs with replicated center point experiments have excellent statistical properties and offer posssibilities to analyze the fitted model by various tests. Sometimes, however, the experimenter is not explicitly interested in the statistical properties of the fitted model, or in analyzing in detail the roles played by the variables. The objective may simply be to visualize the response surface to see where favourable settings of the experimental variables are likely to be found. In such cases, a design with a minimum or with a limited number of experiments would be desirable. If the response surface predicts an optimum and the experimental results are good enough, the goal will have been reached with a minimum of effort.

In this section, some designs are given by which it is possible to fit a quadratic response surface model with fewer runs than what is needed for the composite designs. However, there is always a price to pay for being lazy, and with a reduced number of experiments, the quality of the predictions from the model becomes poorer.

In some cases it would be inconvenient to use more than three levels for the variation of the experimental variables. In the composite rotatable designs, five levels are used. In this section, some examples of three-level response surface designs are also given.

It is evident that there is an infinite number of ways by which experimental points can be distributed in the domain to fit models by regression methods. More or less ingenious designs have been suggested for response surface modelling. It is beyond the scope of this book to review details of these aspects. Thorough discussions of different types of response surface designs are given in the books by Myers[11], and Box and Draper.[12] Below, three alternatives to the composite designs are given: *Equiradial designs, Box-Behnken designs, and Hoke designs.*

7.1. Equiradial designs

The equiradial designs are useful when *two* experimental variables are studied. The coded experimental settings are evenly distributed on the periphery of the unit circle, and with at least one experiment at the center point. Without the center point the **X'X** matrix will be singular. The designs are defined by the following relations:

There should be at least one experiment at the center point, $x_1 = x_2 = 0$, and m experiment on the periphery, defined by

$$x_1 = \cos[\Theta] \quad x_2 = \sin[\Theta]$$

where Θ is defined as $\Theta = j \cdot 2\pi/m$, and with $j = 0, 1, 2,..., (m-1)$, see Fig. 12.16.

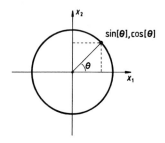

Fig.12.16: Design points on the unit circle.

A quadratic model in two variables contains six parameters, and the smallest design which can be used must therefore contain six experiments. This will correspond to an equiradial design in which the experiments on the periphery ($m = 5$) define a regular pentagon, see Fig. 12.17a.

The most commonly used equiradial design is with $m = 6$, which defines a regular hexagon, see Fig. 12.17b. Such a distribution of experimental points is also obtained when a regular simplex with two variables has reached an optimum domain and have encircled the optimum point. It is therefore possible to establish a response surface model from the simplex experiments and use the model to locate the optimum.

Equiradial design for m = 5 (pentagon)

Exp no	x_1	x_2
1	1.000	0
2	0.309	0.951
3	−0.809	0.588
4	−0.809	−0.588
5	0.309	−0.951
6	0	0

Equiradial design for m = 6 (hexagon)

Exp no	x_1	x_2
1	1.000	0
2	0.500	0.866
3	−0.500	0.866
4	−1.000	0
5	−0.500	−0.866
6	0.500	−0.866
7	0	0

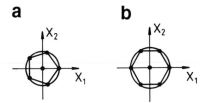

Fig. 12.17: Equiradial designs: *(a)* Pentagon design: *(b)* Hexagon design.

An advantage of the hexagonal distribution of the experimental points is that it is easy to displace the design into an adjacent experimental domain by adding a few complementary runs. This is useful if it should be found from the first design that an optimum is likely to be found outside, but in the vicinity of, the first explored domain. An example on this is shown below.

An equiradial design with $m = 8$ defines an octagon. This is also the distribution which is obtained by a central composite rotatable design.

The equiradial designs are rotatable. It is possible to adjust the number of center point experiments to achieve uniform precision as well as near-orthogonal properties. Uniform precision for the pentagon and the hexagon is obtained with three center point experiments. Orthogonal properties are obtained with five center point experiments for the pentagon , and with six center point experiments for the hexagon.

7.2. Example: Response surface by an equiradial design for determination of the optimum conditions for the synthesis of 4-(N,N-dimethylamino)acetophenone[13]

A procedure for the synthesis of 4-(N,N-dimethylamino)acetophenone by heating 4-chloroacetophenone in the presence of aqueous dimethylamine was described in a Japanese patent.[14]

A yield of 77 % was reported in the patent. However, it was suspected that the experimental conditions in the procedure described were not optimized with regard to yield. It was therefore interesting to explore the experimental domain around the conditions given in the patent to see whether or not the yield could be improved. An equiradial design (hexagon) was used to establish a response surface model. The experimental variables and their range of variation are summarized in Table 12.10, and the experimental design and yields obtained are given in Table 12.11.

The response surface as determined from the first seven experiments, indicated that the optimum conditions were likely to be found just ouside the explored domain, see Fig. 12.9.

Table 12.10: Variables and their variation in the synthesis of 4-(N,N-dimethylamino)acetophenone

Variables	Center point	Unit variation of x_i
x_1: Amount of Me_2NH/ketone (mol/mol)	3.00	1.00
x_2: Reaction temperature[a] (°C)	230	20

[a] The experiments were run in sealed vessels.

Table 12.11: Design and yields obtained in the synthesis of 4-(N,N-dimethylamino)acetophenone

Exp no	Variables		Yield (%)
	x_1	x_2	y
1	0	0	87
2	1.00	0	93
3	0.50	0.87	88
4	−0.50	0.87	74
5	−1.00	0	59
6	−0.50	−0.87	63
7	0.50	−0.87	80
2 − 6 = 8	1.50	0.87	90
2 − 4 = 9	1.50	−0.97	85
2 − 5 = 10	2.00	0	91
11	1.22	0.22	89
12	1.22	0.22	92
13	1.22	0.22	91[a]
14	1.22	0.22	92[a]

[a] Isolated yield from scale-up experiments in a steel autoclave.

To cover the optimum, the design was augmented by three complementary runs, *Exp no 8 −10*. This results in the domain being displaced and in a new hexagon which covers the expected optimum being formed, see Fig. 12.9. *Experiment no 2* is to be the center point in the new hexagon, and *Exp no 4, 5, 6* are to be replaced by the complementary runs. The variable settings in these experiments are computed from the settings in the first design. This is done by subtracting the coordinates of the point that is to be replaced from the coordinates of the new center point. These subtractions are shown in Table 12.11.

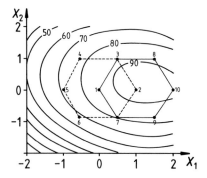

Fig 12.18: Distribution of experimental points and contour plot of the response surface in the synthesis of 4-(N,N-dimethylamino)acetophenone.

From the ten experiments, the response surface model was recalculated:

$$y = 84.7 + 17.0\,x_1 + 5.5\,x_2 - 1.7\,x_1x_2 - 7.1\,x_1^2 - 8.8\,x_2^2 + e$$

A three-dimensional plot of the response surface is shown in Fig. 12.10. The model predicts a maximum yield $y^{\text{Pred.}} = 95.2\ \%$ for $x_1 = 1.17$ and $x_2 = 0.20$.

The confirmatory runs, *Exp no 11 − 14* were conducted close to the predicted optimum and afforded satisfactory results. The yields were considerably improved compared to the original procedure given in the patent. The response surface model predicted a yield of 78.5 % for the conditions given in the patent. This is quite close to the reported yield of 77 %.

7.3. Incomplete three-level factorial designs

In the introduction to this chapter, it was said that complete three-level factorial designs with more than two variables would give too many runs to be convenient for response surface modelling. It is, however, possible to select a limited number of runs from such designs to obtain incomplete 3^k designs which can be used to fit quadratic models.

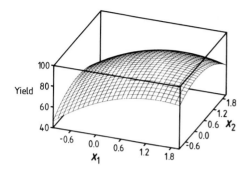

Fig. 12.19: Three-dimensional plot of the response surface in the synthesis of
4-(N,N-dimethylamino)acetophenone.

Box-Behnken designs[15]

In the Box-Behnken designs, the k variables are varied two at a time by 2^2
designs, while maintaining the remaining $(k - 2)$ variables fixed at their middle
(zero) level. The overall design can therefore be regarded as being composed of
intersecting 2^2 designs, augmented by at least one center point. In the original paper
by Box and Behnken[15] such designs composed of intersecting 2^2 and 2^3 designs are
described for many variables. It is rare that more than four variables need to be
considered in response surface modelling for synthesis optimization. Therefore, only
the designs for three and four variables are shown here. The distribution of the
experimental points for the design in three variables is shown in Fig. 12.20. With four
center points, the design is orthogonal. It is not rotatable, however. The design is
rotatable; it is actually a rotated central composite design. With seven center point
experiments a uniform precision is obtained, and with twelve center point
experiments, the design is orthogonal.

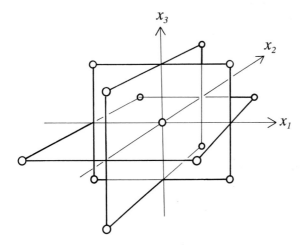

Fig 12.20: Distribution of experimental points in a Box-Behnken design.

Box-Behnken design for three variables

Exp no	x_1	x_2	x_3
1	−1	−1	0
2	1	−1	0
3	−1	1	0
4	1	1	0
5	−1	0	−1
6	1	0	−1
7	−1	0	1
8	1	0	1
9	0	−1	0
10	0	−1	0
11	0	1	0
12	0	1	0
13	0	0	0
.	.	.	.
.	.	.	.
N	0	0	0

Box—Behnken design for four variables

Exp no	x_1	x_2	x_3	x_4
1	−1	−1	0	0
2	1	−1	0	0
3	−1	1	0	0
4	1	1	0	0
5	0	0	−1	−1
6	0	0	1	−1
7	0	0	−1	1
8	0	0	1	1
9	−1	0	−1	0
10	1	0	−1	0
11	−1	0	1	0
12	1	0	1	0
13	0	−1	0	−1
14	0	1	0	−1
15	0	−1	0	1
16	0	1	0	1
17	−1	0	0	−1
18	1	0	0	−1
19	−1	0	0	1
20	1	0	0	1
21	0	−1	−1	0
22	0	1	−1	0
23	0	−1	1	0
24	0	1	1	0
25	0	0	0	0
.	.	.	.	.
.	.	.	.	.
N	0	0	0	0

Hoke designs[16]

The Hoke designs are similar to the central composite designs. The difference is that the axial points are placed at $\pm \alpha = \pm 1$. This means that only three levels of the experimental variables are used, which can be convenient sometimes. Apart from this, the Hoke designs do not offer any advantages over the central composite designs. Hoke designs are not rotatable. They can be brought to be orthogonal by chosing a proper number, N_0, of center point experiments. With k variables, N_0 should be: $k = 2$, $N_0 = 1$; $k = 3$, $N_0 = 6$; $k = 4$, $N_0 = 12$.

8. Optimization when there is more than one response variable

In the sections above it was discussed how response surface models can be used to optimize a single response. It is, however, common that there is more than one response of interest. Examples of this are:

* We may wish to determine conditions which maximize the yield of a desired product, but also that these conditions should minimize the yield of byproducts. It is sometimes the case that one of the byproducts is detrimental to subsequent steps (e.g. a catalyst poison) and that its formation must be completely suppressed.

* We may wish to maximize both the yield and the selectivity of a new reaction.

* We are developing an industrial synthesis, and a number of criteria must be fulfilled: a high yield, several quality specifications, as well as a low overall cost.

Sometimes it is possible to find experimental conditions which satisfy all the specified criteria; sometimes there are conflicts between certain criteria and a compromise solution must be found. It is evident that such problems can be difficult to solve. A special branch of mathematics, "Optimization theory" is devoted to this type of problem. In this area it is assumed that the *object function* (the "theoretical" response function) is perfectly known, and that the final solution can be reached by using mathematical and numerical methods. [17] From the discussions in Chapter 3, it is evident that mathematical optimization theory is difficult to apply in the area of organic synthesis, especially when new ideas are explored. Conclusions must be drawn from observations in suitably designed experiments. The response surface models thus obtained are local and approximate, and definitely not perfectly known. Nevertheless, we shall see that we can use experimentally determined models to find solutions to the problems sketched above.

When there are relatively few response variables to consider, it is possible to map each response separately by a response surface model. These models can then be evaluated simultaneously to determine suitable settings of the experimental variables. Aspects of this are discussed below.

When there are many response variables to consider, simultaneous evaluation of response surface models from each response becomes cumbersome. In such cases, a considerable simplification can often be achieved by multivariate analysis of the *response matrix*. For such purposes, *principal components analysis* and/or multivariate correlation by PLS are useful. These methods are discussed in Chapters 15 and 17.

8.1. Optimization by response surface modelling when there is more than one response variable

The methods discussed in this section are useful when there are only a few different responses to consider. The technique is to establish separate models for each response.

In simple cases, these models can be evaluated by visual inspection of the contour plots. For such evaluations, it is convenient to plot the isocontours for the different responses in the same plot. To obtain the contour plots, it will be most helpful to have access to a suitable computer program. There are a number of programs available on the market, both for main-frame computer and for microcomputers.[9]

In more complicated cases, canonical analysis of the different models can be useful to explore the experimental conditions, especially when ridge systems occur. Another technique to optimize several responses simultaneously is to weigh them together into one single criterion, a *Desirability function*, which can be used as a criterion for the optimization. This technique is not recommended as a *general* method for optimization of many responses. It can, however, be of value as a tool for evaluation of simultaneous mapping by response surface models. Desirability functions are discussed below.

Often there is one principal response which is the most important one, and the other responses can be regarded as constraints imposed on the possible solution, e.g. impurities should be below a certain level, or that there is a maximum allowable cost.

When new synthetic methods are being developed, the problems usually faced belong to the categories *(a) − (c)* below.

(a) The most simple case is when the response surface for the principal response is described by a constant (or nearly constant) ridge. In such cases, any conditions which give an acceptable response along the ridge will be satisfactory. The problem is then reduced to finding conditions along the ridge under which the constraining responses will also be acceptable. A similar approach can be used when the principal response is described by an attenuated maximum surface. A solution is likely to be found along the canonical axes corresponding to numerically small negative eigenvalues. In these directions the principal response will be least sensitive to variations of the experimental settings.

(b) An approach similar to *(a)* is used when the principal response is described by a rising ridge. The steps to be taken are: (1) Explore the ridge to

determine an optimum setting of the experimental variables, and analyse if these conditions also fulfill the requirements imposed by the constraints.

(2) Explore the canonical directions from the ridge which correspond to small eigenvalues if the constraints cannot be met on the ridge.

This will still give an acceptable level of the principal response, but the search off the ridge will be in directions where the principal response is least sensitive to variations in the experimental conditions.

(c) It is rather common that two responses are of equal interest, e.g. yield *and* selectivity. The problem is to find experimental conditions which give acceptable results for both. A solution can often be found by a simple visual evaluation of the contour plots.

An optimization problem solved by graphic evaluation of contour plots is shown in the following example.

8.2 Example: Suppression of a by-product in the synthesis of an enamine [18]

Synthesis of the morpholine enamine from pinacolone (3,3-dimethyl-2-butanone) by the modified $TiCl_4$-method discussed above was complicated by considerable self-condensation of the ketone.

The byproduct, 2,2,3,6,6-pentamethyl-3-hepten-5-one, was difficult to separate from the enamine by distillation. When the conditions optimized for enamine synthesis from methyl isobutyl ketone were applied to the synthesis from pinacolone, ca. 30 % of the byproduct was formed. A response surface study was undertaken to overcome the problem and to determine how the experimental conditions should be adjusted to achieve a maximum yield of the enamine, with a concomitant suppression of the formation of the byproduct.

The variables are specified in Table 12.12. The experimental design and yields obtained are given in Table 12.13. Two responses were measured: The yield (%) of enamine, y_1, and the yield of by-product, y_2. The reaction was monitored by GLC and the yields obtained after 4 h are given here. The derived response D is a desirability value which will be discussed later. The experimental design is orthogonal, but not rotatable.

The response surface models obtained by fitting quadratic models to y_1 and y_2 are:

$$y_1 = 58.54 + 5.36\ x_1 + 8.64\ x_2 + 5.25\ x_3 + 1.80\ x_1x_2 + 0.87\ x_1x_3 + 0.85\ x_2x_3 - \\ -\ 0.61\ x_1^2 - 1.26\ x_2^2 - 0.04\ x_3^2 + e$$

$$y_2 = 12.92 - 4.65\ x_1 + 4.64\ x_2 - 1.61\ x_3 - 1.41\ x_1x_2 - 0.39\ x_1x_3 - 0.61\ x_2x_3 + + + \\ 1.26\ x_1^2 + 0.41x_2^2 + 0.46x_3^2 + e$$

The estimated 95 % confidence limits of the parameters are:
For y_1: The constant term, ±1.91; the linear terms, ± 1.17; and the second order terms, ± 1.43.
For y_2: The constant term, ± 1.96; the linear terms, ± 1.20; and the second order terms, ± 1.47.

Taking these confidence limits into account, it is seen that the variations in response are largely described by the linear terms and the interaction term for x_1x_2. The coefficients of the other terms are not significant.

Table 12.12: Variables and experimental domain in the study of byproduct formation

Variables	−1.414	−1	0	1	1.414
x_1: Morpholine/ketone (mol/mol)	3.00	3.59	5.00	6.41	7.00
x_2: TiCl$_4$/ketone (mol/mol)	0.50	0.57	0.75	0.93	1.00
x_3: Reaction temperature (°C)	52	60	80	100	108

308

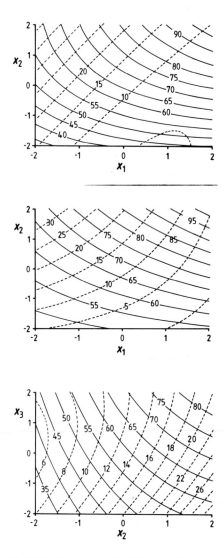

Fig.12.21: Isocontour plot. The variable not varied in the respective plot is at its zero level. Solid lines show the yield of enamine. The dashed lines show the by-product.

Table 12.13: Experimental design and responses in the study of byproduct formation

Exp no[*]	Variables			Responses		
	x_1	x_2	x_3	y_1	y	D
1	−1	−1	−1	41.6	14.6	0.12
2	1	−1	−1	45.1	6.7	0.345
3	−1	1	−1	51.7	26.2	$0.217 \cdot 10^{-5}$
4	1	1	−1	64.7	17.7	0.057
5	−1	−1	1	47.8	11.9	0.219
6	1	−1	1	57.1	7.5	0.439
7	−1	1	1	63.0	26.1	$0.331 \cdot 10^{-5}$
8	1	1	1	77.8	11.0	0.402
9	1.414	0	0	66.7	8.1	0.487
10	−1.414	0	0	49.5	22.2	0.001
11	0	1.414	0	70.4	18.9	0.032
12	0	−1.414	0	43.9	8.0	0.304
13	0	0	1.414	66.4	9.8	0.414
14	0	0	−1.414	52.4	17.3	0.055
15	0	0	0	56.5	13.8	0.186
16	0	0	0	60.0	12.3	0.267
17	0	0	0	58.6	12.6	0.247
18	0	0	0	57.2	13.6	0.197

[*] The experiments were run in random order.

The following conclusions can be drawn directly form the mathematical expressions of the models: An increase in the yield of enamine, y_1, can be expected by increasing the settings of all three variables. A decrease in the yield of byproduct, y_2, can be expected by increasing x_1 and x_3, and decreasing x_2. Fig. 12.22 shows the contour plots projected in the planes spanned by the experimental variable axes. These projections lead to the same conclusions.

To further analyze the experimental results, a reduced models including only the significant terms (linear terms and the x_1x_2 interaction term) was fitted to each response.[8] A projection of the isocontours when x_1 and x_2 are varied and with x_3 set to its upper level 1.414, is shown in Fig. 12.22. The models show that y_1 increases and that y_2 decreases when the temperature is increased.

---

[8] The predictions from the models were less noisy after removal of the insignificant model terms.

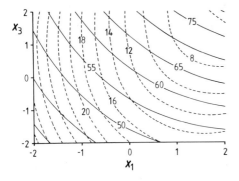

Fig. 12.22: Isocontours of the yield of enamine and by-product with x_3 set at its upper level.

The plot in Fig. 12.22 further supports the conclusions. It is seen that a large amount of morpholine should be used, (x_1 should be at its upper level). The amount of by-product is largely controlled by the amount of titatium tetrachloride, x_2, under these conditions. The level of x_2 will therefore be subject to compromise judgements. To determine a suitable level of x_2, a projection of the reduced models with x_3 set to 2.00 is shown in Fig. 12.23. The temperature setting $x_3 = 2.00$ is an extrapolation outside the explored domain and as such not strictly allowable. It corresponds to a reflux temperature (120 °C, petroleum ether) which was considered convenient and worth trying.

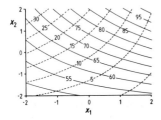

Fig. 12.23: Isocontours of the yield of enamine and by-product with $x_3 = 2.00$.

An experimental run under slightly extrapolated conditions ($x_1 = 2.0$, $x_2 = 1.7$, and $x_3 = 2.0$) afforded yields of $y_1 = 85$ %, and $y_2 = 8$ %. This was an improvement to what was obtained in the experiments in Table 12.13.

This result inspired us to further increase the levels of the variables so that they reached outside the explored domain. The conditions which were finally adopted for preparative applications of the reaction were to use a molar ratio of morpholine/ketone = 10.0 (x_1 = 3.7), and a molar ratio of TiCl$_4$/ketone = 1.0 (x_2 = 1.6) at reflux temperature. These conditions afforded a 93−94 % yield of the enamine and less than 3 % yield of the by-product after one hour of reaction. This was a significant improvement compared to the earlier results. The excess of morpholine is recovered by distillation and can be reused. To use an excess of morpholine was therefore not considered as a drawback of the procedure.

This example furnishes an illustration of the step-wise approach to experimental studies: The experimental domain of the experiments in Table 12.13 was not arbitrarily chosen. The optimum conditions for previously studied enamine syntheses were well within this domain, but pinacolone is a different case as it is sterically hindered and the rate of enamine formation was rather slow. To prevent the formation of the byproduct, a large excess of reagents was necessary. The direction for changing the experimental conditions was determined from the initial response surface study.

The next step was to study if even more hindered ketones, like camphor, could be converted to enamine by an excess of reagents. Formation of enamines from camphor by traditional methods affords very poor yields. By the present method, molar scale synthesis from camphor afforded yields of enamines in the range 78−81 %.[19]

An important and general conclusion from the above results is, that it can be very misleading if the scope of a reaction is determined by using different substrates under some kind of "standardized conditions". To make fair comparisons, it may be necessary to adjust the experimental conditions for each substrate under study. Aspects of this are discussed in Chapter 16.

8.3. On the use of desirability functions

The problem of many responses of approximately equal interest is often met in industrial synthesis. Cost, quality specifications, overall yield as well as pollution due to waste are necessary to control, and it is sufficient that one of the responses fails to meet the requirements for an overall result to be poor. In such cases it is possible to use a technique by which all responses are weighed together into one criterion which is then optimized. First, it is necessary to clearly define, *what* the desired result is for each response. The measured value of the response is then scaled to a dimensionless measure of the desirability, d_i, of the response y_i. The scaling is done so that d_i is in the interval $[-1 \leq d_i \leq 1]$. An overall desirability, D, can then be defined as the *geometric mean* of all individual desirabilities d_i, $i = 1,2,..., m$.

$$D = (d_1 \cdot d_2 \cdot ... \cdot d_m)^{1/m}$$

It is seen, that a high value of D is obtained only if all individual desirabilities d_i are high. If one d_i is poor, this is enough to give an overall poor desirability. If all d_i are equal, then $D = d_i$. The geometric mean is therefore seen to meet the requirements of an overall desirability function. The problem is then to define the individual desirabilities.

It may be required that a response y_i should be in a specified interval, $[a \leq y_i \leq b]$, while other responses should be either below or above a certain level: $y_i \leq a$ or $y_i \geq b$. One way to scale the responses into desirability values would be to assign the value $d_i = 1$ when the specifications are met, and the value $d_i = 0$ when they are not, see Fig. 12.24.

Such scaling of d_i as shown in Fig. 12.24 would give two possible overall results: success or failure. At best, this can be used for classification of the experimental conditions as being useful or useless.[9] Thus, a discrete scaling like this cannot be used for optimization.

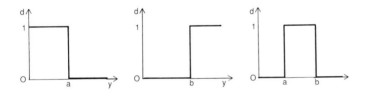

Fig 12.24: Discrete desirability functions.

A smooth and continuous function to define d_i is necessary, and such functions can be obtained by mathematical transformation of the measured response. One way to do this has been suggested by Harrington [20], and this procedure is given here. Other ways of scaling have been given by Derringer and Suich.[21] See also a discussion in the book by Box and Draper.[22]

The scaling is done so that d_i is assigned an arbitrary value for a given value of the response. The following assignments of d_i have been suggested.

---

[9] A far better way to achieve a classification is to use the SIMCA method which is briefly discussed in Chapter 15.

1.00 – 0.80   Excellent
0.80 – 0.63   Good
0.63 – 0.37   Acceptable but mediocre
0.37 – 0.20   Borderline acceptance
0.20 – 0   Poor

For two-sided specifications, $a \leq y_i \leq b$, it is possible to use an exponetial tranformation

$$d_i = \exp[-|q_i|^\alpha]$$

with $q_i = [2y_i - (a + b)] / (b - a)$ and α is a positive number (not necessarily an integer).

This transformation will change the discrete square function into a symmetric peak. The exponent α is computed by assigning a desirability value to the most desirable response.

$$\alpha = [\ln \ln (1/d_i)] / \ln q_i$$

For one-sided specifications, $y_i \leq a$ or $y_i \geq b$, another exponential transformation can be used to change the discrete step function into a sigmoid curve.

$$d_i = \exp[-\exp-(c_0 + c_1 y_i)$$

The scaling parameters, c_0 and c_1, are determined by assigning desirability values to different levels of y_i

$$\ln \ln(1/d_i) = c_0 + c_1 y_i$$

Graphs showing typical desirability functions for one-sided and two-sided specifications are shown in Fig.12.25.

Drawbacks of the method and suggestions as when to it could be used

Though it is *possible* to weigh together several responses into an overall desirabilty function, it is not a method which is recommended for general use. By squeezing all responses into one criterion, multidimensional information is lost. There is no possibility whatsoever to reverse the transformation back to the individual *responses*. It is therefore not possible to understand the chemical phenomena involved from the D value. The function D should therefore *not be used* as an experimental response for optimization if the aim is to do better chemistry. The rather esoteric mathematical tranformations also precludes any possibility of having a feeling for what the scaled value really represents. A large portion of subjectivity is also introduced when the scalings are defined.

314

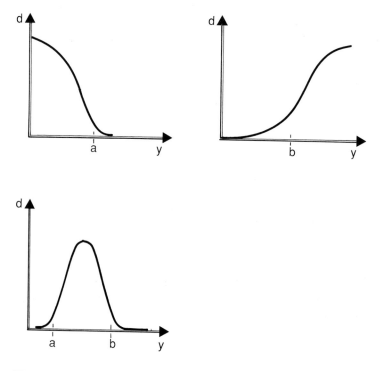

Fig. 12.25: Desirability functions: *(a)* One-sided specification, *(b)* two-sided specification.

There is, however, one application when the overall desirability D can be rather safely used, namely as a tool in conjunction with response surface modelling. In this context, it can be used to explore the joint modelling of several responses so that a near-optimum region can be located by simulations against the response surface models. The search for conditions which incrase D can be effected either by simplex techniques or by the method of steepest ascent. For the steepest ascent, a linear model for D is first determined from the experiments in the design used to establish the response surface models. The settings which increase D can be translated back into the individual responses by using the response surface models. Thus, it is possible to establish immediately whether the simulated reponse values correspond to suitable experimental conditions. Such results must, of course, be verified by experimental runs.

To give an example on how a desirability function can be defined, the enamine experiment in Table 12.13 is used. The two responses, y_1 and y_2, can be scaled into the indvidual desirabilities d_1 and d_2 as following:

For the enamine, an excellent result would be $y_1 = 90\ \%$ which thus is assigned the value $d_1 = 0.80$. A poor result would be if the yield drops to 50 %. A value of $d_1 = 0.3$ is assigned for $y_1 = 50$.

For the byproduct, a good result would be if $y_2 \leq 1\ \%$, which is assigned the value $d_2 = 0.80$. A poor result is when the yield of byproducd exceeds 10 %, and a value $d_2 = 0.3$ is assigned for $y_2 = 10\ \%$. These assignments give:

$$d_1 = \exp[-\exp(2.2956 - 0.04214\,y_1)]$$

$$d_2 = \exp[-\exp(-1.6872 + 0.1873\,y_2)]$$

$$D = (d_1 \cdot d_2)^{1/2}$$

The values of D computed from the observed yields are given in Table 12.13. A linear model for D will be

$$D = 0.227 + 0.135\,x_1 - 0.085\,x_2 + 0.088\,x_3 + e$$

It is seen that an improved overall result can be expected by increasing x_1 and x_3, and by decreasing x_2. These were also the conclusions previously drawn directly from the projections of the response surfaces.

References

1. G.E.P. Box and K.B. Wilson
   J. Roy. Statist. Soc. B 13 (1951) 1.

2. *(a)* G.E.P. Box
   Biometrics 10 (1954) 16;

   (b) G.E.P. Box and P.V. Youle
   Biometrics 11 (1955) 287.

3. R. Carlson, unpublished

4. G.E.P. Box and J.S. Hunter
   Ann. Math. Statist. 28 (1957) 195.

5. R.M. Myers
   Response Surface Methodology
   Allyn and Bacon, Boston 1971, pp. 139–165.

6. W.A. White and H. Weingarten
   J. Org. Chem. 32 (1967) 231.

7. R. Carlson, Å. Nilsson and M. Strömqvist
   Acta Chem. Scand. B 37 (1983) 7.

8. *(a)* Å. Nilsson
   Diss. Umeå University, Umeå 1984;

316

(b) R. Carlson and Å. Nilsson
Acta Chem. Scand. B 38 (1984) 49.

9.   (a) *MODDE*
Umetri AB, P.O.Box 1456, S-901 24 Umeå, Sweden

(b) *NEMROD*
LPRAI
Att. Prof R. Phan-Tan-Luu
Université d'Aix-Marseille III
Av. Escadrille Normandie Niemen
F-13397 Marseille Cedex 13, France

For other programs, see
C.J. Nachtsheim
J. Qual. Technol. 19 (1987) 132.

10. L. Hansson and R. Carlson
Acta Chem. Scand. 43 (1989) 1888.

11. R.M. Myers
Response Surface Methodology
Allyn and Bacon, Boston 1971, pp. 139–165.

12. G.E.P. Box and N.R. Draper
Empirical Model-Building and Response Surfaces
Wiley, New York 1987.

13. T. Lundstedt, P. Thorén and R. Carlson
Acta Chem. Scand. B 39 (1984) 523.

14. K. Yamamoto, k. Nitta, N. Yagi, Y. Imazato and J. Oisu
Jpn. Pat. 79 132 542, Mitsui Toatso Chemical Inc.
[*Chem. Abstr. 92* (1979) 163 712].

15. G.E.P. Box and D.W. Behnken
Technometrics 2 (1960) 455.

16. T.A. Hoke
Technometrics 17 (1974) 375.

17. For an example, see
D.E. Gill, W. Murray and M.H. Wright
Practical Optimization
Academic Press, London 1981.

18. R. Carlson, L. Hansson and T. Lundstedt
Acta Chem. Scand. B 40 (1986) 444.

19. R. Carlson and Å Nilsson
Acta Chem. Scand. B 39 (1986) 181.

20. E.C. Harrington
Ind. Qual. Control 21 (1965) 494.

21. G.C. Derringer and R Suich
J. Qual. Technol. 12 (1980) 214.

22. G.E.P. Box and N.R. Draper
Empirical Model-Building and Response Surfaces
Wiley, New York 1987, pp. 373–375.

Suggestions for further reading

G.E.P. Box and N.R. Draper
Empirical Model-Building and Response Surfaces
Wiley, New York 1987.

G.E.P. Box, W.G. Hunter and J.S. Hunter
Statistics for Experimenters
Wiley,New York 1978.

Appendix 12A: Transformation of the model matrix of an orthogonal design to obtain a diagonal dispersion matrix

A least squares fit of a response surface model is computed by the relation:

$$(X'X)^{-1}X'y = b$$

These calculations are easily carried out by hand if the the experiments have been laid out so that $(X'X)^{-1}$ is a diagonal matrix. A design to which this applies, is called an orthogonal design.

Some experimental designs for quadratic models have intrinsic orthogonal properties. Orthogonality implies that the columns of the model matrix X are orthogonal to each other. However, this is not true of X directly. The columns of the squared variables contain elements which are either positive or zero. As a consequence, the scalar product of the square columns to the constant column I will not be zero. Neither will the scalar product of the square columns to each other be zero.

For a design with intrinsic orthogonal properties, a diagonal dispersion matrix can be obtained by fitting the model after a transformation to a form in which the square variables are replaced by $(x_i^2 - \bar{x}_i^2)$:

$$y = b'_0 + \Sigma\, b_i\, x_i + \Sigma\Sigma\, b_{ij}x_ix_j + \Sigma\Sigma\, b_{ii}(x_i^2 - \bar{x}_i^2) + e$$

The consequence for the model matrix is that the average of each squared variable, x_i^2, is subtracted from the elements of the square columns in the original model matrix. The constant term, b_0, in the transformed model is related to the intercept term, b_0, in the original model by the relation

$$b_0 = b'_0 + \Sigma\, \bar{x}_i^2\, b_{ii}$$

These principles are illustrated by a central composite rotatable design in two variables. This design will be orthogonal with eight experiments at the center point.

The original model matrix **X** is

I	x_1	x_2	x_1^2	x_2^2	$x_1 x_2$
1	−1	−1	1	1	1
1	1	−1	1	1	−1
1	−1	1	1	1	−1
1	1	1	1	1	1
1	−1.414	0	2	0	0
1	1.414	0	2	0	0
1	0	−1.414	0	2	0
1	0	1.414	0	2	0
1	0	0	0	0	0
1	0	0	0	0	0
1	0	0	0	0	0
1	0	0	0	0	0
1	0	0	0	0	0
1	0	0	0	0	0
1	0	0	0	0	0
1	0	0	0	0	0

The matrix **X'X** is not a diagonal matrix.

	I	x_1	x_2	x_1^2	x_2^2	$x_1 x_2$
I	16	0	0	8	8	0
x_1	0	8	0	0	0	0
x_2	0	0	8	0	0	0
x_1^2	8	0	0	12	4	0
x_2^2	8	0	0	4	12	0
$x_1 x_2$	0	0	0	0	0	4

The average $\bar{x}_i$ of x_i^2 is $(1 + 1 + 1 + 1 + 2 + 2 + 0 + ... + 0)/16 = 0.5$.
Substracting these averages from the elements of the square column in **X** gives the transformed model matrix **Z**:

I	x_1	x_2	$x_1^2 - 0.5$	$x_2^2 - 0.5$	$x_1 x_2$
1	−1	−1	0.5	0.5	1
1	1	−1	0.5	0.5	−1
1	−1	1	0.5	0.5	−1
1	1	1	0.5	0.5	1
1	−1.414	0	1.5	−0.5	0
1	1.414	0	1.5	−0.5	0
1	0	−1.414	−0.5	1.5	0
1	0	1.414	−0.5	1.5	0
1	0	0	−0.5	−0.5	0
1	0	0	−0.5	−0.5	0
1	0	0	−0.5	−0.5	0
1	0	0	−0.5	−0.5	0
1	0	0	−0.5	−0.5	0
1	0	0	−0.5	−0.5	0
1	0	0	−0.5	−0.5	0
1	0	0	−0.5	−0.5	0

Computation of $Z'Z$ gives:

$$
Z'Z \;=\;
\begin{bmatrix}
16 & 0 & 0 & 0 & 0 & 0 \\
0 & 8 & 0 & 0 & 0 & 0 \\
0 & 0 & 8 & 0 & 0 & 0 \\
0 & 0 & 0 & 8 & 0 & 0 \\
0 & 0 & 0 & 0 & 8 & 0 \\
0 & 0 & 0 & 0 & 0 & 8
\end{bmatrix}
$$

The dispersion matrix is therefore

$$
(Z'Z)^{-1} \;=\;
\begin{bmatrix}
1/16 & 0 & 0 & 0 & 0 & 0 \\
0 & 1/8 & 0 & 0 & 0 & 0 \\
0 & 0 & 1/8 & 0 & 0 & 0 \\
0 & 0 & 0 & 1/8 & 0 & 0 \\
0 & 0 & 0 & 0 & 1/8 & 0 \\
0 & 0 & 0 & 0 & 0 & 1/8
\end{bmatrix}
$$

Appendix 12B: Transformation of the response variables to improve the model fit

To allow a statistical analysis of the model by the t-distribution or the F-distribution it is assumed that the experimental errors are normally and independently distributed in the experimental domain. It is also assumed that the error variance is constant.

When a model is fitted to experimental data, there will always be deviations between the observed responses and the responses calculated from the model. With an adequate model, these deviations should be nothing but a manifestation of the experimental error. Under these circumstances, it can be assumed that the residuals should also be normally and independently distributed and have a constant variance. Therefore, any indications that these assumptions are violated should jeopardize the model.

Indications of this can be obtained by plotting the residuals in different ways, to make sure that they do not show any abnormal pattern. Sometimes it is found that the plot of the residuals against the predicted response has a funnel-like pattern, which shows that the size of the residuals are dependent on the magnitude of the responses. A situation when this is often encountered is when integrated chromatographic peaks are used to determine concentrations in the sample. When such patterns are observed it is a clear indication that the assumption of a constant variance is violated. This obstacle can sometimes be removed by a mathematical transformation of the response variable y.

The transformations are made to counteract violations of the underlying assumptions when models are fitted to experimental data, so that the model can be evaluated by statistical tests. This is totally different to the subjective scaling involved in the definitions of desirability functions.

The units of measurement of an *observed* response is dependent on the method used to analyze the experiment. Some analytical methods may be afflicted by a non-constant error variance, though the "true" response has a constant variance. For instance, the enatioselectivity of a reaction can be described as an enatiomeric excess, *ee*. However, this quantity can be determined through a variety of methods: by measuring the optic rotation. The observed value will then depend on the nature of the solvent and the wavelength used; by integration of NMR spectra in the presence of chiral shift reagents; by measuring the peak areas in chromatograms obtained using chiral columns.

For such purposes, a family of power transformations are useful. By these, the transformed response, Y, is obtained by raising the original response to a power λ. The common nomenclature is to use the Greek letter λ for the exponent. This should not be confused with the eigenvalues from the canonical analysis.

$$Y = y^\lambda$$

Some standard transformation corresponding to different values of λ are:

λ =	1	no transformation
λ =	0.5	square root
λ =	0	natural logarithm[10]
λ =	−0.5	inverse square root
λ =	−1	reciprocal

A suitable value of λ can be determined from the transformations described by Box and Cox.[1] A brief description of this procedure is given below.

Box-Cox transformation

The model fit can be judged from the residual sum of squares, RSS. An improvement of the model fit will show up as a reduction of RSS. The scaling procedure by Box and Cox makes it possible to compare the different RSS independently of the scaling of the original response.

The transfomations to $Y(\lambda)$ for a given value of λ are defined by

$$Y(\lambda) = (y^\lambda - 1)/[\lambda \cdot \mathring{y}^{(\lambda-1)}] \quad \text{if } \lambda \neq 0$$

$$Y(\lambda) = \mathring{y} \cdot \ln y \quad \text{if } \lambda = 0$$

where $\mathring{y}$ is the geometric mean of the observed responses, $y_1,...y_n$.

$$\mathring{y} = (y_1 \cdot y_2 \cdot ... \cdot y_n)^{1/n} \quad \text{(It is assumed that } y_i > 0)$$

Computation of $\mathring{y}$ is easily made by the following relation:

$$\ln \mathring{y} = 1/n \cdot (\Sigma \ln y_i)$$

Let RSS_λ denote the residual sum of squares from the model fitting obtained with the scaled response $Y(\lambda)$. The optimum value of λ will give the smallest RSS. This value, λ_{Opt}, can be determined by plotting RSS_λ against λ and determining the minimum point. As RSS_λ can vary in magnitude it is convenient to plot $\ln(RSS_\lambda)$ against λ. There is another advantage: $\ln(RSS_\lambda)$ is a log-likelihood function and the difference $\ln(RSS_\lambda) - \ln(RSS_{\lambda Opt})$ will have a χ^2 distribution with one degree of fredom.

$$\ln(RSS_\lambda) - \ln(RSS_\lambda{}^{Opt}) = (1/r) \cdot \chi^2{}_{(\alpha;1)} \quad r \text{ is the degrees of freedom of RSS}$$

An approximate $(1 - \alpha)$ confidence interval for λ can therefore be determined from the plot by the following procedure:

Draw a horizontal line through the minimum point. Draw a second line parallel to this line so that the difference between the line corresponds to $1/r \cdot \chi^2{}_{(\alpha;1)}$. The λ coordinates of the intersections of this line with the curve gives the confidence limits for λ.

---

[10] $y^\lambda = e^\lambda \ln y$. A Taylor expansion gives
$e^{\lambda \ln y} = 1 + \lambda \ln y + 1/2! \cdot (\lambda \ln y)^2 + 1/3! \cdot (\lambda \ln y)^3 + ...$
which can be written
$(y^\lambda - 1)/\lambda = \ln y + \lambda/2! \cdot (\ln y)^2 + \lambda/3! \cdot (\ln y)^3 + ...$
From this follows: $(y^\lambda - 1)/\lambda \rightarrow \ln y$ when $\lambda \rightarrow 0$
It is seen that this gives the Box-Cox tranformation for $\lambda = 0$.

To determine the optimum value of λ, the following steps are taken:

(1)   Determine the geometric mean, $\hat{y}$, of the responses.

(2)   Compute the values of the transformed responses $Y(\lambda)$ for a series of λ in the interval $-1 \leq \lambda \leq 1$; preferably by a step variation of 0.1 or 0.2.

(3)   Fit the model to the transformed responses and determine the residual sum of squares, RSS_λ for the each λ.

(4)   Plot $\ln(RSS_\lambda)$ against λ, and determine the value of λ which gives the minimum RSS.

(5)   Determine an approximate $(1 - \alpha)$ confidence interval of λ as was described above.

(6)   Select a suitable value of λ for the transformation. Choose a standard transformation which is closest to λ. Fit the model to the transformed response and proceed by analysing the fit of the new model. Plot residuals etc.

The above procedure is illustrated by the example of enamine synthesis given in Table 12.5. Experiment *No 11* has been deleted from the calculation. This experiment is an outlier. The geometric mean $\hat{y} = 83.13146$ an this value is used to compute the transformed responses summarized in Table 12B.1

The residual sum of squares (6 degrees of freedom) obtained after fitting a quadratic response surface model to the responses in Table 12B.1 are summarized in Table 12B.2. A plot of $\ln(RSS)$ against λ is shown in Fig. 12B.1.

Table 12B.1: Transformed response values from the enamine example

Exp no	Transformed responses $Y(\lambda)$ for different λ values										
	−1.0	−0.8	−0.6	−0.4	−0.2	0	0.2	0.4	0.6	0.8	1.0
1	6822.12	3458.38	1820.48	1003.51	583.980	360.924	237.437	165.891	122.359	94.509	75.8
2	6812.96	3449.45	1811.77	995.02	575.706	352.859	229.575	158.227	114.888	87.226	68.7
3	6842.09	3478.52	1840.78	1024.00	604.648	381.782	258.490	187.146	143.822	116.187	97.7
4	6840.42	3476.79	1839.00	1022.16	602.760	379.822	256.466	185.056	141.665	113.960	95.4
5	6790.00	3427.76	1791.27	975.65	557.397	335.551	213.209	142.751	100.250	73.379	55.6
6	6827.11	3463.31	1825.36	1008.34	588.762	365.658	242.123	170.529	126.951	99.055	80.3
7	6830.71	3466.92	1828.96	1011.95	592.365	369.261	245.725	174.131	130.551	102.655	83.9
8	6834.20	3470.44	1832.51	1015.52	495.974	372.901	249.396	177.834	134.287	106.423	87.7
9	6840.20	3476.56	1838.76	1021.91	602.498	379.563	256.199	184.781	141.382	113.668	95.1
10	6817.96	3454.30	1816.47	999.59	580.137	357.160	233.748	162.276	118.817	91.039	72.4
11	6840.72	3477.10	1839.32	1022.48	603.084	380.166	256.821	185.422	142.041	114.348	95.8
12	6835.75	3472.01	1834.11	1017.15	597.620	374.571	251.092	179.551	136.034	108.196	89.5

324

Table 12B.2: Residual sum of squares obtained for different values of λ

λ	RSS	ln(RSS)
−1.0	207.352	5.334
−0.8	195.546	5.276
−0.6	185.584	5.224
−0.4	177.759	5.180
−0.2	171.590	5.145
0.0	166.810	5.117
0.2	63.344	5.096
0.4	61.116	5.082
0.6	59.885	5.075
0.8	59.659	5.073
1.0	60.297	5.077

The minimum value of ln(RSS) is obtained for $\lambda = 0.752$. This is fairly close to 1.0 (no transformation). It is therefore not indicated that the original response should be transformed to improve the fit.

The intersecting horizontal line to determine the confidence limits of λ is not shown in Fig.12B.1. The value of χ^2 for $\alpha = 5 \%$ and one degree of freedom is 3.84.

$$\ln(RSS_\lambda) - \ln(RSS_{0.752}) = 1/r\chi^2 = 3.84/6 = 0.64$$

This value is far beyond the observed range of variation of ln(RSS) for the values of λ studied.

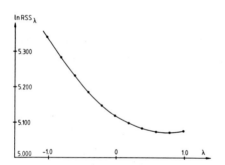

Fig. 12B.1: Lambda plot in the Box-Cox transformation of the responses in the synthesis of enamine.

Reference

1. G.E.P. Box and D.R. Cox
    J. Roy. Statist. Soc. B 26 (1964) 211, 244.

Chapter 13

Summary of strategies to explore the experimental space

1. Benefits of a step-wise strategy

When the experimental conditions for a chemical reaction are studied with a view to developing a synthetic method, there are many problems which must be solved. The process may start at a moment when an idea is born by a sudden flash of inspiration, and may last until a series of confirmatory runs has been accomplished and found to give an optimum and reproducible result. In the preceding chapters it was discussed how experiments can be designed to provide answers to questions which may arise during this process. At the beginning, little is known and the questions are rather qualitative. A screening experiment will give a clearer picture and develop the questions for the next step. Towards the end, when precise information is needed, response surface modelling and canonical analysis of the model will give detailed information and also give clues to an understanding at the mechanistic level.

This chapter shows how the various tools described in the preceding chapters can be integrated into a step-wise strategy. The benefits of such a step-wise approach to problem-solving are obvious: Knowledge is accumulated as the project is progressing and new experiments can be planned in the light of what has been gained, which gives an adaptive and flexible strategy. A well-designed screening experiment can accomodate many variables and cover a large number of different experimental conditions. If it should be found that the reaction under study does not give any promising results, it is unlikely that it can be developed into something useful. Poor ideas can therefore be abandoned at an early stage.

2. Flow sheet to define a strategy

The different steps are numbered. At each step there are several points which should be considered. Breakpoints are marked with Q.

1: Define the problem.

(a): State the objective clearly. Define a criterion of success.

(b): Analyze the problem and answer the questions:
   (i):   What is already known?
   (ii):   What is not known?
   (iii):   What do I need to know?

2: Go through the experimental procedure.

(a): Determine the critical steps.

(b): Pay attention to practical constraints.

3: Define response(s) and variables.

(a): Determine a suitable response to measure. If possible, use a method by which the reaction mixture can be analyzed directly.

(b): Evaluate *1* and *2* and determine which experimental variables must be included.
   (i):   If there are discrete variables which are difficult to change in a randomized manner, consider the possibility of a design divided into blocks.

4: Run a pilot experiment to check if it is reproducible.

Q: (a): If the answer is *yes*, then proceed to *5*.

(b): If the answer is *no*, something is out of control. Go back to *1* and reconsider the problem.

5: Design a screening experiment.

(a): Assign a tentative experimental domain
   (i):   With totally new procedures, this may be difficult. To ensure that a sufficient range of variation has been chosen, run a small pilot experiment with the settings of the variables at the extremes of the tentative domain.

(b): Suggest a response surface model for the screening experiment.

    (i):   Use a linear model if there are many variables to consider

    (ii):   Use directly a second-order interaction model, if it is suspected that there are several strong interaction effects.

(c): Select a design to fit the model chosen:

    (i):   If possible, use a two-level factorial or fractional factorial design.

    (ii):   If there is a constraint on the number of possible runs, consider a Plackett-Burman design if a fractional factorial design cannot be used.

    (iii):   If there are constraints on the possible settings of the experimental variables so that a two-level design cannot be used, use a D-optimal design.

    (iv):   If there are discrete variables which are difficult to vary in randomized order, run the experiments in blocks.

(d): Run the experiments in random order.

6: Evaluate the screening experiment and determine the significant variables.

    (a): Evaluate the model.

        (i):   Check the model fit by ANOVA.

        (ii):   Plot residuals

        (iii):   Draw projections of the response surface to assist the interpretation.

    (b): Identify the significant variables.

        (i):   If the experimental error is known, evaluate the significance of the estimated effects from their confidence limits.

        (ii):   Use a Normal probability plot to identify significant variables if the experimental error is not known.

    (c): If there are any ambiguities due to confounding of interaction effects with main effects, analyze the alias pattern and consider a complementary run.

        (i):   Use a complementary fraction of a factorial design.

        (ii):   Use a fold-over design.

    (d): Run a confirmatory experiment under the best conditions predicted from the model.

Q (a): If this experiment gives a satisfactory result which fulfils the criterion of success, then repeat this experiment to make sure that it is reproducible. The goal has been reached. Stop and write a report.

    (b): If the experimental variables are not found to be significant and the experimental results are inferior to what is satisfactory, go back to *1* and reconsider the problem. If this does not improve the result, then stop and turn to a more promising project.

    (c) If there are significant variables, but the result is below the level of satisfaction, then proceed to *7*.

7: Determine a better experimental domain.

    (a): Fix the settings of the discrete variables at their most favourable levels.

    (b): Adjust the settings of the continuous experimental variables so that the explored domain is left in a direction which gives an improvement.

        (i):   Use the method of the steepest ascent. To determine the direction of the search path the linear coefficients from the screening experiment can be used.

        (ii):   Use a simplex method with the significant variables.

Q (a) If the best result obtained during the search is satisfactory and fulfils the criterion of success, then repeat this experiment to confirm the reproducibility. Stop and write a report.

    (b): If the results are promising, but not satisfactory, and the best settings are far from the first explored domain, there is a risk that

the influence of the variables may have been changed. To avoid false conclusions, go to 5 and consider a new screening of the variables in the improved domain.

(c): If the result is promising but not satisfactory, and experimental conditions for the best result obtained during the search are close to the explored domain, then proceed to *8.*

8: Determine a quadratic response surface model to map the optimum domain.

(a): Analyze the model fit.
- *(i):* Use ANOVA to analyze the regression.
- *(ii):* Plot the residuals.
- *(iii):* If necessary, transform the response variable.

(b): Use the model to localize the optimum conditions.
- *(i):* Visualize the shape of the response surface by projections and make an intuitive interpretation from the plots.
- *(ii):* Make a canonical analysis to determine the nature of the stationary point, and to explore ridges. Use this analysis to achieve a feed-back to an understanding of underlying mechanisms.
- *(iii):* Consider auxilliary responses, which may impose constraints on the solution. Map these responses by separate response surface models and evaluate the models simultaneously to cope with the constraints.

(c): Predict the optimum conditions from the model.

9: **Confirm the conclusions by experiments.**

References

Suggestions to further reading

G.E.P. Box, W.G. Hunter and J.S. Hunter
Statistics for Experimenters
Wiley, New York 1978.

G.E.P. Box and N.R. Draper
Empirical Model-Building and Response Surfaces
Wiley, New York 1987.

S.M. Deming and S.L. Morgan
Experimental Design. A Chemometric Approach
Elsevier, Amsterdam 1987.

Chapter 14

The Reaction Space

1.1. What is the reaction space

Up to now we have discussed how to design experiments to explore the reaction conditions in the *experimental space* with a view to answering the questions "Which variables are important?" and "How should the conditions be adjusted to achieve an optimum performance?". For such experiments, it is assumed that the reaction system (substrate, reagents and solvent) is defined and that the number of alternative choices with regard to discrete variations is limited. In the chapters to follow, we shall discuss what to do when this is not the case, and when the problem is first to determine a suitable reaction system which is worth optimizing with regard to the experimental conditions.

We shall define *the reaction space* as the union of all possible variations of the nature of the substrate, of the reagent(s), and of the solvent, see Fig. 14.1. The reaction space will thus contain all possible *discrete* variations of the reaction system.

Several common problems in organic synthesis are rooted in the reaction space. Some examples are:

*   How should suitable model substrates be selected to make it possible to develop the experimental conditions for key steps in the synthesis of complex natural products so that precious material is not wasted by inefficient transformations?

*   How should test systems be chosen with a view to assessing the utility of new reactions, i.e. to determine their scope and limitations with regard to variations in substrates as well as in reagent(s) and solvent?

*   How sensitive is the reaction to variation in solvent composition? A broad scope with regard to solvent variation is likely to open up possibilities to develop efficient experimental procedures for consecutive transformations by one-pot procedures.

*   Which is the optimum reagent for a given tranformation when there are several possible alternatives?

For almost any given reaction, the number of all possible combinations of substrates, reagents and solvents will be overwhelmingly large. The problem is therefore to select *representative* test systems for the exploration of the roles played by the discrete variations.

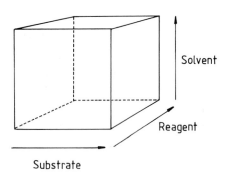

Fig.14.1:   The reaction space is the union of all possible discrete variations of substrates, reagents and solvents.

For the moment, it is sufficient to regard the "axes" of the reaction space, Fig. 14.1, just as "variations". In the next chapter, the concept of *principal properties* is developed. By the principal properties we can obtain a quantitative description of the gradual change of the properties of the reaction systems which is induced by the discrete variations. Quantification of the "axes" will thus make it possible to explore the reaction space systematically.

When a plausible reaction mechanism is known, it is sometimes possible to predict how reagents and solvent should be combined to obtain a suitable reaction system for which the experimental conditions can be adjusted towards optimum performance.

However, with new reactions, mechanistic details are still obscure and it is not likely that such details will be revealed *before* the utility of the reaction has been demonstrated. A consequence of this is, that any conclusions as to the utility of new reactions must be inferred from experimental observations without access to detailed mechanistic knowledge. Thus, it is important that the selected test systems should cover the entire range of interest of the variations.

We shall see that the principles for experimental design developed in the preceeding chapters can be extended with minor modifications also for designs in the reaction space.

1.2. A multivariate design is necessary

The most common approach to the study of discrete variations in the reaction space is to consider one type of variation at a time. For example: The scope of a reaction with regard to substrate variation is often studied by means of a series of test substrates while maintaining the reagent(s) and solvent fixed; the influence of solvent variation is studied with one test substrate and with a given reagent, etc.

Such a study will, however, describe one-dimensional excursions through the reaction space and like any other one-variable-at-a-time study it cannot provide any information whatsoever on possible interaction effects.

A moment of reflection reveals that such interactions are not only possible, but are likely to be present in any variation of the reaction systems. Such interactions can give rise both to compensating and to amplifying effects:

* A sluggish substrate can be brought to reaction by using a more aggressive reagent.

* Substrate and reagents will be differently solvated if the solvent is changed. This is is likely to provoke a change in the reactivity pattern.

* A chemoselective tranformation can often be achieved by adjusting the reactivity of the reagent to fit the reactivity of the substrate.

The ability to recognize such interaction effects is therefore very important when a new reaction is elaborated, otherwise the potential of the reaction for preparative use may be overlooked. It is therefore necessary to use multivariate strategies to explore the reaction space so that the joint influence of varying the substrate, the reagent(s), and the solvent can be evaluated. This can be accomplished by multivariate designs in the principal properties. These designs will define sub-sets of test systems which can furnish the desired information. The principles are discussed in Chapter 16.

1.3. Interdependencies between the reaction space and the experimental space

If in a text-book on the art of baking it was claimed that all types of bread, e.g. loaves of black bread, sponge-cakes and ginger biscuits, should be baked at the same oven temperature and for an equally long time, that book would probably not be a best-seller. This is in essence the concept of "standardized conditions".

The joy of cooking and baking is not very different from the joy of organic synthesis. Unfortunately, in organic synthesis the use of standardized conditions is the rule, not the exception when reactions are studied. It is therefore even more unfortunate that results obtained under such conditions enter into review articles and monographs, thus forming a basis of common knowledge on synthetic reactions.

Of course, using standardized conditions *may* give valuable information as to the scope of reactions. For example, if a reaction is studied with a series of substrates and *all substrates* give excellent results under standardized conditions, the conclusion that the reaction is of wide scope and gives excellent results is obviously valid.

But, if some substrates should give inferior results, it is not a valid conclusion that the reaction is poor for these substrates. It can well be that the reaction conditions should be adjusted for these substrates. One example will illustrate this.[1]

The synthesis of enamines by the modified titanium tetrachloride method was discussed in Chapter 12. The final yield and the rate of enamine formation depend on the molar ratios of TiCl$_4$/substrate and amine/substrate. The optimum conditions with regard to these variables were determined by response surface technique and/or simplex technique for a series of carbonyl compounds. The results obtained for the morpholine enamines are summarized in Fig. 14.2. It is seen that the more crowded substrates require an excess of the reagents. The use of standardized conditions would have led to the wrong conclusions as to the utility of the method. For instance, when the optimum conditions for synthesis of the morpholine enamine from methyl isobutyl ketone were applied to diisopropyl ketone a yield of 12 % was obtained after 4 h. Under optimized conditions yields > 70 % could be obtained.

The above example points to a problem when the scope of a reactions is to be determined. To make fair comparisons, the experimental conditions for each reaction system must be adjusted to an optimum performance. This would obviously be a rather cumbersome process, especially when a large number of test systems is considered. Fortunately, there is a remedy. This remedy is PLS modelling[1] [2] by which it is possible to obtain quantitative models which relate the properties of the reaction system to the variation in optimum experimental conditions. Such models make it possible to *predict* the optimum conditions for *new* systems. As PLS models can also be used to relate variations in the reaction space to variations in the observed response, it is possible to determine *which* properties of the reaction system are essential for achieving the desired result, e.g. a selective tranformation.

---

[1] PLS is a method by which blocks of multivariate data sets (tables) can be quantitatively related to each other. PLS is an acronym: Partial Least Squares correlation in latent variables, or Projections to Latent Structures. The PLS method is described in detail in Chapter 17.

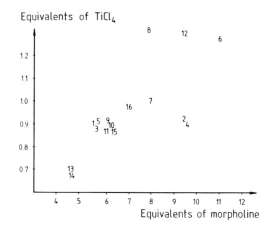

Fig. 14.2: Variation in the conditions affording the maximum yield in the synthesis of morpholine enamines from a series of carbonyl compounds: (1) methyl isopropyl ketone, (2) pinacolone, (3) methyl isobutyl ketone, (4) methyl neopentyl ketone, (5) diethyl ketone, (6) diisopropyl ketone, (7) ethyl isobutyl ketone, (8) diisobutyl ketone, (9) isobutyraldehyde, (10) isovaleraldehyde, (11) cyclohexanone, (12) camphor, (13) acetophenone, (14) *p*-methoxyacetophenone, (15) *p*-nitroacetophenone, (16) isobutyrophenone. The optimum conditions were established by response surface technique or by the simplex method.

PLS modelling is discussed in Chapter 17. Here it is sufficient to say that a suitably designed small subset of test systems can be used to establish a first PLS model. This can then be used to predict the outcome for new systems. These predictions can be validated by experimental runs and the observed results can be used to update and refine the PLS model. This will permit a step-wise approach to the exploration of the reaction space.

The questions posed to the experimental space can be linked to a suitable experimental design to answer the *Which?* and *How?* questions. With new reactions, the answers to these questions will give a solid basis for further studies aimed at seeking the answer also to the *Why* questions. At this stage in the study of a new reaction it is known whether or not the reaction is useful for synthetic purposes. Studies of reaction mechanisms (the *Why?* questions) can therefore be focused on reactions which are known to be practically useful and hence worth learning more about.

336

References

1. *(a)* Carlson, R., Nilsson, Å. and Strömqvist, M.
   Acta Chem. Scand. B 37 (1983) 7;

   (b) Carlson, R. and Nilsson, Å.
   Acta Chem. Scand. B 38 (1984) 49.

2. Wold, S., Albano, C., Dunn III, W.J., Edlund, U, Esbensen, K., Geladi, P., Hellberg, S., Johansson, E., Lindberg, W. and Sjöström, M. in Kowalski, B (Ed.)
   Proc. NATO Adv. Study in Chemometrics, Cosenza, Italy. September 1983. Riedel Publ. Co., Dordrecht 1984.

Chapter 15

Principal properties

When studies on synthetic methods are presented in the literature, the papers sometimes give the impression that the authors have used the "what could be found on the shelf" strategy to select their test systems. Of course, this *may* furnish valuable information on the scope and limitations of the method presented, but sometimes the information is highly biassed due to too narrow a span of important properties of the reaction system.

In this chapter, a strategy is presented by which *all* pertinent properties can be jointly considered prior to selecting the test systems. The strategy is based on a statistical analysis of data on the properties of the potential test candidates. It is realized that this may give quite large amounts of data. The computations involved to analyze the data must therefore be carried out on a computer. Several programs are commercially available, both for personal computers and for mainframe computers. A list of programs is given at the end of this chapter.

1.   Molecular properties

1.1. Observables, descriptors

When a molecule interacts with its surroundings, or when it takes part in a chemical reaction, it is the properties on the molecular level which determine its chemical behaviour. Explanations of this are usually expressed in terms of intrinsic properties which relate to the electronic distribution over the molecular framework as well as to conformational and steric effects. However, such intrinsic properties cannot be measured directly. What *can* be measured are macroscopic, observable, manifestations of the intrinsic properties. These observed properties are then related back to the molecular level by physical-chemical models. For instance, the rather fuzzy concept of *nucleophilocity* is usually understood as the ability of a molecule to transfer electron density to an electron deficient site. A number of observable properties can be related to this ability, e.g. ionization potential, basicity as measured by pK, proton affinity, refractive index, n-π transitions, parameters related to rates in standardized test systems, e.g. the Swain-Scott n parameter [1].

However, one cannot say that the nucleophilic properties, in a philosophical sense, are *explained* by, for instance, the refractive index. There is not a cause-effect relation between these properties. The measured properties are related to nucleophilicity because they depend on the same *intrinsic molecular property*, viz. the polarizability of electrons.

Instrumental methods in chemistry have dramatically increased the availability of measurable properties. Any molecule can be characterized by many different kinds of data. Examples are provided by: *Physical measures*, e.g. melting point, boiling point, dipole moment, refractive index; *structural data*, e.g. bond lengths, bond angles, van der Waals radii; *thermodynamic data*, e.g. heat of formation, heat of vaporization, ioniziation energy, standard entropy; *chemical properties*, e.g. pK, lipophilicity (log P), proton affinity, relative rate measurements; *chromatographic data*, e.g. retention in HPLC, GLC, TLC; *spectroscopic data*, e.g. UV, IR, NMR, ESCA.

Data which can be determined by observations of the properties of a compound and which can be used to characterize the compound, will be called *observables*. It is highly probable that several of the observables of a molecule are related to the chemical behaviour of that molecule when it undergoes a given chemical reaction. It is also probable that some observables are not at all related to the phenomena involved in the reaction under study, e.g. it is not to be expected that the acute toxicity of a compound towards guinea-pigs will have any strong relation to the rate of the reaction in a nucleophilic displacement. This example shows an important principle:

When we have a problem for which we have to consider the properties of a molecule, we can make certain *a priori* assumptions as to the relevance of the observables. Those observables which we believe to be relevant to our problem, will hereafter be called *descriptors*.

1.2 Selection of test items

In a situation where we wish to select a series of test compounds, e.g. for determining the scope and limitations of a reaction, we are facing a problem which involves a discrete variation between the test objects. To cover the possible variation of interest in order to determine as broad a scope as possible, we have to consider all relevant properties of the reaction system. Due to possible interaction effects, all these properties must be taken into account simultaneously. This is not a trivial problem. To take all pertinent descriptors simultaneously into consideration, the problem must be approached by multivariate methods.

In the following sections, the methods of *principal components analysis* (PCA) and *factor analysis* (FA) are described. Principal components analysis is a special case of factor analysis. It is discussed how these methods can be applied to the above problem. Fortunately, these methods are not difficult to understand. A great advantage is that the results obtained by these methods cases can usually be evaluated graphically by visual interpretation of various plots. The systematic variation of the properties for a set of compounds is graphically displayed and it is therefore easy to see how a selection which spreads the properties should be made.

In a situation where test compounds are selected, e.g. for studies on the scope of a reaction, or when the objective is to determine a suitable reagent for a given transformation, it is likely that the possible test candidates are in some way *similar* to each other. This will have several consquences:

(1) Potential substrates, for instance, will belong to a class of compounds which has the essential functional group(s) in common. Differences between items in this class will be reflected by differences in the observable properties of these compounds, and compounds which are similar to each other will have one or several properties which are similar. (Perhaps the reverse is more true: compounds which have similar properties are considered to be similar.) Compounds which are different to each other can be expected to have properties which are different.

(2) Observables which depend on the same intrinsic properties can be expected to be correlated to each other in their variations over the class. For example, melting points and boiling points are correlated since both depend on the strength of the intermolecular forces. As this property, at least in part, is dependent on van der Waal's forces, there will also be a correlation between the molar mass and the melting point/boiling point. It is also reasonable to assume that observables which depend on *different* intrinsic properties are only weakly correlated to each other.

The problem of selecting suitable test objects which span a sufficient range of variation in the intrinsic properties can therefore be stated as follows: From the set of all potential test candidates, a subset should be selected in such a way that a sufficient spread in all pertinent properties is obtained in the test set. These properties are specified by the descriptors.

1.3. Data tables of descriptors

Assume that the set of potential test compounds contains N different compounds. Assume also that each compound has been characterized by K different descriptors.

The same descriptors have been used to characterize all compounds. The data can be summarized in a table, a $(N \times K)$ matrix in which the columns correspond to the K descriptors, and the rows correspond to the N different compounds, see Fig. 15.1.

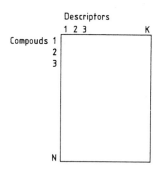

Fig. 15.1: Data matrix of K descriptors for N compounds.

However, if the objective is to select a small subset of compounds which span a large range of variation in all property descriptors, it is very difficult to make a selection of this kind by visual inspection of such tables. An example of a data table showing the variation of ten properties over a set of 28 Lewis acids is given in Table 15.1.[2] The data in Table 15.1 were compiled from standard reference handbooks. The following descriptors were used: 1, dipole moment (vapour phase) (D); 2, negative of standard enthalpy of formation (kJ mol^{-1}); 3, negative of standard Gibbs energy of formation (kJ mol^{-1}); 4, standard entropy of formation (J mol^{-1} K^{-1}); 5, heat capacity (J mol^{-1} K^{-1}); 6, mean bond length (Å); 7, mean bond energy (kcal mol^{-1}); 8, dielectric constant; 9, ionization potential (eV); 10, magnetic and diamagnetic susceptibility (10^{-6} cgs).

It is necessary to resist the temptation to analyze tables like the one below by considering one variable at a time. The risk of finding spurious correlations between the variables increases rapidly with an increasing number of variables. If we wish to be sure at a probability level of 95 % that a descriptor variable is correlated to a chemical penomenon, this also means that we accept a risk of 5 % that this variable is by *pure chance* correlated to the phenomenon. This risk with k variables will be

Table 15.1: Lewis acid characterized by ten property descriptors

Acids	Descriptors									
	1	2	3	4	5	6	7	8	9	10
1 AlCl$_3$	2.1	704.2	628.8	110.47	91.84	2.26	102	15.8	12.01	-
2 BF$_3$	0	1137	1120.33	254.12	50.46	1.265	154	3.9	15.5	-
3 MoS$_2$	-	235.1	225.9	62.59	63.55	-	-	-	-	-
4 SnCl$_4$	0	511.3	440.1	258.6	165.3	2.43	7.6	2.87	-	-115
5 SO$_2$	1.63	320.5	-	237.6	-	1.4321	119	15.4	12.34	-18.2
6 POCl$_3$	2.4	519.1	520.8	222.46	138,78	1.95	122	13.3	11.89	-57.8
7 Me$_3$B	-	143.1	32.1	238.9	-	1.56	89	-	10.69	-
8 Me$_3$Al	-	136.4	99	209.41	155.6	-	61	2.9	9.76	-
9 Me$_2$SnCl$_2$	3.56	336.4	-	-	-	2.37	-	-	10.43	-
10 TiO$_2$	-	913.4	853.9	56.3	9.96	1.97	160	48	10.2	0
11 ZnCl$_2$	2.12	415.05	369.39	11.46	71.34	2.32	96	-	12.9	-65
12 TiCl$_3$	-	720.9	653.5	139.7	97.2	2.138	110	-	-	1110
13 TiCl$_4$	0	804.2	737.2	252.3	145.2	2.19	181	2.8	11.76	-54
14 VCl$_4$	0	576.8	503.27	242.44	-	2.03	92	-	-	1130
15 CrCl$_2$	-	395.2	356.1	114.5	71.1	2.12	91	-	9.97	6890
16 MnCl$_2$	-	481.3	440.3	118.2	72.9	2.32	98.8	-	11.02	14250
17 FeCl$_2$	-	341.79	302.3	117.95	76.65	2.38	95	-	10.34	14750
18 FeCl$_3$	1.28	404.6	398.3	146.4	128	2.32	81	-	-	-
19 CoCl$_2$	-	325.2	282.2	106.5	78.5	2.53	86	-	10.6	12660
20 NiCl$_2$	3.32	305.33	259.03	97.65	71.76	-	87	-	11.23	6145
21 CuCl$_2$	-	220.1	175.1	108.07	71.88	-	91.5	-	-	1080
22 GaCl$_3$	0.85	524.17	454.36	172	-	2.208	78.7	-	11.96	-63
23 GeCl$_4$	0	543.4	-	347.15	29.21	2.1	81	2.43	11.68	-72
24 AsCl$_3$	1.53	335.24	294.7	233.2	-	2.16	70	1.59	11.7	-79.9
25 BCl$_3$	0.61	427.2	387.4	206.3	106.7	1.75	109	0	11.62	-59.9
26 SiCl$_4$	0	601.54	569.32	328.6	145.17	2.019	95.3	2.4	12.06	-88.3
27 SbCl$_3$	3.9	381.75	324.1	186	104.87	2.325	74	33	10.75	-86.7
28 PCl$_3$	7.8	314.7	272.3	217.1	-	1.95	78.5	3.43	9.91	-

$1 - 0.95^k$, i.e. with two variables $1 - 0.92^2 \approx 10$ %, with ten variables $1 - 0.95^{10} \approx 40$ %, and with 60 variables this risk has increased to > 95 %.[1]

Analysis of data tables with several variables must be made by means of multivariate methods which can also take the covariances of the variables into account. For this, *Factor Analysis* (FA), and *Principal Components Analysis* (PCA) are

---

[1] An example will illustrate this: Environmental pollution is a matter of great concern today. Instruments which permit automatic monitoring of several pollutants are finding increased use. Chemical analysis of samples of soil, air and water by sensitive analytical methods can detect hundreds of different contaminants. This implies that environmental samples can be characterized by a large number of variables. By improper analysis of such data, i.e. by considering one variable at a time, the probability is very high that at least some variable *by pure chance* is correlated to the frequency of any imaginable disease och medical disorder. News media are not alway aware of the risks basing their alarm reports on spurious correlations.

suitable tools. In this chapter is decribed how these tools can be employed to derive the concept of *Principal properties*. In the next chapter is discussed how the principal properties can be used for designing experiments aimed at a systematic exploration of the reaction space.

2. Geometric description of principal components analysis

The principles behind principal components analysis are most easily explained by means of a geometrical description of the method. From such a description it will then be evident how a principal components (PC) model can be used to simplify the problem of which test compounds should be selected. The data of the Lewis acids in Table 15.1 will be used to give an example of such selections, after the general presentation of the method which follows.

2.1. Descriptor space

Assume that a set of N componds has been characterized by K variables. For simplicity we can asssume that $K = 10$ as in the Lewis acid example. Let each of the ten descriptor variables define a coordinate axis. The ten descriptors will thus define a ten-dimensional space. Such a space is difficult to imagine, and to illustrate the general principles a three-dimensional space will be used. Geometric concepts like *points*, *lines*, *planes*, *angles*, *distances*, will have the same meanings in higher dimensions as they have in the two- and three-dimensional spaces. We can therefore have an intuitive comprehension of geometries in higher dimensions as analogs to their two- and three-dimensional counterparts.

In the multidimensional descriptor space, each compound can be described by a *point*, with the coordinates along the descriptor axes equal to the measured values of the corresponding property descriptor, see Fig. 15.2.

A series of N different compounds, characterized by the same descriptors, will then define a swarm of ponts in the descriptor space, Fig. 15.3

If the compounds should happen to be very similar to each other, there will only be minor variations in the values of the different property descriptors. The data points would then be close to each other in the descriptor space, and the average values of the descriptors would give an adequate approximation of the distribution of the data points in the descriptor space. This would correspond to a point with the coordinates defined by the average values, Fig. 15.4.

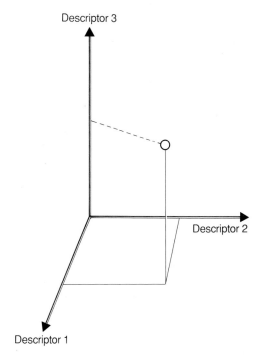

Fig. 15.2: Data point in the multidimensional descriptor space.

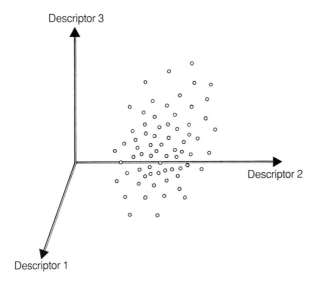

Fig. 15.3: Data points from a series of compounds.

In most cases, however, the compounds are different to each other and there is a spread in their properties. The average values will not give a sufficient description of the data. If the compounds are different to each other, the distribution of the data points will have an extension in the descriptor space. This extension will be outside the range of random error variation around the average point.

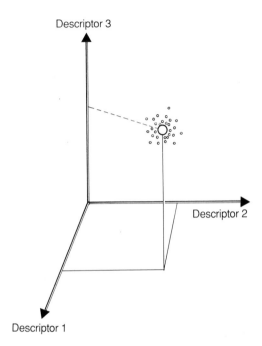

Fig. 15.4: Average point of the data in the descriptor space.

The next step is to determine the direction through the swarm of points along which the data show the largest variation in their distribution in the space. This will be the direction of the first principal component vector, denoted p_1. If this vector is anchored in the average point, it is possible to make a perpendicular projection of all the points in the space on this vector. A projected point will then have a coordinate, measured from the average point, along the first principal component vector, p_1. The distribution of data points in the p_1 direction of the descriptor space can thus be described by their coordinates along p_1. If all descriptors should happen to be correlated to each other, the swarm of points would describe a linear structure in the descriptor space. As such a linear structure will be picked up by the first principal component, the *systematic variation* of the properties over the set of compounds would be described by a single measure, viz. the coordinates along the first principal component vector, see Fig. 15.5. These coordinates are called the principal components *scores*, and will be denoted by the letter t. Compound "i" will have the score t_{1i} along the first component.

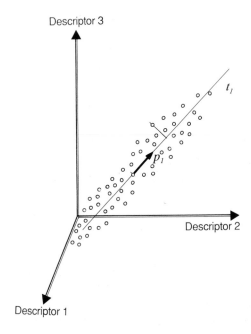

Fig.15.5: Projections of data points along the first principal component vector, p_1. The score of compound "i" is dentoted by t_{1i}.

The extent to which a descriptor contributes to a systematic variation along the principal component determines how much the direction of the *p* vector is tilted towards the corresponding descriptor axis. The direction of a principal component vector p_i is given by the cosine of the angle Θ_{ik} between the axis of descriptor "k" and p_i. This value is called the *loading* of the descriptor variable to the principal component. If a descriptor variable "k" does not contribute at all, i.e. the loading is zero, the corresponding descriptor axis is perpendicular to p_i, $\cos\Theta_{ik} = 0$. With K different descriptors there will be K different loadings to each component vector p_i. The loadings are usually denoted by p_{ik}, where the index "i" assigns the component, and the index "k" assigns the descriptor variable. The principal component vector p_i can thus be written

$$p_i = [p_{i1} \, p_{i2} \, \cdots \, p_{ik}]'$$

The principal component vectors are therefore often called *loading vectors*.

After projection on the first principal component vector it is possible to determine how well this component describes the variation in the data. The distance between the original data point and its projection on the component vector will be a measure of the variation which is not described by the first principal component. This distance is a residual, e_i, between the original data and the model. It is therefore possible to determine the residual sum of squares, Σe_i, and then use the F distribution to compare this sum to the sum of squared deviations between the original data point and the average point. A significant F indicates that the component provides a significant description of the systematic variation. The ratio between these two sums of square describes how large a part of the original variation (sums of squares) which is described by the first principal component. This proportion is usually expressed as a percentage of explained variance. For this, a correction for the number of degrees of freedom must be made.

The next step in the analysis is to determine whether there is more systematic variation which is not described by the first principal component. A second component vector, p_2, is determined so that p_2 describes the direction through the swarm of points with the second largest variation. The direction of p_2 is determined so that it is orthogonal to p_1, and anchored so that it passes the average point. It is then possible to project the data points on p_2 and determine the corresponding scores, t_2. The result of this is that the swarm of points is projected down to the plane spanned by the principal components, and that the projection is made so that a maximum of the systematic variation is portrayed by the projected points., see Fig. 15.6. This means that as much as possible of the "shape" of the swarm of points is preserved in the projection.

Compounds which are similar to each other will have similar values of their descriptors, and the corresponding data points will be close to each other in the descriptor space. Such points will also be close to each other in the projection. Furthermore, compounds which are different to each other will be projected apart from each other. The projection will therefore portray similarities and dissimilarities of the compounds.

If there is still systematic variation left after fitting the two first principal components, it is possible to determine a third component, orthogonal to the first two. This can then be repeated until all systematic variation has been exhausted. Principal components modelling will thus constitute a projection of the data in the K dimensional descriptor space down to a lower dimensional (hyper)plane spanned by the principal component vectors. The projections are made so that the systematic variations of the original data are described by the coordinates along the principal component vectors.

2.2 Principal properties

Here comes the important point: The variations of measured macroscopic molecular properties can reasonably be assumed to be reflections of variations of the intrinsic properties at the molecular level. Descriptors which depend on *the same* molecular property are most likely to be correlated to each other. The principal components describe the systematic variation of the descriptors over the set of compounds. Descriptors which are correlated to each other will be described by *the same* principal component. The principal component vectors are mutually orthogonal, and different component will therefore describe independent and uncorrelated variations of the descriptors. Hence, different components will portray a variation in the data due to different intrinsic properties. These intrinsic properties, which manifest themselves as a variation of the macroscopic descriptors, are called the *principal properties*.[3]

2.3 Plots

Score plots

The variation of the principal properties over the set of compounds is measured by the corresponding scores, t_{ij}. The important result of this is that we can now describe the systematic variation using fewer variables (the scores) than the K original descriptors. This is a considerable simplification without loss of systematic information.

A graphic illustration of how the principal properties vary over the set of compounds is obtained by plotting the scores against each other, see Fig. 15.7.

Such score plots provide a tool by which it is possible to select test compounds to ensure a desired variation of the intrinsic (principal) molecular properties.

As principal components modelling is based on projections it can take any number of descriptors into account. Projections to principal components models offer a means of simultaneously considering the joint variation of *all* property descriptors. By means of the score values it is therefore possible to quantify the "axes" of the reaction space to describe the variation of substrates, reagent(s) and solvents. When there are more than one principal component to consider for each type of variation, the "axes" will be multidimensional. Quantification of the axes of the reaction space offers an opportunity to pay attention to all available back-ground information prior to selecting test compounds.

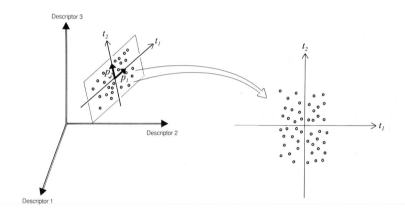

Fig. 15.6: Projection of the swarm of data points down to the plane spanned by the two
            first principal components. A plot of the scores against each other portrays the
            systematic variation.

Loading plots

The contribution of the original descriptors to the principal components can also be displayed graphically by plotting the loadings of the components against each other. Such plots are called *loading plots* and are complementary to the score plots in such a way that the loading value p_{ik} (component "i", descriptor "k") shows the contribution of descriptor "k" to the variation in t_i. Descriptors which have a strong contribution to the variation of t_i will be plotted far from the origin and close to the p_i axis in the loading plot. Descriptors with only minor influence are projected close to the origin in the loading plot. Descriptors which contribute to several components are projected in the quadrants of the loading plots. This occurs when a descriptor is related to several fundamental (principal) properties at the molecular level. For instance, the melting point can depend on van der Waals forces, hydrogen bonding, dipole-dipole attractions etc.

2.4. Example: Selection of Lewis acid catalysts in screening experiments

The problem

Assume that you are about to test a new reaction for which electrophilic catalysis is strongly believed to be beneficial. For this, addition of a Lewis acid catalyst would be worth testing. As there are many potentially useful Lewis acids to chose among, the problem is to find a good one. If the reaction bears a strong resemblance to previously known reactions, it is often possible to make an educated guess of which catalyst to use. If the reaction is totally new, an educated guess is not possible and a suitable catalyst must be determined by experimental observations. In this situation you will face the problem of which catalysts should be tested. Assume also that you have access to a large number of possible catalysts on your shelf, but that it is out of question to test all of them. Then the problem is to select a limited number of test catalysts to cover a range of their molecular properties with a view to finding a catalyst which has promising properties for future development of the method. This problem can be seen as a screening problem and the selected test items will define a screening experiment. The winning candidate from the screening experiment is not necessarily the best choice. It is probable that more experiments will be needed to establish the best choice, but it is reasonable to assume that this catalyst will have properties which are similar to the properties of the winning candidate in the screening experiment. Such a rational and systematic search can be accomplished by analyzing the properties of Lewis acids, and to illustrate the principles the data in Table 15.1 is used.

Projection to determine the principal properties

A projection of the data in Table 15.1 down to the plane spanned by the first two principal components accounted for 54 % of the total variance of the descriptors. The score plot and the loading plot are shown in Fig. 15.7.

a   **b**

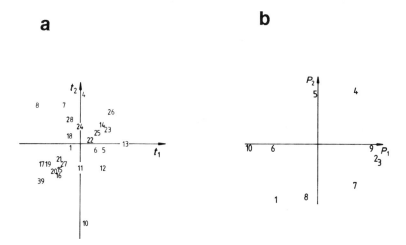

Fig. 15.7: *(a)* Score plot of the Lewis acids; *(b)* Loading plot of the descriptors.

Lewis acids which differ with respect to their descriptors are projected far from each other in the score plot. A selection of test candidates with different properties can therefore be accomplished by selecting Lewis acids which are plotted far from each other in the score plot.

Experiments

The experiments described here, were to test the idea of using the pricipal properties as a criterion for selection.[2] The reactions used were three well-known Lewis acid catalyzed reactions:

(A) Alkylation of silyl enol ethers with *t*-butyl chloride, for which titanium tetrachloride is known to be a suitable catalyst.[4]

(B) Diels-Alder reaction of furan and acentylenedicarboxylic ester, for which aluminium trichloride is known to be useful.[5]

(C) Friedel-Crafts acylation, for which alumnium trichloride is preferred.[6]

$\text{cyclohexenone-SiMe}_3$ enol ether + t-Bu-Cl $\xrightarrow[\text{2/ H}_2\text{O}]{\text{1/ Lewis acid}}$ 2-t-Bu-cyclohexanone

furan + $\text{MeOC-C}\equiv\text{C-COMe}$ $\xrightarrow{\text{Lewis acid}}$ bicyclic diester $-\text{CO}_2\text{Me}$, CO_2Me

(aryl-CH$_2$CH$_2$-CO-Cl) $\xrightarrow{\text{Lewis acid}}$ indanone

The catalysts for testing are encircled in Fig. 15.8. The selection was based solely on the distribution of the points in the score plot, and the selection was made with a view to covering a large variation. Boron trifluoride (2), trimethylborane (7), and trimethylalane (8) are gaseous and difficult to apportion and were excluded from the screening experiment for this reason. Titanium dioxide was considered as an outlier due to the rather extreme value of descriptor 8 and was also excluded.

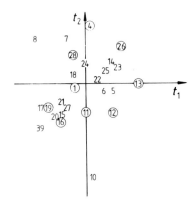

Fig.15.8:   Selected Lewis acid catalysts in the screening experiments

The experimental results obtained in these reactions are summarized in Table 15.2. The results fully confirm the choice of catalysts reported in the literature as the preferred ones. It is interesting to note that ferric chloride (18) has been reported to be a superior catalyst in certain Friedel-Crafts reactions.[7]. Ferric chloride is projected close to aluminium chloride in the score plot.

Table 15.2: Screening experiments with selected Lewis acids

Reaction[a]	Lewis Acid	Maximum yield (%)	Rate[b] t_{50} (min)
A	$AlCl_3$	39	1.9
	$CoCl_2$	0	-
	$MnCl_2$	0	-
	PCl_3	0	-
	$SiCl_4$	0	-
	$SnCl_4$	38	22.1
	$TiCl_3$	0	-
	$TiCl_4$	45	3.1
	$ZnCl_2$	44	378
B	$AlCl_3$	50	0.1
	$CoCl_2$	0	-
	$MnCl_2$	0	-
	PCl_3	0	-
	$SiCl_4$	0	-
	$SnCl_4$	2.5	20.2
	$TiCl_3$	13	1.5
	$TiCl_4$	26	7.0
	$ZnCl_2$	0	-
C	$AlCl_3$	94.1	0.8
	$CoCl_2$	3.5	630
	$MnCl_2$	6.6	71.2
	PCl_3	0	-
	$SiCl_4$	0	-
	$SnCl_4$	0	-
	$TiCl_3$	25.7	36.0
	$TiCl_4$	55.0	750
	$ZnCl_2$	5.0	1800

[a] A: Alkylation of silyl enol ether; B, Diels-Alder reaction; C, Friedel-Crafts reaction.
[b] It was difficult to measure the initial rates. As a rough estimate of the kinetics, t_{50}, is given. This quantity is defined as the time necessary to obtain 50 % of the final yield.

The conclusion from the result in Table 15.2 was that score plots which show the principal properties can be used for selecting test items in screening experiments.

Strategies for selecting test systems to cope with other problems in the development of synthetic procedures are discussed further in the Chapter 16.

3. Mathematical description of Factor Analysis and Principal Components Analysis

A table containing the values of K descriptors (columns) for N different compounds (rows), displays two types of variations:

Vertically, the table shows the between-compound variation of the descriptors.

Horizontally, the table shows the within-compound variations of the descriptors.

Factor analysis and Principal Components Analysis partition these variations into two entities, *scores* and *loadings*, which can be evaluated.

3.1. Scaling of the descriptor variables

The units of measurements can be very different for the descriptors, and their measured values can vary in magnitudes, e.g. melting point, refractive index, wave number in the infrared spectrum. For this reason, it is necessary to scale the values of the descriptors so that their variations can be compared to each other.[8] This implies that the axes of the descriptor space should be adjusted. Otherwise the "shape" of the swarm of data points may be very strange. One common way to scale values of the descriptors is to divide them by their standard deviation over the set of compounds. This will give each descriptor a variance equal to one. This procedure is usually referred to as *autoscaling*, or *variance scaling*, or *scaling to unit variance*. It is the standard procedure in a first attempt to determine a model, especially if the descriptor variables are of different kinds. Scaling to unit variance will give each descriptor variable an equal chance to influence the model. With this procedure, no prior assumptions as to the relevance of the different descriptors are made. The relevance of a descriptor can be evaluated from loading plots and from analysis of residuals after fitting a first model. Irrelevant descriptors can then be deleted in the final analysis of the principal properties.

There are certain situations where scaling to unit variance is not the preferred procedure and where no scaling at all is better. If all descriptor variables are of the same kind and measured in the same unit, e.g. intensities of spectral absorbtion at different frequencies or peak heights in chromatographic profiles, it is sometimes unnecessary to scale the variables. Autoscaling such variables would exaggerate minor variations. Another case, when scaling may be unnecessary is when a variable is almost constant over the set of compounds and its variation is only slightly above the noise level of its measurement.

The principal component score is a linear combination of the descriptors. Chemical phenomena are rarely purely linear. In a limited domain of variation we can regard a principal components model as a local linearization by a Taylor expansion. As such it is likely to apply for classes of *similar* compounds.

If the values of some descriptor vary in magnitudes over the set of compounds it is difficult to assume that a linear model will be a good approximation to account for such large variations. In these cases, a better model can often be obtained after a logarithmic transformation of this variable prior to scaling to unit variance.

Another situation where transformation of a descriptor, x_i is indicated is when it is known by some physical model how this descriptor is linked to more fundamental properties of the molecule. This may indicate some kind of mathematical transformation of the descriptor prior to autoscaling, e.g. $\ln x_i$, $\exp[x_i]$, $1/x_i$. One example is spectral data, for which the wavelength, x_i, of an absorption by the reciprocal transformation $1/x_i$ is transformed into a frequency measure which is proportional to the energy of excitation.

3.2. Data matrix used for modelling

Scaling converts the original descriptor matrix to a normalized data matrix, which hereafter will be denoted by $\mathbf{X}$.[2] The columns in $\mathbf{X}$ are the scaled descriptor variables, see Fig. 15.9.

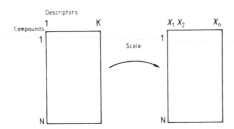

Fig. 15.9: Scaling converts the original descriptor matrix into the normalized data matrix $\mathbf{X}$.

---

[2] The scaled data matrix is usually denoted by X in Factor analysis and Principal components analysis. Unfortunately, the same symbol is used to denote the model matrix in response surface (linear regression) modelling. Hopefully, the reader will excuse this overlap in nomenclature.

3.3 Mathematical derivation of the principal components model. Loading vectors, score vectors

Geometric illustrations are used to show the mathematical aspects of the modelling process. It is shown that these models are related to eigenvalues and eigenvectors. See Fig 15.10 for the definitions.

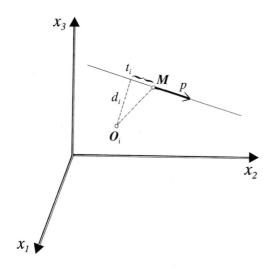

Fig. 15.10: Projection of an object point O_i on a vector p.

Loading vectors

Let M be the point in the descriptor space which corresponds to the average value of each descriptor:

$$M = (\bar{x}_1, \bar{x}_2,..., \bar{x}_K)$$

Let O_i be the data point corresponding to object "*i*":

$$O_i = (x_{1i}, x_{2i}, ..., x_{Ki})$$

Let p be a vector which is anchored in M, and let d_i be the perpendicular distance from O_i to p. This is a projection of O_i on p. Let t_i be the distance from the projected point to the average point, M.

The squared distance from O_i to M is the squared norm $\|O_iM\|^2$ of the vector O_iM.

A consequence of the Pythagorean theorem is

$$\|O_iM\|^2 = d_i^2 + t_i^2$$

For N given (compound) object points, the sum of the squared distances of the corresponding points O_i to M is constant.

$$\Sigma \|O_iM\|^2 = \Sigma d_i^2 + \Sigma t_i^2 = Constant$$

It is therefore seen that a minimization of the sum of squared distances between the object points and the vector p is equivalent to maximizing the sum Σt_i^2.

The value of t_i is the scalar product of the vector O_iM and p.

$$t_i = (O_iM) \cdot p$$

Let $Y = \Sigma t_i^2$

After summation over all points O_i this corresponds to

$$Y = [(X - \bar{X})p]'[(X - \bar{X})p] = p'(X - \bar{X})'(X - \bar{X})p$$

where $\bar{X}$ is the ($N \times K$) matrix in which the elements in each column k is the average of the scaled variable x_k.

If we now introduce the constraint that the vector p should be normalized to unit length, i.e. $p'p = 1$, we can write this constraint

$$C = (1 - p'p) = 0$$

To maximize Y with the constraint C, we can use the method of Lagragne multiplicators to determine the vector p which maximizes

$$L = Y + \lambda\, C$$

$$L = p'(\mathbf{X} - \overline{\mathbf{X}})'(\mathbf{X} - \overline{\mathbf{X}})p + \lambda\, (1 - p'p)$$

Setting $\partial L / \partial p = o$

$$2(\mathbf{X} - \overline{\mathbf{X}})'(\mathbf{X} - \overline{\mathbf{X}})p - 2\,\lambda\, p = o$$

Which gives

$$(\mathbf{X} - \overline{\mathbf{X}})'(\mathbf{X} - \overline{\mathbf{X}})p = \lambda\, p$$

We see that p is an eigenvector to the matrix $(\mathbf{X} - \overline{\mathbf{X}})'(\mathbf{X} - \overline{\mathbf{X}})$.

The matrix $(\mathbf{X} - \overline{\mathbf{X}})'(\mathbf{X} - \overline{\mathbf{X}})$ is a real symmetric square matrix in which all elements are greater than or equal to zero. The rank of a matrix is the number of its non-zero eigenvalues. Imprecision due to an omnipresent error in all experimental measured data will give as result that, in practice, all eigenvalues will be different to zero and
$(\mathbf{X} - \overline{\mathbf{X}})'(\mathbf{X} - \overline{\mathbf{X}})$ will be a full rank matrix. Hence, there will be K different eigenvectors, p_i. Remember, that eigenvectors to different eigenvalues are orthogonal to each other.

When the average centred variable matrix $(\mathbf{X} - \overline{\mathbf{X}})$ is used, the matrix
$(\mathbf{X} - \overline{\mathbf{X}})'(\mathbf{X} - \overline{\mathbf{X}})$ contains the sums of squares and the crossproducts of the variables. Since the means of each variable have been subtracted, the elements in $(\mathbf{X} - \overline{\mathbf{X}})'(\mathbf{X} - \overline{\mathbf{X}})$ are related to the variances and the covariances of the variables over the set of N compounds. The total sum of squares is equal to the sum of the eigenvalues. The variation described by a component is proportional to the sum of squares associated with this component, and this sum of squares is equal to the corresponding eigenvalue. It is usually expressed as a percentage of the total sum of squares and is often called "explained variance", although this entity is not corrected for the number of degrees of freedom. Percent "explained variance" by component "j" is therefore obtained as follows:

Percent "explained variance" $= 100 \cdot \lambda_j / (\Sigma\, \lambda_i)$

When the descriptors have been scaled to unit variance, the matrix
$(\mathbf{X} - \overline{\mathbf{X}})'(\mathbf{X} - \overline{\mathbf{X}})$ is the correlation matrix, and the sum of squares will be equal to K (the number of descriptors).

Score vectors

The values of the scores of the compounds along the eigenvector p_i are computed by multiplying p_i by $(\mathbf{X} - \overline{\mathbf{X}})$. The vector defined by the set of scores of the compounds along p_i is called the corresponding *score vector* and is denoted by t_i, see Fig. 15.11.

$$t_i = (\mathbf{X} - \overline{\mathbf{X}})p_i$$

The eigenvectors p_i are composed of contributions from the descriptor variables. When p_i is normalized to unit length, the corresponding eigenvalue describes how large a proportion of the total variation can be attributed to the original descriptors, assuming that the descriptors reflect the intrinsic molecular properties. Since the eigenvectors are mutually orthogonal they depict *different* intrinsic properties (principal properties). The score value, t_{ij}, of a compound "j" will thus be a measure of how much the principal property "i" influences the properties of "j".

Score vectors to different components are orthogonal to each other, see Appendix 15B.

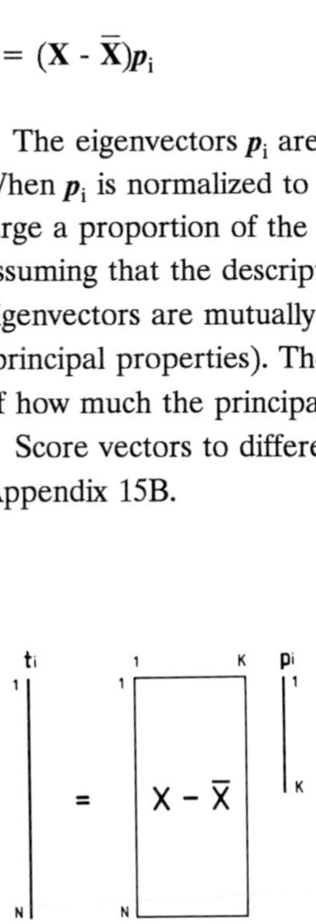

Fig.15.11: Computation of a score vector.

Let $\mathbf{P}$ be the $(K \times K)$ square matrix in which the column vectors are the eigenvectors p_i $(i = 1, 2,..., K)$, and let $\mathbf{T}$ be the rectangular $(N \times K)$ matrix of the corresponding score vectors: It is seen in Fig. 15.12 that

$$\mathbf{T} = (\mathbf{X} - \overline{\mathbf{X}})\mathbf{P}$$

If both sides of the above relation is post-multiplied by **P'** we obtain

$$\mathbf{TP'} = (\mathbf{X} - \overline{\mathbf{X}})\mathbf{PP'}$$

PP' is the unit matrix since **P** is an orthogonal matrix of eigenvectors and $\mathbf{P'} = \mathbf{P}^{-1}$. This gives the following important relation which shows that any matrix can be factorized into two different matrices.

$$(\mathbf{X} - \overline{\mathbf{X}}) = \mathbf{TP'}$$

The matrices **T** and **P** have orthogonal columns. **T** describes the between-compound variation of the descriptors over the set of compounds, and **P** describes the between-descriptor variation of the compounds. Some other aspects of matrix factorization are briefly discussed in Appendix 15B.

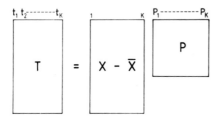

Fig.15.12: Computation of score vectors.

3.4. Factor analysis and principal components

The procedure by which the factorization described above is carried out on a data matrix obtained after centering and scaling each descriptor to unit variance is often called *Factor Analysis*. (The term "Factor analysis" in mathematical statistics has a slightly different meaning which will not be discussed here. For details of this, see [9]). It can be effected by computing the eigenvectors to $(\mathbf{X} - \overline{\mathbf{X}})'(\mathbf{X} - \overline{\mathbf{X}})$. Another, and more efficient method is to use a procedure called *Singular value descomposition* (SVD), see Appendix 15B.

It is often found that some eigenvectors correspond to very small eigenvalues, and will therefore describe a very small part of the systematic variation of the original data. Due to the presence of an experimental error in all experimental data it is often reasonable to assume that the small variation components are mere reflections of the error variation. The large and systematic variation will be portrayed by those components which correspond to large eigenvalues. These components are therefore called the *principal components*.

One criterion often used for selecting the principal components is to use only those components which correspond to eigenvalues greater that one.[9] Another criterion is to use as many components as necessary to describe a specified amount, say 80 %, of the total sum of squares. We will not use any of these criteria to determine the principal components. Instead, we shall use a criterion based on *cross validation* throughout this book. The cross validation criterion determines the optimum number of components to ensure a maximum reliabililty in the prediction by the model. This criterion is discussed below, after a discussion of how a principal component model can be determined by a step-wise procedure.

3.5. Step-wise procedure for determining principal components

A problem when a matrix **X** is factorized by computing the eigenvalues and eigenvectors to the corresponding correlation matrix is that it is assumed that **X** is complete, i.e. there are no missing data. This is not a realistic assumption when property descriptors are compiled from literature sources. It is often found that certain descriptors are not available for more than a limited number of compounds; they have simply not been measured for more compounds. Another problem is that literature data are not consistent, since different authors have used different methods to determine the data. A consequence of these problems is that, in practice, there will always be missing data when descriptors are compiled for a large number of compounds. Fortunately, there are procedures by which the principal components can be determined through iterative methods which can tolerate missing data, provided that the "holes" in the data matrix due to missing data are not too many and fairly randomly distributed over the descriptor matrix. The amount of missing data must be less than about 25 % of the total amount of data. One algorithm for computing the principal components is the NIPALS algorithm, suggested by Fisher [10] and further devloped by Wold [11]. It is briefly described in Appendix 15C. By the NIPALS algorithm the principal components are determined one at a time in decreasing order according to how much variance they describe, until all systematic variation in the descriptor space is described by the model. This furnishes a stepwise method which can be summarized as follows:

(1) A standard procedure is to scale each descriptor to unit variance. Transform the descriptors, if necessary, prior to scaling; see discussion on

scaling above. This transforms the original descriptor matrix into the normalized data matrix $\mathbf{X}$.

(2) Compute the mean, $\bar{x}_i$, of each descriptor "i" over the set of compounds and subtract this value from the value of "i". This yields the scaled and mean centred descriptor matrix $(\mathbf{X} - \bar{\mathbf{X}})$.

Another way of describing this is to say that $\mathbf{X}$ can be decomposed into the average matrix $\bar{\mathbf{X}}$ and a matrix of residuals $\mathbf{E} = (\mathbf{X} - \bar{\mathbf{X}})$.

$$\mathbf{X} = \bar{\mathbf{X}} + \mathbf{E}$$

The average matrix $\bar{\mathbf{X}}$ can be written as the product of a $(N \times 1)$ column vector $\mathbf{1}$ of ones and a $(1 \times K)$ row vector $\bar{\mathbf{x}} = [\bar{x}_1 \ \bar{x}_2 \ ... \ \bar{x}_K]$

$$\bar{\mathbf{X}} = \mathbf{1}\bar{\mathbf{x}}$$

The next step is to analyze the residual matrix $\mathbf{E}$ to see if it contains systematic variation which can be described by a principal component.

(3) Determine the first principal component $\mathbf{p}_1$ by means of the NIPALS algorithm. Multiplication of $\mathbf{p}_1$ by $\mathbf{E}$ gives the first score vector $\mathbf{t}_1$. The matrix $\mathbf{M}_1$ obtained by the multiplication $\mathbf{t}_1\mathbf{p}_1'$ describe how much variation in $\mathbf{E}$ is described by the first component. We can therefore write the original matrix $\mathbf{X}$ as a sum of three matrices

$$\mathbf{X} = \bar{\mathbf{X}} + \mathbf{M}_1 + \mathbf{E}$$

where $\mathbf{E}$ is the the residual matrix obtained after removing the averages and the variation described by the first component from $\mathbf{X}$.

$$\mathbf{E} = \mathbf{X} - \mathbf{1}\bar{\mathbf{x}} - \mathbf{t}_1\mathbf{p}_1'$$

(4) The next step is to extract a second component using $\mathbf{E}$ as the starting matrix. This is repeated until all systematic varation in $\mathbf{X}$ has been exhausted. The overall result after extracting the principal components can be summarized as shown in Fig 15.13. The principal components model can be written in matrix notation as

$$\mathbf{X} = \mathbf{1}\bar{\mathbf{x}} + \mathbf{TP'} + \mathbf{E}$$

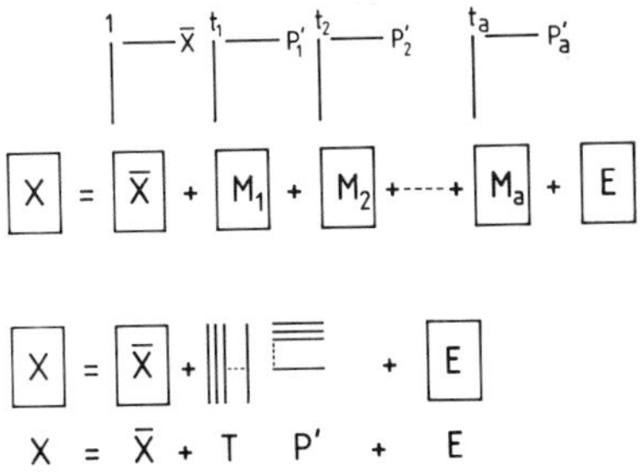

$$X = \bar{X} + M_1 + M_2 + \cdots + M_a + E$$

$$X = \bar{X} + T\,P' + E$$

Fig.15.13: Decomposition of a data matrix **X** into principal components. The matrix **X** is expressed as a sum of the average matrix, the component matrices, and the residual matrix **E** as shown in the upper part of the figure. The matrix **X** can also be expressed as the sum of the average matrix, the matrix obtained by multiplying the transposed loading matrix **P'** by the score matrix **T**, and the residual matrix as is shown in the lower part of the figure.

To describe the *systematic* variation in **X**, it is very likely that only a few components are necessary. The remaining components, not included in the model, will then describe a non-systematic variation. Such non-systematic variation may arise if some irrelevant descriptor which does not contribute to a systematic variation has been introduced in the data set. A non-systematic variation of the same order of magnitude as the experimental error is alway found. If we have prior knowledge of the error variance we can use this to determine the significant components. However, the precision of the descriptors are generally not known. We must therefore resort to other principles to determine which components are significant.

3.6. Cross validation

A significant[3] component should describe systematic variation above the noise level. The principal components model can be used to predict the values of the descriptors. Since a model can never tell the whole story, there will always be a deviation between the observed and the predicted values. We can then determine

---

[3] This means "statistically significant". To be of chemical significance, a component must also be statistically significant. The reverse is not always true. There can be statistically significant differences, which are uninteresting from a chemical point of view. Aldehydes and ketones, for instance, are significantly different to each other in many respects, but in their reaction with lithium aluminum hydride they do not show any pronounced chemoselectivity.

the variance of the prediction error and compare this to the residual variance after fitting the principal components model to the data. When the components are peeled off one at a time as was described above, it is possible to make predictions after addition of each component to the model. To be significant the new component should describe a variation above the noise level, and the errors of prediction should be less than what is obtained without this component in the model. This prevents the inclusion of insignificant components in the model, and ensures that the model affords reliable predictions. This is, in essence, the principles behind the cross validation criterion.[12] These principles in combination with the NIPALS algorithm to determine the optimum number of components in Principal Components Analysis were first described by Wold.[13] This reference also includes a thorough discussion on the computational details of cross validation in principal components analysis. Only a few aspects of this are given here.

Assume that a principal components model with A components $(p_1, p_2,..., p_A)$ has been determined. The model can be used in two ways: (1) For a new compound "r" the corresponding score values can be determined by projecting the descriptors of "r" down to the hyperplane spanned by the components. (2) It is then possible to predict the original descriptors of "r" from the scores and the loading vectors. If the model is good, the predicted value, $\hat{x}_{ir}$, of a descriptor should be close to the observed value, x_{ir}. The difference, $\hat{x}_{ir} - x_{ir} = f_{ir}$, is the prediction error. (The letter f will be used to denote the errors of prediction in the cross validation procedure)

The score values of "r" are computed by multiplication of the loading matrix $\mathbf{P}$ by the centred and scaled descriptors of "r".

$$[t_{1r}\ t_{2r}\ ...\ t_{Ar}] = [x_{1r}\ x_{2r}\ ...\ x_{Ar}]\mathbf{P}$$

The predicted values of the descriptors are obtained by multiplication of the transposed loading matrix $\mathbf{P}'$ by the scores.

$$[\hat{x}_{1r}\ \hat{x}_{2r}\ ...\ \hat{x}_{Kr}] = [t_{1r}\ t_{2r}\ ...\ t_{Ar}]\mathbf{P}'$$

The NIPALS algorithm can tolerate missing data. It is therefore possible to compute a principal components model if data are left out from the data matrix during the modelling process. This can be used to determine whether or not a new component is significant by examining how well the expanded model with the new component can predict left-out data, as compared to the model without the new component. If the new component does not improve the predictions, it is considered not to be significant. The cross validation procedure can be summarized as follows:

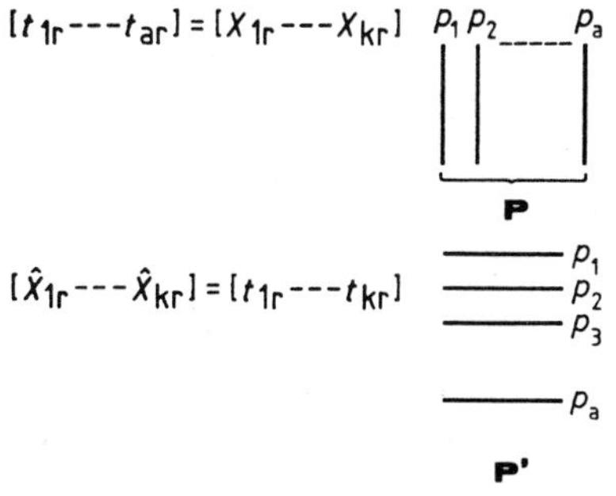

$$[t_{1r}\text{---}t_{ar}] = [X_{1r}\text{---}X_{kr}] \quad \begin{matrix} P_1\,P_2 \underline{\quad\quad} P_a \\ | \; | \quad\quad\; | \\ | \; | \quad\quad\; | \\ \underline{} \\ \textbf{P} \end{matrix}$$

$$[\hat{x}_{1r}\text{---}\hat{x}_{kr}] = [t_{1r}\text{---}t_{kr}] \quad \begin{matrix} \underline{}\,P_1 \\ \underline{}\,P_2 \\ \underline{}\,P_3 \\ \underline{}\,P_a \\ \textbf{P'} \end{matrix}$$

Fig.15.14: The upper part shows the computation of scores for a new compound. The lower part shows the prediction of descriptors from the scores.

After fitting A principal components, the variation not described by the model is given by the residual matrix $\mathbf{E}$.

$$\mathbf{E} = \mathbf{X} - \mathbf{1}\overline{\mathbf{x}} - \mathbf{t}_1\mathbf{p}_1' - ... - \mathbf{t}_A\mathbf{p}_A'$$

From $\mathbf{E}$ compute the residual sum of squares, RSS $= \Sigma\Sigma\,e_{ij}$, as the sum of the squared elements of the residual matrix.

The next step, is to leave out part of the data from the residual matrix and compute the next component from the truncated data matrix. The left-out data can then be predicted from the expanded model and the error of prediction $f_{ir} = \hat{x}_{ir} - x_{ir}$ is determined. This is repeated over and over again, until all elements in $\mathbf{E}$ have been left out once and only once. The prediction error sum of squares, PRESS $= \Sigma\Sigma\,f_{ij}$, is computed from the estimated errors of prediction. If it should be found that PRESS exceeds the residual sum of squares RSS calculated from the smaller model, the new component does improve the prediction and is considered to be insignificant. A ratio PRESS/RSS > 1 implies that the new component predicts more noise than it explains the variation.

Sometimes one component is found to be insignificant whereas that the next one is highly significant. This occurs when the data points describe a flat and almost circular (pancake-shaped) swarm of points in the descriptor space. In such cases the direction of the first component will not be well-defined, but inclusion of a second component will account for this feature. Both components are significant in such cases. To ensure that a correct number of significant components has been found, two successive components should be insignificant according to the cross validation criterion.

3.7. Residuals

When a swarm of data points in the K-dimensional descriptor space is projected down to the hyperplane defined by a A-dimensional principal components model, there will always be deviations, residuals, between the original data point and its projected point on the model, see Fig. 15.15. These residuals can provide additional information.

Relevance of descriptor variables: The perpendicular distance from a data point O_i to the principal component p_j is given by the residual d_{ij}, see Fig. 15.15. The sum of squared deviations with respect to p_j over all data points is $\Sigma\ d_{ij}$ ($(i=1$ to $N)$. This sum of squares is the variation *not* accounted for by p_j. After correction for the number of degrees of freedom this can be expressed as "unexplained variance".

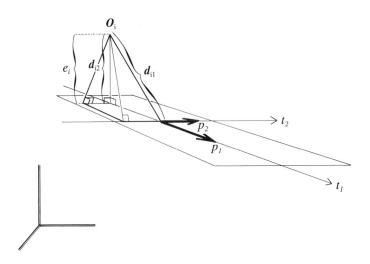

Fig. 15.15: Residual between an object point O_i and the PC model.

The sum of squares $\Sigma\, d_{ij}$ over all objects and all components is the residual sum of squares not accounted for by the model. This can be partitioned into components, Σc_{ik}^2, which show how much the "unexplained variance" for each descriptor, "k", contributes to the total sum of squares, see Fig. 15.16. It is seen in Fig. 15.16 that

$$c_{i1}^2 + c_{i2}^2 + c_{i3}^2 = d_{ij}^2$$

If all objects "i" are taken into account, the above expression can be generalized to K descriptors and the total sum of square deviations with respect to the component "j" is partitioned according to

$$\Sigma d_{ij}^2 = \Sigma c_{i1}^2 + \Sigma c_{i2}^2 + \ldots + \Sigma c_{iK}^2$$

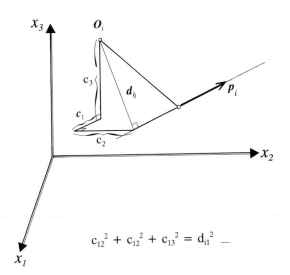

$$c_{12}^2 + c_{12}^2 + c_{13}^2 = d_{i1}^2 \quad -$$

Fig. 15.16: Partition of the residual sum of squares into variance components of each descriptor.

This partitioning of "unexplained variance" offers a means of determining the relevance of each descriptior in the principal components model. Descriptor variables with large "unexplained variance" contribute only to a small or negligible extent to the description of the systematic variation. Cross validation ensures a correct number of components. Taken together, these techniques make it possible to analyze the relevance of all considered descriptors for a given problem.

Deviations of object points: The perpendicular deviation, e_i, between a data point O_i and its projected point on the model is shown in Fig. 15.15. The sum of squared deviations is the residual sum of squares, RSS, after fitting the model to the data points. Dividing RSS by the number of degrees of freedom[4] after fitting a A-dimensional model to the N data points gives the residual variance, $s_0{}^2$.

$$s_0{}^2 = \Sigma \ e_i{}^2 \ / \ [(N - A - 1)(K - A)]$$

This gives an estimate of the residual standard deviation, s_0, which makes it possible to use the t distribution and assign confidence limits around the model. From these confidence limits it is then possible to determine whether or not a compound O_i fits to the model by comparing its residual, e_i, to s_0.

A principal components model can be used to determine the principal properties for selecting test compounds in synthetic experiments. If a test compound which has a large residual when projected down to the model would be selected, it would not be a typical compound. If an atypical compound is included in the set of test compounds when the general applicability of a synthetic method is investigated, the results may be misleading. Test compounds which are in some respect atypical are, of course, of interest when the limitations of a method is studied.

Outliers: When data for a set of objects are used to determine a principal components model, there is always a risk that an object is in some respect atypical. Such objects are easily detected, they appear as outlying isolated points in the score plots. As principal components modelling is a least squares method, the model is fitted also to minimize the deviations to such outlying points. This can have a bad influence and tilt the orientation of the principal component vector if the aberrant point is outside the vector which fits well to the remaining points. It can also reinforce the model of the aberrant point is in the direction of the extension of the

---

[4] With N objects and K descriptors there are $N*K$ initial degrees of freedom. Computing and subtracting the averages consumes K degrees of freedom. Each p vector consumes K degrees of freedom and with A model dimensions, $A*K$ degrees of freedom is consumed. Each score vector takes N degrees of freedom, and the scores consume $N*A$ degrees of freedom. However, normalizing each component vector is a constraint which gives back A degrees of freedom, and the constraint that the p vectors should be orthogonal ($p'p$ is a unit matrix with A^2 matrix element) gives back $A*A$ degrees of freedom. The number of degrees of freedom for the residual variance is therefore: $N*K - A*K - K - N*A + A + A^2 = (N - A - 1)(K - A)$.

remaining points, see Fig. 15.17. The ability of an outlier to tilt the model is sometimes called leverage.[5]

A data point suspected to be an outlier should always be considered and the risk of a bad influence on the model should not be overlooked. This risk can easily be eliminated when principal properties are determined since usually only a few principal components (two or three) are necessary. Outliers are then easily detected from the score plots. The best in such cases is to delete the suspected outlier from the modelling process and refit the model to the remaining compounds. If the model thus obtained should be very different both with regard to the scores and the loading (examine the plots), and if the "explained variance" is highly improved, the suspected object was probably an outlier with a large leverage effect. If we wish to determine the scores of the deleted object, these can be obtained by projecting the deleted object down to the model. If it is still an outlier as determined from the confidence limits around the model, the suspected object is an outlier which is atypical. It is then often possible to give an *ad hoc* explanation why this object is an outlier.

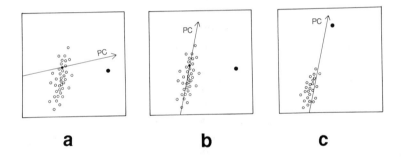

a   **b**   **c**

Fig.15.17: Outliers in PC modelling: *(a)* The outlier has a great and bad influence on the model; *(b)* Removal of the aberrant point improves the model; *(c)* The outlier reinforces the model.

---

[5] It is evident that both atypical objects (compounds) and atypical (irrelevant) descriptor variables can influence the properties of the model. A general analysis of which objects and variables will have an effect on the properties of the model can be obtained from the leverage matrices: $T(T'T)^{-1}T'$ (objects), and PP' (variables). The diagonal elements in these matrices will have values between zero and one. A value close to zero indicates small effects due to leverage, and a value close to one indicates a large leverage effect.

It is difficult to detect outliers through inspection of large data matrices. Two examples illustrate this: When principal properties were determined for a set of ketones[14], 2-furanone (which is a lactone) was included in the set of compounds to check the sensitivity of the model for detecting outliers. As expected it was a clear outlier. When a set of aldehydes was analyzed[14], hydroxy substituted aromatic aldehydes were projected as clear ouliers. The ability of these compounds to strong hydrogen bonding sorted them out as different to the remaining, more lipophilic aldehydes.

4.   Some general aspects on the use of principal components analysis

The very useful metod of principal components analysis has found applications in so many different scientific and technical disciplines that it is not possible to give an overview. Some general references to principal components analysis and factor analysis are given in the reference list. Here we shall only briefly discuss three other general applications of PCA.

4.1. Detection of outliers

It is advisable always to use principal components analysis to check the consistency when large data tables are analysed. Two or three components are sufficient to detect clear outliers. Outliers are almost always found, and in most cases they are artifacts arising from blunder-type errors committed in the compilation of the data or errors committed in the preparation of the data, e.g. typing errors when data are transferred, aberrant observations, or errors committed in analysis or experimental procedures. Erroneous data of this kind are almost impossible to detect by visual inspection of the data table, but are easily detected from the score plots.

Outliers due to real differences in data will also be detected by plotting the first two score vectors against each other. It depends on the specific problem whether or not they should be included in the final analysis. Nevertheless, they can be *detected* at an early stage of the invesitgation.

4.2. Overview of the data

Correcting the input data by removal of inconsistencies due to error and aberrant objects will give a data matrix which, hopefully, contains relevant data for the problem under study. A principal components model with two or three components gives an opportunity for visual interpretation of the general features of the structure of the data. Often, two or three components are sufficient.

Objects: The score plots describe the large and systematic variation between the objects. Often, clear groupings can be discerned in the score plots. This indicates that there are subgroups in the set of objects and that objects which belong to such a subgroup are more similar to each other than to members of other subgroups. Such groupings can then be used for classification of new objects by the SIMCA method, see below.

Variables: The loading plots from such an overview analysis reveal which descriptors are correlated to each other and thus are associated with the same properties of the objects. Such loadings are either projected close to each other (positive correlation) or opposite (180 °) to each other with respect to the origin (negative correlation). Descriptors which influence only one component will appear in the loading plot close to the corresponding axis. Descriptors which have a strong influence on the model will be projected far from the origin, and descriptors with negligible or minor influence will appear close to the origin in the loading plot.

4.3. Classification by the SIMCA method[15]

It is often the case that several variables are used to characterize a set of objects and that these can be divided into natural and disjunct subgroups. It can reasonably be assumed that members of such subgroups are similar to each other, but are more or less different to members of other subgroups. Such subgroups is hereafter called *classes*. Examples of such classes are: Analytical samples of different origins; different batches of raw material; different substitution patterns in chemical compounds, e.g. *cis /trans* isomers; clinical data of healthy and ill persons.

We may wish to use an *a priori* known classification of the objects to determine to which class a *new* object belongs. We may also wish to know in which respect the classes are similar to each other and in which respect they are different.

It is seen that a number of common problems can be regarded as classification problems.

In the SIMCA[6] method for classification,[15,16] separate principal components models are determined for *each class*. The idea behind this is that classes are fairly homogeneous and that the objects in a class are similar to each other, and it is very likely that a principal components model with few components is sufficient to describe the variation within a class. When a new object is projected down to the

---

[6] SIMCA is an acronym for: Statistical Isolinear Multiple Components Analysis, or Soft Independent Modelling of Class Analogy, or SIMilarographic Computer Analysis, or SIMple Classification Algorithm, or ...

372

class models, it is possible to use the confidence limits around the models to determine to which class the new object fits, see Fig. 15.18.

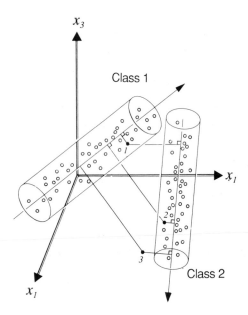

Fig.15.18: Classification of new objects in SIMCA analysis: Object *1* belongs to Class 2, Object *2* belongs to Class 1, and Object *3* is outside the confidence limits of both models and can be of another kind.

It is not within the scope of this book to go into any details on the SIMCA technique. Readers who wish to have more of technical details should consult the literature given in the reference list.[15–17]

Examples of the SIMCA technique in organic chemistry: A very early example of the use of the SIMCA technique is the classification based on ^{13}C NMR data of substituted norbornanes to determine if they are *endo* or *exo* substituted.[18]

Another interesting example is a study on rate data in solvolysis reactions.[19]. It was found that *exo* substituted norbornane systems were classified to belong to the same class as substrates which were known to give rise to stabilized carbocations. The existence of non-classical carbonium ions has been subject for much debate. This study supported the interpretation that there are some involvement of extra stabilization (non-classical ions) in the solvolysis of *exo*-substituted bornyl derivatives.

I am not aware of any applications of the SIMCA technique to organic synthesis. There are certain types of problems where SIMCA definitely would be an appropriate tool. Some examples are:

* In what way are different batches of starting material influencing the result of a synthetic reaction? Useful descriptors would be analytical data for the different batches, and the classes would be the different batches.

* There is a strong suspicion that two competing reaction mechanisms are operating, and that this is determining the selectivity of the reaction. Which mechanism will dominate is dependent on the structure of the substrate. To predict whether or not the reaction can be assumed to be selective for a new substrate it would be valuable to analyze which features of the substrate structure favour the respective mechanism.

* A synthetic reaction which has been successfully applied on many occasions is found to fail with certain combinations of substrates and solvents. The problem is to clarify why it fails.
  As there may be many causes of a failure it is not very probable that a useful class model can be obtained for the failure systems. The successful systems might be described by a class model based on pertinent descriptors of the substrates and the solvents. Analysis of the principal components model will then make it possible to determine in which respect the successful systems are similar to each other. Such an analysis may also furnish information on how to attain an optimum performance of the reaction.

5. Some examples of principal properties in organic synthesis

In this section, we shall present some examples of principal properties which can be used to select test systems for exploring synthetic reactions. In the next chapter, strategies for selection will be discussed in more detail.

5.1. Solvents[20]

Most organic reactions are carried out in solution. The choice of a good solvent is therefore often critical. A problem is that there are so many different solvents to choose among. If the reaction under study bears a strong resemblance to other well-known reactions, or when the reaction mechanism is fairly well known, it is often possible to make an educated guess as to a suitable solvent. However, with a totally new reaction, this is not possible, and to determine a suitable solvent it is necessary to run experiments. Solvents are commonly classified according to their properties: Polar−apolar, hydrogen bond acceptors−hydrogen bond donors, Brönsted acid−base properties, Lewis acid−base propeties (EPA−EPD) etc. Such classifications make it possible to select a type of solvent which we believe will fit the reaction. Often our choice is based on assumptions as to the reaction mechanism. We should, however, always be aware of that speculations in reaction mechanisms may be wrong. If a selection of a test solvent was based on a hypothesized reaction mechanism which later on was found to be false and if this erroneous assumption led to a rather narrow selection of tested solvents, and if the reaction failed in the solvent selected, there is a risk that a promising new method has been overlooked. A more dispersed set of test solvent would have made it possible to detect suitable solvent and would also have revealed that the mechanistic assumptions were wrong.

It is common to use "solvent polarity" as a criterion for the section. This is quite a vague concept.[7] It is normally used about the ability of the solvent to interact with charged species in solution. Often the dielectric constant or the dipole moment is used as a measure of "polarity". If a reaction was assumed to proceed by an ionic mechanism and consequently only polar solvents were tested, this would be an example of too narrow a choice of test solvents. If it should be found later on that the critical step of the reaction was homolytic, it is evident that the "polarity" of the solvent was not the most critical property.

A solvent can play many roles in a reaction:

* It can interact with the reactants and the intermediates and hence influence the energetics of the system. This determines the detailed reaction mechanism. A change of solvent will modify these interactions and may alter the reaction mechanism.

* It can act as a heat transport medium.

* It can serve merely as a diluent.

---

[7] "The characterization of a solvent by means of "polarity" is an unsolved problem since the term "polarity" has until now, not been precisely defined", C. Reichardt.[21]

In view of these many roles, it is evident that all attempts at deriving single and unique scales of solvent descriptors accounting for *all* solvent "effects" should be regarded with scepticism.

To account for the involvment of a solvent in a chemical reaction it is therefore necessary to use multivariate methods and in this context the principal properties are useful. Many attempts have been made to derive various "scales" of solvent properties to account for solvent-related phenomena. There are ca. 30 different empirical solvent scales described in the book by Reichardt[22] but few of these descriptors are available for a sufficiently large number of solvent to be practically useful as selection criteria. The principal properties described here were determined from the following property descriptors:

1, melting point (*mp*); *2*, boiling point (*bp*); *3*, dielectric constant (*de*); *4*, dipole moment (*my*); *5*, refractive index (n_D); *6*, the empirical polarity parameter E_T^N described by Reichardt[23] (*ET*); *7*, density (*d*); *8*, lipophilicity as measured by log P (log *P*); *9*, water solubility measured as log (mol L^{-1}) (*aq*). The symbols given within parentheses are used to identify the descriptors in the loading plot.

Descriptor *6*, E_T^N is the Reichardt-Dimroth $E_T(30)$ parameter[24] expressed in SI units. $E_T(30)$ defined as the transition energy (kcal) at 25 ° C of the long wave absorption band of standard betain dyes when dissolved in the solvent. Descriptor *8*, log P is the logarithm of the equilibrium constant of the distribution of the solvent between 1-octanol and water at 25 °C.

The data of 103 solvents are summarized in Appendix 15A, Table 15A.1. Data for solvents 1–82 were taken from the the first edition of the book by Reichardt[22a] and these data were also used in the first determination of the principal properties of organic solvents.[20] The numbering of the solvent in [20] was the same as in the book by Reichardt. To make it possible to compare the results given here to the previous results, the same numbering has been kept for the first 82 solvents. The augmented data set used here has been compiled from the second edition of the book by Reichardt[22b] and from other sources, (see Appendix 15A, Table 15A.1).

Principal components analysis

Principal components analysis of the data in Table 15A.1 showed that a two-component model was significant according to cross validation and accounted for 59.9 % of the total variance (32.9 % by PC1 and 27.0 % by PC2). The score plot is shown in Fig. 15.19a and the loading plot in Fig. 15.19b.

From the loading plot, Fig. 15.19b, it is seen that the principal properties of a solvent can be interpreted as "polarity" (PC1), and "polarizability (PC2). The first component is mainly described by the typical polarity descriptors, while the second component is described by the refractive index and by the density. It is seen that the refractive index is almost orthogonal to the polarity measures. Melting point and boiling point reflect the strength of the intramolecular forces, and as expected they contribute to both components. The intermolecular forces depend on polar attractions and hydrogen bonding (PC1) as well as weak dispersion forces (PC2). "Polarity" and "polarizability" are therfore the principal properties of a solvent and when test solvents are selected both properties must be considered and the selected solvents should span a sufficient range of both. From a practical point of view, such a selection can be made from a plot in which the refractive index is plotted against the dielectric constant for the potential test solvents. Such a plot would be very similar to the score plot in Fig. 15.19a. How such a selection could be accomplished to cope with different types of problems is discussed in the next chapter.

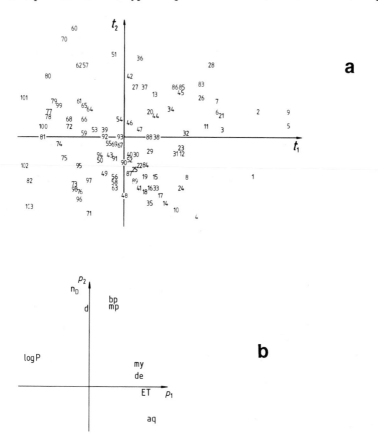

Fig.15.19: Principal properties of organic solvents: *(a)* score plot, *(b)* loading plot.

5.2. Aldehydes and ketones[25]

The carbonyl group is by far the most important single functional group in organic chemistry. The majority of known synthetic methods utilizes manipulations of carbonyl compounds in one way or another. For this reason, synthesis of complex molecules often involves key steps in which a carbonyl group plays a role. To develop such key reactions with a view to attaining an optimum result, studies are often carried out on more simple model compounds, often using commercially available aldehydes and ketones. To ensure that the important features of the key reaction can be grasped from such model studies it is desirable that the test systems are selected to cover a sufficient span in the principal properties.

It is possible to envisage several factors which govern the behaviour of a carbonyl compound, e.g. the electronic distribution over the carbonyl group; the kinetic and thermodynamic acidities of the α-hydrogen(s); stereoelectronic effects; steric hindrance towards attacking reagents; solvent coordination (hydrogen bonding, coordination to Lewis base solvents). Such aspects have been the subject of many elegant studies, but a problem which is always encountered when descriptors are compiled is that data related to interesting properies are only available for a limited number of the potential test candidates. To be of practical use, the set of possible test candidates must be quite large so that a sufficient range of variation of the principal properties is spanned. This will, however, limit the choice of descriptors. The descriptors given below were used to characterize a set of 113 commercially available aldehydes, and a set of 79 available ketones. The data are summarized in Appendix 15A, Tables 15A.2 and 15A.3.

1, molar mass (*mm*); *2*, melting point (*mp*); *3*, refractive index (n_D); *4*, density (*d*); *5*, boiling point (*bp*); *6*, wave number of infrared absorption (*IR*); *7*, molar volume (*mv*). The symbols given within parentheses are used to identify the descriptors in the loading plot.

The selection of these descriptors was based on an initial study of a smaller set of ketones for which two more descriptors were available: *8*, the ionization potential and *9*, the ^{13}C NMR chemical shift of the carbonyl carbon. The score plot obtained with these two extra descriptors in the data set was almost identical to the plot obtained from a model without these descriptors included in the model. It was therefore assumed that they were not necessary for obtaining an overall picture, see Ref.[25] for details.

Principal components analysis

Principal components analysis of the set of 113 aldhydes (Table 15A.2) afforded two significant components according to cross validation. The model accounted for

78 % of the total variance. For the set of 79 ketones (Table 15A.3), two significant components accounted for 88 % of the total variance. Fig. 15.20 shows the score and loading plots of the aldehydes, and Fig. 15.21, the corresponding ones for the ketones. The identification numbers are given in the tables in Appendix 15A.

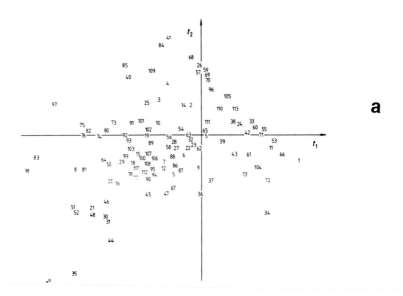

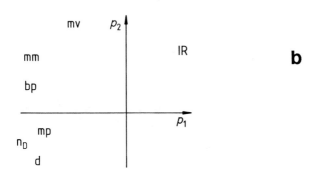

Fig. 15.20: Principal properties of aldehydes: *(a)* Score plot; *(b)* Loading plot.

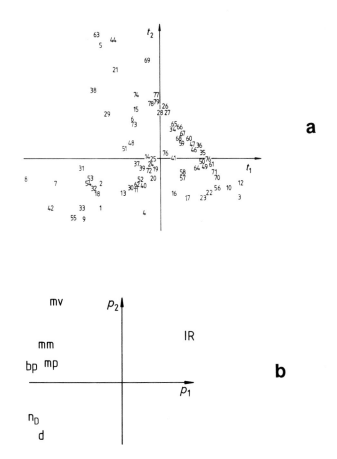

Fig.15.21: Principal properties of ketones: *(a)* Score plot; *(b)* Loading plot.

5.3. Amines[26]

A rough estimate based on pharmaceutical literature is that more than 95 % of all commercially available pharmaceuticals are amine derivatives. Hence, the chemistry of the amino group is of great practical importance.

Descriptors

The chemical behaviour of amines depends on at least four factors: Basicity, nucleophilicity, solvation, and steric environment of the nitrogen atom. As missing data always are a problem, the number of available descriptors is limited. Only easily measured descriptors are commonly available. Data compiled for 126 amines are summarized in Appendix 15A, Table 15A.4. The following descriptors are included in the data set:

1, molar mass (*mm*); *2*, melting point (*mp*); *3*, refractive index (n_D); *4*, density (*d*); *5*, boiling point (*bp*); *6*, molar volume (*mv*); *7*, pK_a (*pK*) The symbols given within parentheses are used to identify the descriptors in the loading plot.

The selection of descriptors was based on the following considerations:

Basicity: The only basicity descriptor generally available for a large number of amines is pK_a (measured in water).

A PCA study on a limited set of amines[27] for which also proton affinity and gas phase basicity were available as descriptors produced almost the same result as did pK alone.[26]

Nucleophilicity: Good descriptors of this property are rare. Consistent kinetic data from nucleophilic reactions are available only for a limited number of amines. Since nucleophilicity related to the properties of the lone pair electrons on the nitrogen, indirect measures related to the lone pair must be used. The base properties are involved, so pK_a is in part a descriptor of nucleophilicity. Also the polarizability is part of the picture, and for this property the refractive index is a generally available descriptor. Other descriptors which can be used in this context are the ionization potential and the frontier orbital energies. Unfortunately, these descriptors are not available for a sufficiently large number of amines.

Solvation: When the solvent is changed, the solvation of the substrate and the reagent(s) may be altered. This may modify the reactivity pattern. An example is when strongly associated hydrogen bonding solvents and weakly associated solvents are to be included among the test solvents. When the problem is to consider both the variation of an amine and the solvent in a reaction, it is possible to cope with the situation by using a multivariate design based on the principal properties of both amines and solvents. Such designs are disussed in the next chapter.

A quite unexpected result reported in Ref.[26] was that hydrogen bonding does not have a strong influence on the principal properties as measured from the descriptors in Table 15A.4. Analysis by the SIMCA method with one class of amines defined for primary and secondary amines, and one class defined for tertiary amines, afforded the unexpected result that these class models were almost identical and completely overlapping. Thus, properties like melting point, boiling point and density evidently depend on weak dispersion forces and not on hydrogen bonding. Weak dispersion forces are responsible for the solvation in aprotic non-polar organic solvents commonly used in organic synthesis. It has previously been shown that in the absence of strong interactions, several properties in the liquid state can be adequately described by principal components models.[28]

Steric environment of the amine nitrogen atom: The descriptors used to determine the principal properties of amines do not take steric effects into account. It is therefore suggested that steric factors should be considered as a separate additional criterion when test items are selected.

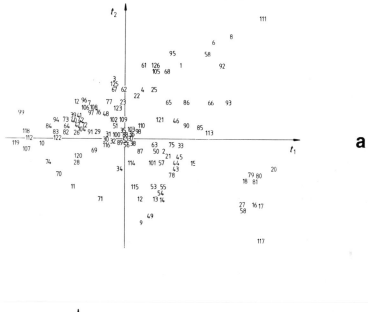

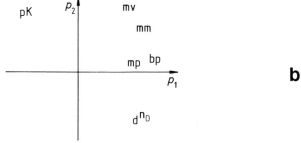

Fig.15.22: Principal properties of amines. *(a)* Score plot; *(b)* Loading plot.

Principal components analysis

A two-component model was significant (cross validation) and accounted for 85 % of the total variance. The score plot and loading plot are shown in Fig.15.22.

5.4. Lewis acids[2]

In the introduction to this chapter, the idea of the principal properties was illustrated by a small set of 28 Lewis acis. However, to be of practical use, the set of potential test candidates must be much larger. A set af 116 Lewis acids characterized by 20 descriptors is summarized in Appendix 15A, Table 15A.5. The following descriptors are included in the data set:

1, coordinate bond energy; *2*, negative of standard enthalpy of formation; *3*, negative of standard Gibbs energy of formation; *4*, standard entropy of formation; *5*, heat capacity; *6*, mean bond length; *7*, melting point, *8*, boiling point; *9*, density; *10*, standard enthalpy of formation of M^{n+} species; *11*, lattice energy; *12*, mean bond energy; *13*, covalent bond energy; *14*, ionic bond energy; *15*, partial charge on central atom; *16*, partial charge on ligand atom; *17*, ionozation potential (gas phase); *18*, magnetic susceptibility; *19*, dipole moment (gas phase); *20*, atomic electronegativity of central atom (in different oxidation states).

Principal components analysis

A two-component model was significant and accounted for 65 % of the total variance. The score plot and loading plot are shown in Fig. 15.23.

5.5. Other examples of principal properties

Principal components models have been used to determine principal properties for classes of compounds in applications to other areas than organic synthesis. Some examples are given below. These examples will not be further discussed here.

Aminoacids, where principal properties have been used to establish quantitative structure-activity relations (QSAR) for peptides.[29] In this context, a more elaborate concept, *dedicated principal properties*[30] has been suggested where the modelling process is carried out in several steps to achieve maximum predictability of the QSAR models. We do not go into details on this.

Environmental hazardous chemicals, where principal properties of homogeneous subgroups have been determined and used to assess toxicity.[31]

Substituents on aromatic rings, where the objective is to obtain orthogonal measures which could be used for design of test set in QSAR modelling.[32]

Eluents for chromatography, where principal properties of solvents have been used to optimize separation in TLC and HPLC.[33]

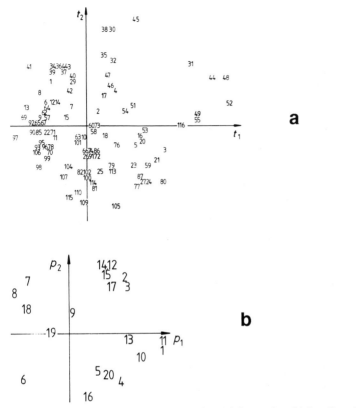

Fig. 15.23: Principal properties of Lewis acids: *(a)* Score plot; *(b)* Loading plot.

6. Summary

This chapter introduces the concept of principal properties. To comprehend the principles, a geometrical description as projections of points in a space down to a plane is sufficient. To understand the modelling process in detail, a mathematical description is necessary. The message conveyed can be summarized as follows:

When a synthetic method is developed, it is often of interest to make experimental studies of discrete variations of the reaction system, e.g. different substrates, various reagents, different solvents. The objectives for this may be to establish the scope and limitations of the method, or to find a better combination of reagents and solvents. A problem is that there are often a large number of test candidates to choose among.

All reactions under study have their background history. There may be a reasonable mechanism suggested, or at least there are some hypotheses at the

384

mechanistic level. The reaction may be quite similar to other methods for which a lot of details are well known. It is therefore always possible to conceive a number of "factors" which might influence the result. A moment of reflection will reveal that several of these "factors" can be linked to observable macroscopic properties which can be used to characterize the potential test candidates. An evaluation of these observables in relation to the problem under study will suggest a set of *descriptors* which are reasonably relevant. Compilation of descriptor data from literature sources for the set of potential test candidates will generally give a large table of data, a data matrix. Principal components analysis of these data will portray the *systematic* variation of the properties over the set of compounds. Descriptors which vary in a concerted manner can reasonably be assumed to depend on the same intrinsic molecular property. As the principal components vectors are mutually orthogonal, they will account for *different* and *independent* intrinsic properties. These properties are therefore called *the principal properties*. Plots of score vectors against each other give a graphic display of the variation of the principal properties over the set of potential test candidates. Such plots can therefore be used to select test items which span a sufficient range of variation of the principal properties. The loading plots show how the descriptors contribute to describe the principal properties. It is sometimes possible to interpret the principal properties in terms of "nucleophilicity", "polarity" etc., i.e. to concepts usually used to describe different "effects" in organic chemistry.

References

1. C.G. Swain and C.B. Scott
    J. Am. Chem. Soc. 75 (1953) 141.

2. R. Carlson, T. Lundstedt, Å. Nordal and M. Prochazka
    Acta Chem. Scand. B 40 (1986) 522.

3. S. Hellberg, M. Sjöström, B. Skagerberg and S. Wold
    J. Med. Chem. 30 (1987) 1126.

4. M. Reetz, W.F. Maier, H. Heimbach, A. Giannis and G. Anastassious
    Chem. Ber. 113 (1980) 3734.

5. A.W. McCulloch, D.G. Smith and A.G. McInnes
    Can. J. Chem. 51 (1973) 4125.

6. H.L. Martin and L.F. Fieser
    Org. Synth. Coll. Vol. 2 (1943) 569.

7. J.-F. Scuotto, D. Mathieu, R. Gallo, R. Phan-Tan-Luu, J. Metzger and M. Desbois
    Bull. Soc. Chim. Belg. 94 (1985) 897.

8. S. Wold, K. Esbensen and P. Gealdi
   Intell. Lab. Syst. 2 (1987) 37.

9. For an example, see
   K. Göreskog, J. Klovan and R. Reyment
   Geological Factor Analysis
   Elsevier, Amsterdam 1976.

10. R. Fisher and W. McKenzie
    J. Agr. Sci. 13 (1923) 311.

11. H. Wold in F. David (Ed.)
    Research Paper in Statistics
    Wiley, New York 1966, pp. 411–444.

12. *(a)* M. Stone
    J. Roy. Statist. Soc. B 36 (1974) 111;

    (b) S. Geisser
    J. Am. Statist. Assoc. 70 (1975) 320.

13. *(a)* S. Wold
    Technometrics 20 (1978) 397.

    (b) H.T. Eastment and W.J. Krzanowski
    Technometrics 24 (1982) 73.

14. R. Carlson, M. Prochazka and T. Lundstedt
    Acta Chem. Scand. B 42 (1988) 145.

15. S. Wold and M. Sjöström
    in B.R. Kowalski (Ed.)
    Chemometrics: Theory and Application
    ACS Symposium Ser. 52 (1977) 243.

16. S. Wold
    Pattern Recognition 8 (1976) 127.

17. *(a)* S. Wold, C. Albano, W. Dunn III, U. Edlund, K. Esbensen, P. Geladi, S. Hellberg,
    E. Johansson, W. Linderg and M.Sjöström
    Multivariate Data Analysis in Chemistry
    in B.R. Kowalski (Ed.)
    Chemometrics: Mathemathics and Statistics in Chemistry
    Reidel, Dordrecht 1984, pp. 17–95;

    (b) S. Wold, C. Albano, W. Dunn III, K. Esbensen, P. Geladi, S. Hellberg,
    E. Johansson, W. Linderg, M. Sjöström, B. Skagerberg, C. Wikström and J. Öhman
    Multivariate Data Analysis. Converting Chemical Data Tables to Plots
    in J. Brandt and I. Ugi (Eds.)
    Computer Applications in Chemical Research and Education
    Dr Alfred Hühtig Verlag, Heidelberg 1989.

18. M. Sjöström and U. Edlund
    K. Magn. Reson. 25 (1977) 285.

19. C. Albano and S. Wold
    J. Chem. Soc. Perkin Trans. II (1980) 1447.

20. R. Carlson, T. Lundstedt and C. Albano
    Acta Chem. Scand. B 39 (1985) 79.

386

21. C. Rechardt
   Liebigs Ann. Chem. (1983) 42.

22. *(a)* C. Rechardt
   Solvent Effects in Organic Chemistry
   Verlag Chemie, Weinheim 1979;

   (b) C. Rechardt
   Solvents and Solvent Effects in Organic Chemistry. Second, completely revised and enlarged edition
   Verlag Chemie, Weinheim 1988.

23. C. Reichardt and E. Harbush-Görnert
   Liebigs Ann. Chem. (1983) 721.

24. K. Dimroth, C. Reichardt, F. Siepman and F. Bohlman
   Liebigs Ann. Chem. 661 (1963) 1.

25. R. Carlson, M.P. Prochazka, and T. Lundstedt
   Acta Chem. Scand. B 42 (1988) 145.

26. R. Carlson, M.P. Prochazka, and T. Lundstedt
   Acta Chem. Scand. B 42 (1988) 157.

27. T. Lundstedt, R. Carlson and R. Shabana
   Acta Chem. Scand. B 41 (1987) 157.

28. R.D. Cramer
   J. Am. Chem. Soc. 102 (1980) 1837, 1849.

29. *(a)* S. Hellberg, M. Sjöström, B. Skagerberg and S. Wold
   J. Med. Chem. 30 (1987) 1126;

   (b) S. Hellberg, M. Sjöström and S. Wold
   Acta Chem. Scand. B 40 (1986) 135.

30. B. Skagerberg, M. Sjöström and S.Wold
   J. Chemometrics 4 (1990) 241.

31. *(a)* J. Jonsson, L. Eriksson, M. Sjöström, S. Wold and M.L. Tosato
   Intell. Lab. Syst. 5 (1989) 169;

   (a) L. Eriksson, J. Jonsson, M. Sjöström and S. Wold
   Intell. Lab. Syst. 7 (1989) 131.

32. B. Skagerberg, D. Boneli, S. Clementi, G. Cruciani and C. Ebert
   Quant. Struct.–Act. Relat. 8 (1989) 32.

33. P. Kaufman
   Diss. Stockholm University, stockholm 1990.

Suggestions to further reading

I.T. Jolliffe
Principal Component Analysis
Springer, New York 1986.

R.G. Brereton
Chemometrics: Application of Mathematics and Statistics to Laboratory Systems
Ellis Horwood, New York 1990.

D.L. Massart, B.G.M. Vanderginste, S.N. Deming, Y. Michotte and L, Kaufman
Chemometrics: a textbook
Elsevier, Amsterdam 1988.

S. Wold, C. Albano, W. Dunn III, U. Edlund, K. Esbensen, P. Geladi, S. Hellberg, E. Johansson,
W. Linderg and M.Sjöström
Multivariate Data Analysis in Chemistry
in B.R. Kowalski (Ed.)
Chemometrics: Mathemathics and Statistics in Chemistry
Reidel, Dordrecht 1984, pp. 17–95.

S. Wold, C. Albano, W. Dunn III, K. Esbensen, P. Geladi, S. Hellberg, E. Johansson, W. Linderg, M.
Sjöström, B. Skagerberg, C. Wikström and J. Öhman
Multivariate Data Analysis. Converting Chemical Data Tables to Plots
in J. Brandt and I. Ugi (Eds.)
Computer Applications in Chemical Research and Education
Dr Alfred Hüthig Verlag, Heidelberg 1989.

Some available programs for principal components analysis and PLS modelling.

The programs use the NIPALS algorithm and can treat data matrices even if there are some data missing.

SIMCA-R (V. 4.3)

Umetri AB
P.O.Box 1456
S-901 24 Umeå, Sweden

MDS, Inc.
371 Highland Ave.
Winchester, MA 01890, USA

SIRIUS

Chemical Institute
Att. O.M. Kvalheim
University of Bergen
N-3007 Bergen, Norway

UNSCRAMBLER II

Camo AS
Jarleveien 4
N-7041 Trondheim, Norway

Appendix 15A: Tables of descriptors and principal component scores

This appendix contains descriptor data and the corresponding principal component scores of the classes of compounds described in the text. The score values are measures of the principal properties.

These data are given to offer an opportunity to the reader to decide whether or not the plots shown in this chapter are applicable to his/her synthetic problem. It might be necessary to include other descriptors and make a new PC model for obtaining the pertinent principal properties in his/her specific case. As it is the author's experience that compilation of descriptor data from the literature can be very time-consuming, the data given here might save some time.

The score values of the Lewis acids given in Table 15A.6 have been taken from the thesis by Prochazka with permission from the author.

M.P. Prochazka
Multivariate Modelling in Synthesis. Determination of Reaction Space Through PC and PLS Modelling. Application to the Fischer Indole Synthesis
Diss. Umeå University, Umeå 1990. ISBN 91-7174-472-X.

Table 15A.1: Descriptors and PC scores of organic solvents.

No	Solvent	Descriptors[a]			
		1	2	3	4
1	Water	0.00	100.0	78.30	5.9
2	Fomamide	2.55	210.5	111.00	11.2
3	1,2-Ethanediol	-12.6	197.15	37.7	7.7
4	Methanol	-97.7	64.5	32.66	5.5
5	N-Methylformamide	-3.8	182.5	182.40	12.9
6	Diethylene glycol	-7.8	245.7	31.69	7.7
7	Triethylene glycol	-4.3	288.0	23.69	10.0
8	2-Methoxyethanol	-85.1	124.6	16.93	6.8
9	N-Methylacetamide	30.6	206.7	191.3	14.2
10	Ethanol	114.5	78.3	24.55	5.8
11	2-Aminoethanol	10.5	170.95	37.72	7.6
12	Acetic acid	16.7	117.9	6.17	5.6
13	Benzylalcohol	-15.3	205.45	13.10	5.5
14	1-Propanol	-126.2	97.15	20.45	5.5
15	1-Butanol	-88.6	117.7	17.51	5.8
16	2-Methyl-1-propanol	-108.0	107.9	17.93	6.0
17	2-Propanol	-88.0	82.2	19.92	5.5
18	2-Butanol	-114.7	99.5	16.56	5.5
19	3-Methyl-1-butanol	-117.2	130.5	15.19	6.1
20	Cyclohexanol	25.25	161.1	15.00	6.2
21	4-Methyl-1,3-dioxol-2-one	-54.5	241.7	64.94	16.5
22	2-Pentanol	–	119.0	13.71	5.5
23	Nitromethane	-28.55	101.2	35.94	11.9
24	Acetonitrile	-43.8	81.6	35.94	11.8
25	3-Pentanol	-75.0	115.3	13.35	5.5
26	Dimethylsulfoxide	18.5	189.0	46.45	13.5
27	Aniline	-6.0	184.4	6.71	5.0
28	Sulfolane	28.45	287.3	43.30	16.0
29	Acetic anhydride	-73.1	140.0	20.70	9.4
30	2-Methyl-2-propanol	25.6	82.3	12.47	5.5
31	N,N-Dimethylformamide	-60.4	153.0	36.71	10.8
32	N,N-Dimethylacetamide	-20.0	166.1	37.78	12.4
33	Propanenitrile	-92.8	97.35	28.86	11.7
34	1-Methyl-pyrrolidine-2-one	-24.4	202.0	32.20	13.6
35	Acetone	-94.7	56.1	20.56	9.0
36	Nitrobenzene	5.8	210.8	34.78	13.3
37	Benzonitrile	-12.75	191.1	25.20	13.4
38	1,2-Diaminoethane	11.3	116.9	12.9	6.3
39	1,2-Dichloroethtane	-35.7	83.5	10.70	6.1

No	Descriptors					Scores	
	5	6	7	8	9	t_1	t_2
1	1.3330	1.000	0.9282	-1.38	1.774	3.70	-1.48
2	1.4475	0.799	1.1334	-1.51	1.401	4.47	0.86
3	1.4318	0.790	1.1088	-1.36	1.060	2.90	0.34
4	1.3284	0.762	0.7914	-0.77	1.393	1.76	-2.98
5	1.4319	0.722	1.0100	-1.30	–	6.00	0.06
6	1.4475	0.713	1.1090	–	–	2.74	1.02
7	1.4558	0.704	1.0682	-1.33	0.852	3.02	1.31
8	1.4021	0.667	0.9650	-0.77	1.103	1.59	-1.37
9	1.4253	0.667	0.9570	-1.05	–	6.38	0.49
10	1.3614	0.654	0.7850	-0.31	1.231	2.21	-2.67
11	1.4545	0.651	1.0180	-1.31	1.216	2.63	0.43
12	1.3719	0.648	1.0492	-0.17	1.002	1.56	-0.47
13	1.5404	0.608	1.0420	1.10	-0.992	0.59	1.73
14	1.3856	0.617	0.8040	0.25	1.126	0.87	-2.32
15	1.3993	0.602	0.8098	0.88	0.125	0.57	-1.46
16	1.3959	0.552	0.7940	0.76	0.176	0.43	-1.78
17	1.3772	0.546	0.7860	0.05	1.117	0.90	-2.16
18	1.3971	0.506	0.8080	0.61	0.176	0.25	-1.88
19	1.4072	0.565	0.8092	1.42	-0.518	0.06	-1.41
20	1.4648	0.500	0.9642	1.23	-0.755	0.61	1.04
21	1.4215	0.491	1.2040	–	–	3.91	0.57
22	1.4064	0.488	0.8100	1.19	-0.612	0.14	-0.99
23	1.3819	0.481	1.1370	-0.35	0.235	1.79	-0.44
24	1.3441	0.460	0.7857	-0.34	1.282	1.93	-1.88
25	1.4104	0.563	0.8210	1.21	-0.234	0.05	-1.10
26	1.4793	0.444	1.1010	-1.35	–	3.12	1.25
27	1.5863	0.420	1.0217	0.90	-0.370	0.26	1.97
28	1.4816	0.410	1.2620	-0.77	–	3.62	2.53
29	1.3904	0.407	1.0820	–	0.070	0.82	-0.58
30	1.3877	0.389	0.7890	0.35	-1.057	0.23	-0.56
31	1.4305	0.404	0.9450	-1.01	1.112	1.98	-0.53
32	1.4384	0.401	0.9370	-0.77	1.110	2.29	0.84
33	1.3658	0.401	0.7820	0.16	–	0.98	-1.92
34	1.4700	0.355	1.0260	–	–	2.09	0.96
35	1.3587	0.355	0.7900	-0.24	1.134	0.85	-2.44
36	1.5562	0.324	1.2040	1.85	-1.778	0.74	2.93
37	1.5282	0.333	1.0100	1.56	-1.013	0.82	1.84
38	1.4568	0.349	0.8990	-2.04	–	1.27	-0.04
39	1.4448	0.225	1.2350	1.48	-1.092	-0.66	0.40

Table 15A.1: (continued).

No	Solvent	Descriptors[a]			
		1	2	3	4
40	2-Methyl-2-butanol	-8.8	102.0	5.78	5.7
41	2-Butanone	-86.7	79.6	18.51	9.2
42	Acetophenone	19.6	202.0	17.39	9.8
43	Dichloromethane	-94.9	39.6	8.93	5.2
44	1,1,3,3-Tetramethylurea	-1.2	175.2	23.60	11.7
45	HMPT	7.2	233.0	29.60	18.5
46	Cyclohexanone	-32.1	155.65	16.10	10.3
47	Pyridine	-41.55	115.25	12.91	7.9
48	Methylacetate	-98.05	56.9	6.68	5.7
49	4-Methyl-2-pentanone	-84.7	117.4	13.11	2.7
50	1,1-Dichloroethane	-97.0	57.3	10.00	6.1
51	Quinoline	-14.85	237.1	8.95	7.3
52	3-Pentanone	-39.0	102.0	17.00	9.4
53	Chloroform	-63.15	61.2	4.81	3.8
54	Triethyleneglycol dimethyl ether	-45.0	216.0	7.50	–
55	Diethyleneglycol dimethyl ether	-64.0	159.8	5.80	6.6
56	1,2-Dimethoxyethane	-69.0	84.5	7.20	5.7
57	1,2-Dichlorobenzene	-17.0	180.5	9.93	7.1
58	Ethylacetate	-83.55	77.1	6.02	6.1
59	Fluorobenzene	-42.2	84.7	5.42	4.9
60	Iodobenzene	-31.35	188.3	4.49	4.7
61	Chlorobenzene	-45.6	131.7	5.62	5.4
62	Bromobenzene	-38.8	155.9	5.40	5.2
63	Tetrahydrofuran	-108.4	66.0	7.58	5.8
64	Methoxybenzene	-37.5	153.6	4.33	4.2
65	Ethoxybenzene	-29.5	169.8	4.22	4.5
66	1,1,1-Trichloroethane	-30.4	74.1	7.25	5.7
67	1,4-Dioxane	11.8	101.3	2.21	1.5
68	Trichloroethene	-86.4	87.2	3.42	2.7
69	Piperidine	-10.5	106.7	5.80	4.0
70	Diphenylether	26.9	258.1	3.69	3.9
71	Diethylether	-116.3	34.4	4.20	3.8
72	Benzene	5.5	80.1	2.27	0.0
73	Diisopropylether	-85.5	68.3	3.88	4.2
74	Toluene	-95.0	110.6	2.38	1.0
75	Dibutylether	-95.2	140.3	3.08	3.9
76	Triethylamine	-114.7	88.9	2.42	2.9
77	1,3,5-Trimethylbenzene	-44.7	164.7	2.28	0.0
78	Carbon disulfide	-111.6	46.2	2.64	0.0

No	Descriptors					Scores	
	5	6	7	8	9	t_1	t_2
40	1.4050	0.321	0.8060	0.89	0.039	0.12	-0.65
41	1.3788	0.327	0.8050	0.29	0.494	0.55	-1.83
42	1.5342	0.306	1.0281	1.58	-1.553	0.32	2.31
43	1.4242	0.309	1.3300	1.25	-0.699	-0.82	-0.63
44	1.4493	0.318	0.9690	0.19	–	1.47	0.74
45	1.4588	0.315	1.0240	0.28	–	2.70	1.52
46	1.4510	0.281	0.9478	0.81	-1.236	0.30	0.49
47	1.5102	0.302	0.9820	0.65	1.094	0.59	0.27
48	1.3614	0.287	0.9330	0.18	0.486	-0.10	-2.11
49	1.3958	0.269	0.7978	1.31	-0.871	-0.84	-1.25
50	1.4164	0.269	1.1760	1.79	-1.259	-1.06	-0.71
51	1.6273	0.269	1.0930	2.03	–	-0.14	3.23
52	1.3923	0.265	0.8138	0.91	-0.270	0.30	-0.87
53	1.4459	0.259	1.4800	1.97	-1.246	-1.26	0.45
54	1.4264	0.253	–	–	–	0.04	0.67
55	1.4078	0.244	–	–	–	-0.51	-0.19
56	1.3796	0.231	0.8269	-0.21	–	-0.37	-1.48
57	1.5515	0.225	1.3050	3.38	-3.029	-1.38	2.84
58	1.3724	0.228	0.9000	0.73	-0.041	-0.37	-1.62
59	1.4684	0.194	1.0230	2.27	-1.797	-1.41	0.30
60	1.6200	0.170	1.8310	3.25	-2.847	-1.76	4.28
61	1.5248	0.188	1.1060	2.84	-2.450	-1.63	0.42
62	1.5568	0.182	1.4950	2.99	-2.570	-1.59	2.80
63	1.4072	0.207	0.8890	0.46	1.142	-0.17	-1.86
64	1.5170	0.198	0.9960	2.11	-1.917	-1.26	1.19
65	1.5074	0.182	0.9670	2.51	-2.332	-1.40	1.36
66	1.4380	0.170	1.3390	2.49	-2.179	-1.46	0.79
67	1.4224	0.164	1.0340	-0.27	1.070	0.10	-0.18
68	1.4773	0.160	1.4640	2.42	-1.995	-1.97	0.81
69	1.4525	0.150	0.8610	0.85	–	-0.10	-0.23
70	1.5763	0.142	1.0750	4.21	-3.947	-2.01	3.76
71	1.3524	0.117	0.7140	0.89	0.076	-1.13	-2.80
72	1.5011	0.111	0.8786	2.13	-1.559	-1.82	0.60
73	1.3681	0.102	0.7241	1.52	-1.982	-1.69	-1.73
74	1.4969	0.099	0.8670	2.73	-1.747	-2.30	-0.11
75	1.3992	0.071	0.7689	3.21	-1.941	-2.06	-0.77
76	1.4010	0.043	0.7275	1.45	0.224	-1.34	-1.90
77	1.4994	0.068	0.8650	3.42	-3.356	-2.73	1.08
78	1.6225	0.065	1.2630	1.94	-1.552	-2.59	1.04

Table 15A.1: (continued).

No	Solvent	Descriptors[a]			
		1	2	3	4
79	Carbon tetrachloride	-22.8	76.6	2.23	0.0
80	Tetrachloroethene	-22.4	121.2	2.30	0.0
81	Cyclohexane	6.7	80.7	2.02	0.0
82	Hexane	-95.3	68.7	1.88	0.0
83	Tetraethylene glycol	-6.2	327.3	19.7	10.8
84	1-Pentanol	-78.2	138.0	13.9	5.7
85	DMEU	8.2	225.5	37.6	13.6
86	DMPU	-20.0	230.0	36.12	14.1
87	2-Pentanone	-76.9	102.3	15.38	9.0
88	Morpholine	-4.8	128.9	7.42	5.2
89	3-Methyl-2-butanone	-92.0	94.2	15.87	9.2
90	3,3-Dimethyl-2-butanone	-49.8	106.3	13.10	9.3
91	2,4-Dimethyl-3-pentanone	-69.0	125.25	17.2	9.1
92	2,6-Dimethyl-4-heptanone	-46.0	168.2	9.91	8.9
93	Diethyleneglycol diethylether	-44.3	188.9	5.70	–
94	Diethylcarbonate	-43.0	126.8	2.82	3.0
95	1,1-Dichloroethene	-122.0	31.6	4.82	4.3
96	*tert*-Butylmethylether	-108.6	55.2	4.50	4.1
97	Diethylamine	-49.8	55.55	3.78	4.0
98	Dipropylether	-123.2	90.1	3.39	4.4
99	1,4-Dimethylbenzene	13.3	138.4	2.27	0.0
100	Tributylamine	-70.0	214.0	–	2.6
101	*cis*-Decaline	-43.0	195.8	2.20	0.0
102	Heptane	-90.6	98.4	1.92	0.0
103	Pentane	-129.7	36.1	1.84	0.0

[a] Descriptors: 1, melting point ($^\circ$ C); 2, boiling point ($^\circ$ C); 3, dielectric constant; 4, dipole moment; 5, refractive index; 6, E_T^N, the normalized Reichardt-Dimroth parameter[1], 8, lipophilicity as measured by log$P$[2], water solubility as measured by log(Mol L^{-1}).

The author wishes to express his gratitude to Dr Erik Johansson, Research Group for Chemometrics, Umeå University, Umeå, fot this compilation of data. Part of Table 15A.1 was also published in [3]

No	Descriptors					Scores	
	5	6	7	8	9	t_1	t_2
79	1.4602	0.052	1.5900	2.83	-2.294	-2.39	1.51
80	1.5057	0.037	1.6230	3.40	-2.895	-2.65	2.44
81	1.4262	0.006	0.7780	3.44	-3.069	-2.78	0.16
82	1.3749	0.009	0.6600	3.98	-3.359	-3.46	-1.50
83	1.4577	0.664	1.1285	1.38	–	2.42	2.11
84	1.4100	0.568	0.8144	1.56	-0.631	0.13	-0.92
85	1.4707	0.364	–	–	–	2.63	1.65
86	1.4881	0.352	–	–	–	2.42	1.60
87	1.3908	0.321	0.8089	0.91	-0.328	0.16	-1.29
88	1.4542	0.312	1.0005	-0.86	1.060	0.95	0.05
89	1.3880	0.315	0.8051	0.56	-0.164	0.23	-1.58
90	1.3952	0.256	0.8012	0.85	-0.706	0.06	-0.89
91	1.3999	0.247	0.8108	1.49	-1.299	-0.30	-0.73
92	1.4122	0.225	0.8053	2.96	-1.731	-0.76	0.12
93	1.4115	0.210	0.9063	–	0.076	0.14	-0.04
94	1.3837	0.194	0.9752	1.21	-0.798	-0.79	-0.54
95	1.4247	0.194	1.2180	1.86	-1.666	-1.80	-0.95
96	1.3690	0.148	0.7405	0.94	–	-1.61	-2.20
97	1.3846	0.145	0.7056	0.58	–	-1.20	-1.61
98	1.3805	0.102	0.7360	2.03	-1.388	-1.74	-1.83
99	1.4958	0.074	0.8611	3.15	-2.806	-2.31	1.31
100	1.4291	0.043	0.7771	4.56	-3.116	-2.84	0.55
101	1.4810	0.015	0.8965	5.08	-5.192	-3.72	1.68
102	1.3876	0.012	0.6838	4.57	-4.046	-3.71	-0.92
103	1.3575	0.009	0.6262	3.39	-3.129	-3.48	-2.38

Table 15A.2: Descriptors and PC scores of aldehydes

No	Aldehyde	Descriptors[a]		
		1	2	3
1	Acetaldehyde	44.05	-125.0	1.3316
2	4-Acetoxybenzaldehyde	164.16	–	1.5379
3	5-Acetoxymethyl-2-furaldehyde	168.5	–	–
4	2-Pentyl-3-phenylpropenal	202.25	-75	1.5571
5	*o*-Anisaldehyde	136.15	38	–
6	*m*-Anisaldehyde	136.15	–	1.5523
7	*p*-Anisaldehyde	136.15	-1	1.5713
8	9-Anthracenecarbaldehyde	206.24	104.5	–
9	Benzalsehyde	106.12	-26	1.5454
10	4-Butoxybenzaldehyde	178.23	–	1.53
11	Butyraldehyde	72.11	-96	1.379
12	*trans*-Cinnamaldehyde	132.16	–	1.6221
13	2-Butenal	70.09	–	1.4365
14	2,4-Dimethyl-2,6-heptadienal	166.27	–	1.4676
15	2,3-Dimethoxybenzaldehyde	166.18	50	–
16	2,4-Dimethoxybenzaldehyde	166.18	70.5	–
17	2,5-Dimethoxybenzaldehyde	166.18	50.5	–
18	3,4-Dimethoxybenzaldehyde	166.18	–	–
19	3,5-Dimethoxybenzaldehyde	166.18	–	–
20	2,4-Dimethxy-3-methylbenzaldehyde	180.21	–	–
21	4,6-Dimethoxysalicylaldehyde	182.18	69	–
22	2,4-Dimethylbenzaldehyde	134.18	-9	1.5492
23	2,5-Dimethylbenzaldehyde	134.18	–	1.5422
24	2,2-Dimethyl-4-pentenal	112.17	–	1.4203
25	Diphenylacetaldehyde	196.25	–	1.5893
26	Dodecanal	184.32	–	1.4344
27	2-Ethoxybenzaldehyde	150.18	–	1.5422
28	4-Ethoxybenzaldehyde	150.18	–	1.5584
29	3-Ethoxy-4-methoxybenzaldehyde	180.21	52	–
30	4-Ethoxy-3-methoxybenzaldehyde	180.21	59.5	–
31	3-Ethoxysalicylaldehyde	166.18	67	–
32	4-Ethylbenzaldehyde	134.18	–	1.538
33	2-Ethylbutyraldehyde	100.16	–	1.4018
34	Formaldehyde 37%/Water	30.03	–	1.3765
35	5-Formylsalicylaldehyde	150.13	108.5	–
36	2-Furaldehyde	96.09	-36	1.5243
37	3-Furaldehyde	96.09	–	1.493
38	Heptanal	114.19	-43	1.4125
39	*trans*-2-Heptenal	112.17	–	1.4473

No	Descriptors				Scores	
	4	5	6	7	t_1	t_2
1	0.788	21.5	1726.2	55.9	4.39	-0.92
2	1.168	–	1763.1	140.55	-0.62	1.09
3	–	–	1732.1	–	-1.91	1.45
4	0.97	288	–	208.51	-1.65	2.20
5	1.127	238	1689.2	120.81	-1.27	-1.29
6	1.119	–	1702.7	121.67	-0.94	-0.71
7	1.119	248	1683.3	121.67	-1.61	-1.03
8	–	–	1669.2	–	-5.78	-1.20
9	1.044	181.5	1702.5	101.65	-0.19	-1.10
10	1.031	285	1694.1	172.87	-2.01	0.67
11	0.817	75	1727.6	88.26	3.10	-0.30
12	1.048	248	1676.7	126.11	-1.98	-1.16
13	0.826	104	1691	82.85	1.85	-1.37
14	0.862	–	1689.7	192.88	-0.70	1.17
15	–	–	1691.7	–	-2.98	-0.60
16	–	–	1672.8	–	-3.96	-1.72
17	–	–	1681.7	–	-3.28	-1.10
18	–	281	1683.1	–	-3.33	-0.88
19	–	–	1703.4	–	-2.55	0.01
20	–	–	–	–	-4.45	-1.71
21	–	–	1645.3	–	-5.16	-2.69
22	0.062	–	1694.5	139.48	-0.78	-0.33
23	0.95	–	1693.6	141.24	-0.62	-0.26
24	0.825	124.5	–	135.96	1.59	0.56
25	1.106	315	1724.1	177.44	-2.60	1.27
26	0.835	–	1728.2	220.74	-0.22	2.70
27	1.074	–	1690.2	139.83	-1.28	-0.42
28	1.08	255	1696.7	139.06	-1.63	-0.25
29	–	–	1676.8	–	-3.82	-1.02
30	–	–	1683.8	–	-4.54	-3.07
31	–	264	1646.4	–	-4.54	-3.07
32	0.979	221	1702.9	137.06	-0.75	-0.10
33	0.814	117	1730.9	123.05	2.13	0.51
34	1.083	–	–	27.73	2.91	-2.94
35	–	–	–	–	-6.03	-5.28
36	1.16	162	1674.5	82.84	-0.25	-2.19
37	1.111	–	–	86.5	0.28	-1.76
38	0.85	153	1727.4	134.34	1.32	0.63
39	0.857	–	1693.9	130.89	0.89	-0.22

398

Table 15A.2: Aldehydes (continued)

No	Aldehyde	Descriptors[a]		
		1	2	3
40	*cis*-9-Hexadecenal	238.42	–	–
41	*cis*-11-Hexadecenal	238.42	–	1.4526
42	Hexanal	110.16	–	1.4035
43	*trans*-2-Hexenal	98.15	–	1.4455
44	2-Hydroxy-4-methoxybenzaldehyde	152.15	42	–
45	2-Hydroxy-3-methoxybenzaldehyde	152.15	4	1.5784
46	3-Hydroxy-4-methoxybenzaldehyde	152.15	114	–
47	5-hydroxymethyl-2-furaldehyde	126.11	33.5	1.5627
48	2-Hydroxy-1-naphtalelecarbaldeyde	172.18	–	–
49	2-Hydroxy-5-nitrobenzaldehyde	167.12	129	–
50	3-Hydroxy-4-nitrobenzaldehyde	167.12	130	–
51	4-Hydroxy-3-nitrobenzaldehyde	167.12	141	–
52	5-Hydroxy-2-nitrobenzaldehyde	167.12	167	–
53	Isobutyraldehyde	72.11	-65	1.3725
54	4-Isopropylbenzaldehyde	148.21	–	1.5298
55	Isovaleraldehyde	86.13	–	1.3882
56	Mesitaldehyde	148.21	14	1.5522
57	7-Methoxy-3,7-dimethyloctanal	186.3	–	1.4374
58	2-Methyl-3-phenylpropenal	146.19	–	1.6045
59	2-Methylundecanal	182.32	–	1.4321
60	2-Methylvaleraldehyde	100.16	–	1.4067
61	*trans*-2-Pentenal	84.12	–	1.4414
62	Phenylacetaldehyde	120.15	-10	1.5293
63	(±)3-Phenylburyraldehyde	148.21	–	1.5179
64	2,3-Diphenylpropenal	208.26	45	–
65	(±)2-Phenypropionaldehyde	134.18	–	1.5176
66	Propionaldehyde	50.08	-81	1.365
67	Salicylaldehyde	122.12	1.5	1.5719
68	Tridecanal	198.35	–	1.484
69	10-Undecenal	168.28	–	1.4427
70	Undecanal	170.3	–	1.4322
71	Valeraldehyde	86.13	-92	1.3942
72	Acrylaldehyde	56.06	-87	1.402
73	3-Benzyloxybenzaldehyde	212.35	57	–
74	4-Benzyloxybenzaldehyde	212.35	73.5	–
75	3-Benzyloxy-4-methoxy-benzaldehyde	242.47	62.5	–
76	4-Benzyloxy-3-methoxy-benzaldehyde	242.47	64	–
77	2-Bromobenzaldehyde	185.03	21.5	1.596

No	Descriptors				Scores	
	4	5	6	7	t_1	t_2
40	–	–	1711.1	–	-3.41	2.47
41	–	–	1728.5	–	-1.57	3.90
42	0.834	131	1727	120.1	1.94	0.37
43	0.846	–	1693.1	116.02	1.26	-0.59
44	–	–	1630.6	–	-4.29	-3.94
45	1.219	–	1630.6	124.82	-2.61	-2.13
46	–	–	1673	–	-4.48	-2.41
47	–	–	1673.8	–	-1.71	-2.14
48	–	–	1645.7	–	-5.19	-3.04
49	–	–	–	–	-7.14	-5.46
50	–	–	1697.7	–	-4.45	-0.94
51	–	–	1686.1	–	-6.01	-2.59
52	–	–	1670.8	–	-6.01	-2.59
53	0.794	63	1738	90.82	3.11	-0.13
54	0.977	235.5	1702.6	151.69	-1.01	0.30
55	0.803	90	1727.6	107.26	2.61	0.09
56	1.005	237	1687.8	147.47	-1.60	-0.25
57	0.877	–	–	212.43	-0.29	2.55
58	1.047	–	1682.3	139.62	-1.79	-0.73
59	0.83	171	1728.9	222.07	0.05	2.66
60	0.808	119.5	1726.3	123.96	2.02	0.43
61	0.86	80.5	–	97.81	1.99	-0.70
62	1.027	195	1723.7	116.66	-0.24	-028
63	0.997	–	–	148.66	-0.66	0.20
64	–	–	1666.7	–	-4.70	-0.79
65	1.011	–	1724.7	132.72	-0.04	0.17
66	0.805	48	1733.3	72.15	3.53	-0.61
67	1.146	197	1664.9	106.56	-1.48	-1.91
68	0.835	–	1728.4	237.54	-0.61	3.10
69	0.81	–	1727.4	207.75	0.12	2.34
70	0.825	–	1728.3	206.42	0.15	2.35
71	0.81	103	1726.2	106.33	2.59	0.09
72	0.839	53	1696.2	66.82	2.84	-1.57
73	–	–	1694.8	–	-4.20	0.62
74	–	–	1687.7	–	-4.75	0.13
75	–	–	1677	–	-5.62	0.42
76	–	–	1678.1	–	-5.61	0.46
77	1.585	230	1697.1	116.74	-3.20	-1.42

Table 15A.2: Aldehydes (continued)

No	Aldehyde	Descriptors[a]		
		1	2	3
78	3-Bromobenzaldehyde	185.03	–	1.5935
79	4-Bromobenzaldehyde	185.03	56.5	–
80	2-Bromo-3-phenylpropenal	211.06	67	–
81	5-Bromosalicylaldehyde	201.02	106.5	–
82	5-Bromocrotonaldehyde	245.08	63.5	–
83	6-Bromoveratraldehyde	245.08	150.5	–
84	4-Butylbenzaldehyde	162.22	–	1.4781
85	3-(4-*t*-Butylphenoxy)benzaldehyde	254.33	–	1.5702
86	2-Chlorobenzaldehyde	140.57	11	1.5658
87	3-Chlorobenzaldehyde	140.57	17.5	1.5645
88	4-Chlorobenzaldehyde	140.57	45.5	–
89	2-Chloro-3-phenylpropenal	166.61	32.5	–
90	2-Chloro-5-nitrobenzaldehyde	185.57	76	–
91	2-Chloro-6-nitrobenzaldehyde	185.57	70	–
92	4-Chloro-3-nitrobenzaldehyde	185.57	64	–
93	5-Chloro-2-nitrobenzaldehyde	185.57	67	–
94	3-Cyanobenzaldehyde	131.13	76.5	–
95	4-Cyanobenzaldehyde	131.13	99	–
96	Decanal	156.27	–	1.428
97	3,5-Dibromosalicylaldehyde	279.93	83	–
98	3,5-Di-*t*-butyl-4-hydroxybenzaldehyde	234.34	188	–
99	2,3-Dichlorobenzaldehyde	175.01	65.5	–
100	2,4-Dichlorobenzaldehyde	175.01	71	–
101	2,6-Dichlorobenzaldehyde	175.01	70.5	–
102	3,4-Dichlorobenzaldehyde	175.01	42.5	–
103	3,5-Dichlorobenzaldehyde	175.01	65	–
104	2-Methylpropenal	70.05	-81	1.416
105	Nonanal	142.24	–	1.424
106	1-Naphtaldehyde	156.18	1.5	1.652
107	2-Naphtaldehyde	156.18	60.5	–
108	Piperonal	150.13	36	–
109	Tridecanal	212.38	24	–
110	Octanal	128.22	14	1.4183
111	*trans*-2-Nonenal	140.23	–	1.4531
112	1,3-Benzenedicarbaldehyde	134.13	89	–
113	2-Ethylhexanal	128.22	–	1.4155

No	Descriptors				Scores	
	4	5	6	7	t_1	t_2
78	1.597	229	1699.2	116.59	-3.25	-1.42
79	–	–	1696.6	–	-3.44	0.05
80	–	–	1698.5	–	-4.53	0.25
81	–	–	1674.6	–	-5.53	-1.07
82	–	–	–	–	-5.73	0.45
83	–	–	1669.7	–	-7.70	-0.60
84	0.926	–	–	255.25	-1.89	3.56
85	0.984	–	1701.5	258.46	-3.62	2.85
86	1.248	212	1697.9	112.64	-1.66	-1.18
87	1.241	213.5	1702	113.27	-1.65	-1.10
88	–	213.5	1704.2	–	-1.80	-0.59
89	–	–	1697.1	–	-3.65	0.01
90	–	–	1690	–	-2.47	-0.18
91	–	–	1709.9	–	-3.29	-1.56
92	–	–	1700.9	–	-3.33	0.62
93	–	–	1697.3	–	-3.48	0.22
94	–	210	1699.7	–	-2.22	-1.34
95	–	–	1706.9	–	-2.62	-1.10
96	0.83	208	1727.9	188.28	0.33	1.98
97	–	–	1682.2	–	-0.68	1.43
98	–	–	1667.9	–	-8.24	-1.26
99	–	–	1686.9	–	-3.66	-0.75
100	–	233	1695.4	–	-3.11	-0.54
101	–	–	1716.2	–	-2.88	0.67
102	–	247.5	1705.2	–	-2.54	0.31
103	–	–	1692.7	–	-3.48	-0.46
104	0.847	69	1695.7	82.75	2.42	-1.23
105	0.827	–	1727.6	171.99	0.92	1.52
106	1.15	–	1689.3	135.81	-2.37	-0.82
107	–	–	1695	–	-2.83	-0.77
108	–	264	1688.1	–	-2.69	-0.93
109	–	–	1728.9	–	-2.51	2.61
110	0.821	171	1727.6	156.18	0.55	0.99
111	0.846	–	1694.2	165.76	0.16	0.63
112	–	–	1695.6	–	-2.84	-1.51
113	0.822	–	–	155.99	1.33	1.13

Table 15A.3: Descriptors and PC scores of ketones

No	Ketones	Descriptors[a]		
		1	2	3
1	1-Acenaphtone	170.21	11	1.682
2	2-Acenaphtone	170.21	54	–
3	Acetone	58.08	-94	1.3585
4	Acetophenone	120.15	19.5	1.5325
5	2-Adamamntanone	150.22	257	–
6	1-Adamantyl methyl ketone	178.28	54	–
7	Anthrone	193.23	153	–
8	Benzanthrone	230.27	169	–
9	Benzophenone	182.22	50	–
10	2-Butanone	72.11	-87	1.3788
11	Butyrophenone	148.21	12	1.5195
12	Cyclobutanone	70.0	–	1.4195
13	Cyclobutyl phenyl ketone	160.22	–	1.547
14	Cyclodecanone	154.25	24	1.482
15	Cyclododecanone	182.31	60	–
16	Cycloheptanone	112.17	–	1.4611
17	Cyclohexanone	98.15	-47	1.4500
18	Cyclohexyl phenyl ketone	188.27	56	–
19	Cyclononanone	140.23	28	1.477
20	Cyclooctanone	126.2	40	–
21	Cyclopentadecanone	224.39	62	–
22	Cyclopentanone	84.12	-51	1.4359
23	Cyclopropyl methyl ketone	84.12	–	1.4241
24	1-Decalone	152.24	–	1.4917
25	2-Decalone	152.24	–	1.49
26	2-Decanone	156.27	4	1.4249
27	3-Decanone	156.27	-4	1.4241
28	4-Decanone	156.27	–	1.4237
29	Decyl phenyl ketone	232.37	35	–
30	3,4-Dimethylacetophenone	148.21	–	1.538
31	2,4-Dimethylbenzophenone	210.28	–	1.592
32	2,5-Dimethylbenzophenone	210.28	35	1.588
33	3,4-Dimethylbenzophenone	210.28	46	–
34	2,6-Dimethyl-4-heptanone	142.24	–	1.4128
35	2,4-Dimethyl-3-pentanone	114.19	-80	1.3986
36	4,4-Dimethyl-2-pentanone	114.19	–	1.4037
37	2,2-Dimethylpropiophenone	162.23	–	1.5084
38	Dodecyl phenyl ketone	260.42	46	–

No	Descriptors				Scores	
	4	5	6	7	t_1	t_2
1	1.12	302	1677.0	151.97	-3.66	-1.77
2	–	300.5	1674.6	–	-3.65	-0.97
3	0.791	56	1715.1	73.43	4.43	-1.45
4	1.03	202	1685.2	116.65	-1.09	-2.00
5	–	–	1720.2	–	-3.67	4.15
6	–	–	1701.4	–	-1.70	1.37
7	–	–	1658.6	–	-6.23	-0.96
8	–	–	1648.0	–	-7.93	-0.78
9	–	–	1659.8	–	-4.53	-2.13
10	0.805	80	1717.5	89.58	3.77	-1.06
11	1.021	221	1687.2	145.16	-1.52	-1.06
12	0.938	99	1783.2	74.63	4.45	-0.95
13	1.05	–	1678.4	152.59	-2.35	-1.19
14	0.958	–	1701.6	161.01	-0.92	-0.04
15	0.906	–	1711.7	201.23	-1.57	1.72
16	0.951	179	1702.1	117.95	0.53	-1.16
17	0.947	155	1714.0	103.64	1.48	-1.39
18	–	–	1667.3	–	-3.86	-1.28
19	o.959	–	1701.6	146.23	-0.54	-0.35
20	0.958	196	1700.6	131.73	-0.50	-0.76
21	0.897	–	1711.6	250.16	-2.77	3.10
22	0.951	130.5	1746.4	88.45	2.57	-1.17
23	0.849	114	1696.7	99.08	2.53	-1.42
24	0.989	–	1710.0	153.93	-0.73	-0.20
25	0.979	–	1714.8	155.51	-0.58	-0.03
26	0.825	211	1718.9	189.42	0.07	1.81
27	0.825	204.5	–	189.42	0.09	1.67
28	0.824	206.5	1714.0	189.65	-0.03	1.74
29	–	–	1688.9	–	-3.30	1.52
30	1.001	243	1681.1	148.07	-1.93	-1.04
31	1.074	–	1622.4	195.79	-4.75	-0.31
32	–	–	1665.1	–	-4.01	-0.98
33	–	–	1656.4	–	-4.79	-1.73
34	0.847	169	1712.7	167.93	0.66	1.00
35	0.806	124	1713.4	141.67	2.32	0.22
36	0.809	127.5	1716.8	141.15	1.96	0.40
37	0.97	220.5	–	167.65	-1.63	-0.14
38	–	–	1689.2	–	-4.13	2.39

404

Table 15A.3: Ketones (continued)

No	Aldehyde	Descriptors[a]		
		1	2	3
39	3-Ethylacetophenone	148.21	–	1.5264
40	4-Ethylacetophenone	148.21	-21	1.5293
41	4-Ethylcyclohexanone	126.2	–	1.4515
42	Flavone	222.4	98	–
43	2-(5*H*)-Furanone[b]	84.07	4.5	1.4692
44	9-Heptadecanone	254.46	52	–
45	2-Heptanone	114.19	-35	1.4095
46	3-Heptanone	114.19	-39	1.4095
47	4-Heptanone	114.19	-33	1.407
48	Heptyl phenyl ketone	190.29	17	1.5077
49	2-Hexanone	100.16	-57	1.4005
50	3-Hexanone	100.16	–	1.4002
51	Hexanophenone	176.26	25.5	1.5105
52	Isobutyrophenone	148.21	–	1.5172
53	2-Methylbenzophenone	196.25	–	1.5958
54	3-Methylbenzophenone	196.25	–	1.597
55	4-Methylbenzophenone	196.25	57	–
56	3-Methyl-2-butanone	86.13	-92	1.388
57	2-Methylcyclohexanone	112.17	–	1.4478
58	4-Methylcyclohexanone	112.17	–	1.4142
59	5-Methyl-3-heptanone	128.22	–	1.4064
60	5-Methyl-3-hexanone	114.19	–	1.4064
61	4-Methyl-2-pentanone	100.16	-80	1.396
62	4-Methylpropiophenone	148.21	7	1.528
63	10-Nonadecanone	282.51	56	–
64	2-Nonanone	142.24	-21	1.421
65	3-Nonanone	142.24	–	1.4204
66	5-Nonanone	142.24	-50	1.419
67	2-Octanone	128.22	-16	1.415
68	3-Octanone	128.22	–	1.415
69	8-Pentadecanone	226.4	42	–
70	2-Pentanone	86.13	-78	1.3897
71	3-Pentanone	86.13	-40	1.3924
72	1-Phenyl-2-butanone	148.21	–	1.5122
73	2-Phenylcyclohexanone	174.24	55	–
74	4-Phenylcyclohexanone	174.24	79	–
75	Pinacolone	100.16	–	1.3964
76	2,2,6-Trimethylcyclohexanone	140.23	–	1.447

No	Descriptors				Scores	
	4	5	6	7	t_1	t_2
39	0.93	–	1685.4	159.37	-1.31	-0.42
40	0.993	–	1682.7	149.25	-1.20	-1.13
41	0.895	193	1717.6	141.0	0.57.	0.07
42	–	–	1646.9	–	-6.54	-1.74
43	1.185	–	–	70.95	–	–
44	–	–	1711.8	–	-2.85	4.24
45	0.82	149.5	1718.1	139.26	1.80	0.34
46	0.818	147.5	1715.4	139.60	1.80	0.29
47	0.817	145	1713.3	139.8	1.75	0.27
48	–	–	–	–	-1.89	0.50
49	0.812	127	1716.7	123.35	2.46	-0.15
50	0.815	123	1714.9	122.90	2.34	-0.16
51	0.958	256.2	1687.1	183.99	-2.26	0.38
52	0.988	217	1684.0	150.01	-1.48	-0.86
53	1.083	–	1664.7	181.21	-4.41	-0.78
54	1.095	–	1660.8	179.22	-4.44	-0.95
55	–	326	1653.9	–	-5.26	-2.09
56	0.85	94.5	1715.7	101.33	3.18	1.01
57	0.924	162.5	1712.6	121.40	1.03	-0.75
58	0.914	170	1717.5	122.72	1.10	0.55
59	0.823	159.5	1714.1	155.80	1.17	0.74
60	0.809	135	1718.3	141.15	1.90	0.43
61	0.800	117.5	1717.0	125.20	2.76	-0.12
62	0.993	238.5	1685.6	149.25	-1.61	-0.90
63	–	–	1705.1	–	-3.94	4.43
64	0.832	–	1718.8	100.10	1.89	-0.33
65	0.821	187.5	1716.8	173.25	0.57	1.32
66	0.826	186.5	1714.8	172.20	0.89	1.09
67	0.819	173	1716.8	156.56	1.15	0.84
68	0.822	167.5	1714.3	155.99	1.10	0.77
69	–	178	1715.7	–	-0.88	3.51
70	0.812	100.5	1717.4	106.07	3.17	-0.65
71	0.814	102	1716.6	105.81	2.88	-0.57
72	0.998	–	1713.0	148.51	-0.74	-0.42
73	–	–	1700.5	–	-1.68	1.20
74	–	–	1709.8	–	-1.59	2.22
75	0.801	106	1708.5	125.04	2.43	-0.20
76	0.904	178.5	1707.1	155.15	0.17	0.22

Table 15A.3: Ketones (continued)

No	Aldehyde	Descriptors[a]		
		1	2	3
77	2-Undecanone	170.3	11	1.428
78	3-Undecanone	170.3	12	1.4291
79	6-Undecanone	170.3	15	1.428

[a] Descriptors: 1, molar mass 10^{-3} kg mol^{-1}); 2, melting point ($^{\circ}$ C); 3, refractive index; 4, density (10^{3} kg m^{-3}); 5, boiling point ($^{\circ}$ C); 6, molar volume (10^{-1} m^{3} mol^{-1}); 7, carbonyl IR absorption (wave number) .

Data in Table 15A.3 are taken from Ref.[4].

[b] 2(5*H*)Furanone is a lactone and not a ketone. It was included in the data set to check the ability of the principal components model to detect an outlier. This was also found to be the case, see [4].

No	Descriptors				Scores	
	4	5	6	7	t_1	t_2
77	0.825	231.5	1719.2	206.42	-0.45	2.31
78	0.827	226	–	205.93	-0.52	2.16
79	0.831	228	1715.2	204.93	-0.53	2.17

Table 15A.4: Descriptors and PC scores of amines

No	Amine	Descriptors[a]		
		1	2	3
1	Undecylamine	171.33	17	1.4388
2	Tripropargylamine	131.18	–	1.4838
3	Tripopylamine	143.27	-94	1.4165
4	Tris(dimethylamino)methane	145.25	–	1.4360
5	Trioctylamine	353.68	–	1.4485
6	Tridecylamine	199.38	31	–
7	Triethylamine	101.19	-7	1.4000
8	1-Tetradecylamine	213.41	41	–
9	Pyridine	79.10	-42	1.5090
10	Propylamine	59.11	-83	1.3889
11	Propargylamine	55.08	–	1.4480
12	2-Picoline	93.13	-70	1.5000
13	3-Picoline	93.13	-19	1.5054
14	4-Picoline	93.13	2	1.5050
15	3-Phenyl-1-propylamine	135.21	–	1.5260
16	2-Phenylpyridine	155.19	–	1.6242
17	3-Phenylpyridine	155.19	–	1.6155
18	4-Phenylpyridine	155.19	–	–
19	4-Phenylmorpholine	163.22	–	–
20	*N*-Phenylbenzylamine	183.25	36.5	–
21	2-Phenylethylamine	121.18	–	1.5225
22	Octylamine	129.25	-2	1.4290
23	2-Amino-2,3,3-trimethylbutane	129.25	–	1.4240
24	Octadecylamine	269.52	51	–
25	Nonylamine	143.27	–	1.4548
26	Neopentylamine	87.17	–	1.4030
27	5-Methylquinoline	143.19	–	1.6135
28	*N*-Methylpropargylamine	69.11	–	1.4332
29	1-Methylpiperidine	99.18	–	1.4378
30	2-Methylpiperidine	99.18	–	1.4459
31	3-Methylpiperidine	99.18	–	1.4470
32	4-Methylpiperidine	99.18	–	1.4458
33	*N*-Methylphenethylamine	135.21	–	1.5162
34	4-Methylmorpholine	101.15	-66	1.4349
35	*N*-Methylcyclohexylamine	113.20	–	1.4546
36	2-Methylcyclohexylamine	113.20	–	1.4565
37	3-Methylcyclohexylamine	113.20	–	1.4525
38	4-Methylcyclohexylamine	113.20	–	1.4531
39	*N*-Methylbutylamine	87.17	–	1.3995

No	Descriptors				Scores	
	4	5	6	7	t_1	t_2
1	0.80	240.0	215.34	10.63	1.31	2.23
2	0.93	–	141.51	–	1.13	-0.47
3	0.75	157.0	190.33	–	-0.61	1.71
4	–	–	–	–	0.13	1.21
5	0.81	366.0	437.18	–	6.35	7.36
6	–	265.0	–	11.00	2.81	2.87
7	0.73	88.8	139.38	10.85	-1.54	0.84
8	–	–	–	10.62	3.03	2.98
9	0.98	115.0	80.88	5.25	0.35	-2.70
10	0.72	48.0	82.21	10.69	-3.08	-0.35
11	0.80	83.0	68.59	8.15	-1.72	-1.59
12	0.94	128.5	98.75	5.96	0.24	-2.02
13	0.96	144.5	97.31	5.70	0.74	-2.10
14	0.96	145.0	97.31	5.99	0.83	-2.03
15	0.95	221.0	142.18	–	1.98	0.86
16	1.09	269.0	142.90	4.77	4.40	-2.07
17	1.08	–	143.44	–	4.25	-2.07
18	–	274.5	–	5.35	3.47	-1.54
19	–	–	–	–	3.10	-0.89
20	1.06	306.5	172.71	–	4.08	-0-76
21	0.96	198.5	125.58	9.88	1.44	-0.86
22	0.78	170.0	165.28	10.65	0.08	1.16
23	0.81	140.0	160.56	--	-0.27	0.92
24	–	–	–	10.60	4.67	4.23
25	0.78	201.0	183.21	10.64	0.50	1.41
26	0.75	–	117.0	9.85	-1.89	0.13
27	1.06	259.0	134.70	5.15	3.90	-2.08
28	0.82	83.0	84.34	–	-1.81	-1.05
29	0.92	106.5	121.54	10.38	-1.04	-0.07
30	0.84	118.59	117.51	10.95	-0.86	0.16
31	0.85	125.5	117.37	11.07	-0.81	-0.13
32	0.84	124.0	118.35	10.78	-0.82	-0.14
33	0.93	203.0	145.39	9.35	1.51	-0.32
34	0.92	115.5	109.95	7.38	-0.48	-1.09
35	0.87	–	139.41	10.72	-0.23	0.00
36	0.86	150.0	132.24	–	-0.03	-0.21
37	0.86	–	132.39	–	-0.09	-0.17
38	0.86	154.0	132.39	–	-0.03	-0.16
39	0.74	91.0	118.44	10.69	-1.88	0.50

Table 15A.2: Amines (continued)

No	Amine	Descriptors[a]		
		1	2	3
40	1-Methylbutylamine	87.17	–	1.4029
41	2-Methylbutylamine	87.17	–	1.4116
42	S(-)2-Methylbutylamine	87.17	–	1.4126
43	2-Methylbenzylamine	121.18	–	1.5435
44	3-Methylbenzylamine	121.18	–	1.5360
45	4-Methylbenzylamine	121.18	–	1.5340
46	N-Isopropylbenzylamine	149.24	–	1.5025
47	Isopentylamine	87.17	–	1.4089
48	Hexylamine	101.19	-23	1.4180
49	Furfurylamine	97.12	-70	1.4900
50	1-Ethynylcyclohexylamine	123.19	–	1.4817
51	1-Ethylpiperidine	113.20	–	1.4440
52	2-Ethylpiperidine	113.20	–	1.4510
53	2-Ethylpyridine	107.16	–	1.4964
54	3-Ethylpyridine	107.16	–	1.5015
55	4-Ethylpyridine	107.16	–	1.5009
56	4-Ethylmorpholine	115.18	-63	1.4415
57	4-Ethyl-2-methylpyridine	121.18	–	1.4874
58	Dodecylamine	185.36	29	–
59	2,6-Diphenylpyridine	231.28	75	–
60	Diphenylamine	169.23	53	–
61	N,N-Dimethyloctylamine	157.29	-57	1.4243
62	1,5-Dimethylhexylamine	129.25	–	1.4209
63	N,N-Dimethylbenzylamine	135.21	-75	1.5011
64	N,N-Diethylmethylamine	87.17	–	1.38837
65	N,N-Diethylcyclohexylamine	155.29	–	1.4562
66	Dicyclohexylamine	181.32	-2	1.4842
67	Dibutylamine	129.25	-62	1.4170
68	Decylamine	157.29	13	1.4360
69	Cyclopentylamine	85.15	–	1.4482
70	Cyclopropylamine	57.09	–	1.4206
71	Cyclobutylamine	71.12	–	1.4316
72	Pentylamine	87.17	-50	1.4110
73	2-Amino-2-methylbutane	87.17	–	1.3996
74	Allylamine	57.09	-88	1.4245
75	N-Allylcyclohexylamine	139.24	–	1.4664
76	2-Amino-3,3-dimethylbutane	101.19	-20	1.4130
77	2-Aminoheptane	115.22	–	1.4175

No	Descriptors				Scores	
	4	5	6	7	t_1	t_2
40	0.74	91.0	118.44	10.65	-1.99	0.32
41	0.74	95.5	118.72	–	-1.84	0.27
42	0.74	–	118.12	–	-1.89	0.25
43	0.98	199.0	124.03	9.19	1.67	-1.11
44	0.97	203.5	125.45	9.45	1.56	-0.94
45	0.95	195.0	127.29	9.54	1.50	-0.85
46	0.89	200.0	167.31	9.69	1.46	0.37
47	0.75	96.0	116.07	10.60	-1.84	0.20
48	0.77	131.5	132.10	10.64	-1.07	0.52
49	1.10	145.5	88.37	–	0.50	-2.62
50	0.91	–	134.94	–	0.83	-0.56
51	0.83	131.0	137.28	10.45	-0.52	0.25
52	0.85	143.0	132.24	–	-0.17	-0.14
53	0.94	149.0	114.37	5.89	0.90	-1.57
54	0.95	166.0	112.34	5.70	1.15	-1.69
55	0.94	168.0	113.76	6.03	1.08	-1.56
56	0.90	139.0	127.27	–	-0.29	-0.41
57	0.92	178.0	131.86	–	0.98	-0.79
58	–	248.0	–	10.67	3.94	-2.25
59	–	–	–	–	5.88	0.22
60	–	302.0	–	0.79	5.12	-4.09
61	0.76	195.0	205.62	–	0.36	2.17
62	0.77	155.0	168.51	–	-0.30	1.30
63	0.90	184.0	150.23	9.03	0.78	-0.20
64	0.72	64.0	121.07	9.75	-2.58	0.29
65	0.85	95.0	182.69	–	1.14	1.06
66	0.91	256.0	199.25	–	2.43	1.08
67	0.77	159.0	168.51	11.25	-0.70	1.38
68	0.79	217.0	199.87	10.64	0.86	1.91
69	0.86	107.0	98.66	10.65	-1.08	-0.69
70	0.82	49.5	69.3	9.10	-2.21	-1.39
71	0.83	82.0	85.38	10.04	-1.70	-0.93
72	0.75	104.0	115.92	10.63	-1.74	0.21
73	0.75	77.0	116.85	–	-2.08	0.29
74	0.76	53.0	–	–	-2.75	-0.94
75	0.96	–	144.74	–	1.26	-0.36
76	0.76	102.5	134.03	–	-1.24	0.56
77	0.77	143.0	150.42	–	-0.73	0.91

Table 15A.42: Amines (continued)

No	Amine	Descriptors[a]		
		1	2	3
78	Benzylamine	107.16	10	1.5424
79	2-Benzylpyridine	169.23	9	1.5785
80	3-Benzylpyridine	169.23	–	1.5815
81	4-Benzylpyridine	169.23	–	1.5818
82	Butylamine	73.14	-49	1.4010
83	sec-Butylamine	73.14	-72	1.3928
84	tert-Butylamine	73.14	-67	1.3790
85	N-Butylbenzylamine	163.27	–	1.5006
86	N-(tert-Butyl)benzylamine	163.27	–	1.4968
87	Cycloheptylamine	113.20	–	1.4724
88	Cyclohexylmethylamine	113.20	–	1.4630
89	Cyclohexylamine	99.18	-17	1.4580
90	N-Cyclohexyl-1,3-propanediamine	56.27	-16	1.4820
91	Diallylamine	97.16	-88	1.4405
92	2,6-Di-tert-butyl-4-methylpyridine	205.35	34.5	1.4767
93	2,6-Di-tert-butylpyridine	191.35	–	1.4739
94	Diethylamine	73.14	-50	1.3861
95	Dihexylamine	185.36	–	1.4320
96	1,3-Dimethylbutylamine	101.19	–	1.4085
97	3,3-Dimethylbutylamine	101.19	–	1.4135
98	N,N-Dimethylcyclohexylamine	127.23	–	1.4535
99	N,N-Dimethylethylamine	73.14	-140	1.3720
100	1,4-Dimethylpiperazine	114.19	–	1.4463
101	2,5-Dimethylpiperazine	114.19	–	–
102	2,6-Dimethylpiperidine	113.20	–	1.4394
103	2,6-Dimethylpiperidine	113.20	–	1.4454
104	1,2-Dimethylpropylamine	87.17	-50	1.4955
105	Dipentylamine	157.29	–	1.4272
106	Dipropylamine	101.19	–	1.4035
107	Ethylamine	45.09	-81	–
108	N-Ethylbutylamine	101.19	–	1.4050
109	2-Ethylbutylamine	101.19	21.5	1.4209
110	N-Ethylcyclohexylamine	127.23	–	1.4525
111	1-Hexadecylamine	241.46	–	–
112	Isopropylamine	59.11	-101	1.3756
113	N-Methylcyclodecylamine	169.31	–	1.4832
114	1-Methylpiperazine	100.17	–	1.4655
115	Morpholine	87.12	-6	1.4541

No	Descriptors				Scores	
	4	5	6	7	t_1	t_2
78	0.98	184.5	109.24	9.33	1.37	-1.43
79	1.05	276.0	160.56	5.13	3.89	-1.26
80	1.04	287.5	162.41	–	4.00	-1.11
81	1.06	287.0	159.5	–	4.07	-1.26
82	0.74	78.0	98.84	10.65	-2.27	-0-08
83	0.72	63.0	101.02	10.63	-2.61	0.01
84	0.70	46.0	105.09	10.69	-2.90	0.23
85	0.91	–	179.22	–	2.12	0.40
86	0.88	–	185.32	10.19	1.58	0.91
87	–	–	–	–	0.23	-0.70
88	0.87	147.0	130.11	11.04	-0.17	0.00
89	0.87	134.0	114.39	10.66	-0.45	-0.36
90	0.92	–	170.41	–	1.53	0.30
91	0.79	111.5	123.46	–	-1.33	-0.05
92	–	233.0	–	–	2.64	2.04
93	0.85	–	224.55	5.02	2.71	1.15
94	0.71	55.0	103.45	11.16	-2.71	0.24
95	0.80	193.5	233.16	–	1.35	2.61
96	0.72	109.0	141.13	–	-1.53	0.91
97	0.75	115.0	134.56	–	-1.33	0.60
98	0.85	158.5	149.86	–	0.26	0.27
99	0.69	37.0	108.36	–	-3.58	0.57
100	0.84	131.0	135.30	–	-0.30	-0.03
101	–	164.5	–	–	0.09	-0.37
102	0.84	128.0	134.07	11.07	-0.62	0.29
103	0.85	144.0	132.71	–	-0.20	-0.09
104	0.76	85.5	115.15	–	-1.87	0.17
105	0.78	202.5	202.45	–	0.75	2.04
106	0.74	107.5	137.11	11.20	-1.65	0.81
107	0.69	19.5	65.44	10.70	-3.82	-0.55
108	0.74	108.0	136.74	–	-1.50	0.76
109	0.78	125.0	130.39	–	-0.70	0.37
110	0.84	165.0	150.75	–	0.29	0.33
111	–	330.0	–	10.61	4.44	3.59
112	0.69	33.5	85.17	10.71	-3.47	-0.17
113	0.98	–	172.41	–	2.33	0.15
114	0.90	138.0	110.93	–	-0.12	-1.00
115	1.00	129.0	87.21	8.33	0.03	-1.70

414

Table 15A.42: Amines (continued)

No	Amine	Descriptors[a]		
		1	2	3
116	Piperidine	85.15	-13	1.4525
117	Tetramethylpyrazine	136.20	85	–
118	Trimethylamine	59.11	-117	–
119	Dimethylamine	45.09	-93	1.3700
120	Pyrrolidine	71.12	–	1.4431
121	Piperazine	86.14	106	1.4460
122	Isobutylamine	73.14	-85	1.3970
123	Heptylamine	115.22	–	1.4243
124	Diisopropylamine	101.19	-61	1.3920
125	Di-*sec*-butylamine	129.24	–	1.4110
126	Diisopentylamine	157.29	–	1.4230

[a] Descriptors: 1, molar mass (10^{-3} kg mol^{-1}); 2, melting point ($^\circ$ C); 3, refractive index; 4, density (10^3 kg m^{-1}); 5, boiling point ($^\circ$ C); 6, molar volume (10^{-6} m^{-3} mol^{-1}), 7, pK_a.
   Data in Table 15A.4 are taken from Ref.[5].

No	Descriptors			Scores		
	4	5	6	7	t_1	t_2
116	0.86	106.0	98.90	11.20	-0.95	-0.59
117	–	190.0	–	2.80	3.17	-3.80
118	0.64	–	92.94	9.80	-3.76	0.05
119	0.68	7.4	66.27	10.73	-3.94	-0.50
120	0.85	87.5	83.47	11.27	-1.64	-0.88
121	–	146.0	–	9.83	0.36	-0.02
122	0.74	68.0	99.51	10.42	-2.55	-0.14
123	0.78	155.0	148.67	10.66	-0.65	0.81
124	0.72	84.0	141.13	10.96	-1.96	0.91
125	0.75	135.0	171.63	–	-0.61	1.49
126	0.77	186.0	204.01	–	0.56	2.11

Table 15A.5: Descriptors of Lewis acids[a]

No	Lewis Acid	Descriptors 1	2	3	4	5	6	7	8	9	10
1	$TiCl_2$	23.6	513.8	464.4	87.4	69.83	2.25	1308.5	475	3.13	585.6
2	$TiCl_3$	51.3	720.9	653.5	139.7	97.15	2.3	440	660	2.64	1220.4
3	$TiCl_4$	94.5	804.2	737.2	252.3	145.18	2.19	-25	136.4	1.73	2220
4	VCl_3	561	–	–	–	–	–	–	–	3.00	1296.5
5	VCl_5	–	570.2	–	235.3	–	2.03	-28	148.5	1.82	2374.9
6	$CrCl_2$	24	395.4	356.1	115.3	71.2	2.09	814	–	2.88	634.7
7	$CrCl_3$	54.3	556.5	486.2	123	91.8	2.38	1152	1300	2.76	1356
8	$MnCl_2$	24	481.3	440.5	118.2	72.93	2.09	650	1190	2.98	602.2
9	$FeCl_2$	25.2	341.8	302.3	118.0	76.65	2.38	677	–	3.16	657
10	$FeCl_3$	54.5	399.5	334	142.3	96.65	–	306	315	2.90	1365.4
11	$CoCl_2$	25.5	312.5	269.9	109.2	78.5	2.53	7.2	1049	3.36	679.8
12	$NiCl_2$	26.2	305.3	259	97.65	71.7	1.82	1001	973	3.55	700.6
13	$CuCl$	7.85	137.2	119.9	86.2	48.5	2.34	429	1490	4.14	260.5
14	$CuCl_2$	26.9	220.1	175.7	108.1	71.9	2.09	493	993	3.99	729.9
15	$ZnCl$	26.8	415.1	369.4	111.5	71.3	2.05	283	732	2.91	665.1
16	BCl_3	–	427.2	387.4	206.1	106.7	1.75	-107.3	12.5	1.35	1785
17	$AlCl_3$	55.6	704.2	628.8	110.5	91.84	2.06	190	182.7	2.44	1310.5
18	$GaCl_3$	57.4	523.4	455.2	135.2	–	2.09	77.9	201.3	2.47	1433
19	$SiCl_2$	–	162.4	176.3	282	51.33	2.00	–	–	–	676.5
20	$SiCl_4$	–	657	617	330.6	90.2	2.09	-70	57.6	1.48	2492.8
21	$GeCl_4$	103.2	504.8	–	347.5	–	2.10	-49.5	84	1.84	2488.6
22	$SnCl_2$	22.7	350	302.1	122.5	–	2.42	246	652	3.94	581.9
23	$SnCl_4$	–	511.3	440.1	258.6	165.3	2.31	-33	114.1	2.23	2227.7
24	PCl_5	–	374.9	305	364.6	112.8	2.03	166.8	162	4.63	4159.3
25	$AsCl_3$	55.9	305	259.4	216.3	–	2.16	-8.5	130.2	2.16	1386.5
26	$SbCl_3$	46.4	382.2	323.7	184.7	108	2.33	73.4	283	3.14	1231
27	$SbCl_5$	–	440.2	350.2	301.3	–	1.97	2.8	79	2.34	3537
28	$AsCl_5$	–	–	–	–	–	–	–	–	–	3990
29	TiF_2	26.1	682.3	694.9	255.6	58.86	1.88	–	–	–	585.6
30	TiF_3	56	1436.2	1360.7	87.9	92.02	1.97	1200	1400	3.40	1220.5
31	TiF_4	101.6	1649.3	1559.3	134.0	114.2	1.92	400	284	2.80	2220
32	VF_3	57.6	–	–	–	–	–	800	–	3.36	1296.5
33	VF_4	–	1404	–	121.4	–	–	325	–	2.98	2374.9
34	CrF_2	21.1	779.9	711.3	89.7	–	1.72	894	1300	4.11	634.7
35	CrF_3	58.1	1113.9	1046	94.0	–	1.90	1100	1200	3.80	1356
36	MnF_2	25.7	795.5	–	92.4	66.78	1.72	856	–	3.98	602.2
37	FeF_2	27.2	711.3	668.6	86.1	68.1	1.99	1000	–	4.09	657
38	FeF_3	58.5	1046.4	840.9	98.4	–	1.92	1000	–	3.52	1365.4

Table 15A.5: Lewis acids

No	Descriptors									
	11	12	13	14	15	16	17	18	19	20
1	2431	120	–	–	–	–	–	570	–	1.54
2	5134	111	–	–	–	–	–	1110	–	–
3	9431	104	–	–	–	–	11.76	-54	0	–
4	5322	101	–	–	–	–	15.80	3030	–	–
5	–	91	–	–	–	–	11.77	113	–	–
6	2455	97	–	–	–	–	9.97	7230	–	1.66
7	5473	86	–	–	–	–	–	6890	–	–
8	2362	96	–	–	–	–	11.03	14350	–	1.55
9	2525	98	–	–	–	–	9.84	14750	–	1.83
10	5364	80	–	–	–	–	–	13450	–	–
11	2709	92	–	–	–	–	10.00	12.7	–	1.88
12	2753	90	–	–	–	–	11.23	6145	–	1.91
13	921	88	34.6	45.2	0.279	-0.279	10.70	-40	–	1.90
14	2774	72	–	–	0.484	-0.242	12.89	1080	–	–
15	2690	78	38.3	40.2	0.328	-0.26	12.90	-65	–	–
16	–	106.1	58.5	45.8	0.357	-0.119	11.62	-59.9	0	2.04
17	5376	101.5	39.5	62.2	0.576	-0.192	12.01	–	1.97	1.61
18	5217	86.8	46	32.7	0.312	-0.104	11.96	-63	–	1.81
19	–	101	–	–	–	–	10.93	–	–	–
20	–	95.6	48.3	45.5	0.440	-0.110	11.80	-88.3	0	1.9
21	–	81.2	–	–	0.264	-0.066	11.68	-72	0	2.01
22	2276	93	–	–	–	-0.160	7.30	-69	–	1.65
23	8355	75.3	41.4	31.6	0.352	-0.088	12.13	-115	0	1.96
24	–	63	–	–	–	-0.060	10.70	-67.8	0	–
25	–	73.8	48.2	18.7	0.183	-0.061	10.55	- 79.9	1.55	2.18
26	5032	75	43.3	28.1	0.294	-0.098	10.20	-86.7	3.80	2.05
27	–	60.4	–	–	–	–	–	-120	–	–
28	–	–	–	–	–	–	–	–	–	–
29	2724	–	–	–	–	–	–	–	–	1.54
30	5644	144	–	–	–	–	–	1300	–	–
31	10012	142	–	–	–	–	–	–	0	–
32	5895	134	–	–	–	–	–	2730	–	–
33	–	–	–	–	–	–	–	–	–	–
34	2778	144	–	–	–	–	10.60	–	–	1.66
35	5958	111	–	–	–	–	–	4370	–	–
36	2644	111	–	–	–	–	11.38	10700	–	1.55
37	2769	117	–	–	–	–	–	9500	–	1.83
38	5870	110	–	–	–	–	–	13760	–	–

Table 15A.5: Descriptors of Lewis acids (continued)

No	Lewis Acid	Descriptors 1	2	3	4	5	6	7	8	9	10
39	CoF_2	27.7	692.9	626.6	82.1	68.9	2.04	1200	1400	4.46	679.8
40	NiF_2	28.5	651.4	604.1	73.6	64.06	1.72	1000	–	4.63	700.6
41	CuF	8.61	192.5	171.6	64.8	44.9	1.75	908	1100	–	260.5
42	CuF_2	28.9	542.7	475.4	86.1	94.14	1.72	950	–	4.23	729.9
43	ZnF_2	28.8	764.4	713.3	73.7	65.65	1.81	872	1500	4.95	665.1
44	BF_3	–	1137	1120.3	254.1	50.46	1.30	-126.7	-99.9	2.99	1785
45	AlF_3	61.2	1504.1	1425	66.4	75.1	1.63	1291	–	2.88	1310.5
46	GaF_3	61.5	–	–	–	–	1.88	800	1000	4.47	1433
47	SiF_2	–	587.9	598.3	256.2	44.5	1.59	–	–	–	676.5
48	SiF_4	–	1614.9	1572.7	282.4	73.6	1.55	-90	–	–	2492.8
49	GeF_4	108.7	1192.5	–	302.9	–	1.68	-32	-36.5	2.46	2488.6
50	SnF_2	25	–	–	–	–	2.06	–	–	–	581.9
51	SnF_4	–	–	–	–	–	1.86	705	–	4.78	2227.7
52	PF_5	–	1595.8	1508.7	300.6	84.8	1.58	-83	-75	5.81	4152.3
53	AsF_3	61.5	821.3	774.2	181.2	126.6	1.71	-8.5	-63	2.67	1386.5
54	SbF_3	53.7	915.5	–	–	–	1.90	292	319	4.38	1231
55	SbF_5	–	–	–	–	–	–	7	149.5	2.99	3537
56	AsF_5	–	–	–	–	–	–	-80	-53	7.71	3990
57	$TiBr_2$	22.9	402	383.2	119.7	77.82	2.40	500	935.8	4.31	585.6
58	$TiBr_3$	50	548.5	523.8	176.6	101.71	2.40	–	794.2	–	1220.5
59	$TiBr_4$	92.9	616.7	589.5	243.5	131.50	2.31	39	230	2.60	2220
60	VBr_3	51.8	447.9	–	142.4	–	–	–	–	–	1297.5
61	VBr_4	–	393.6	–	334.9	–	2.30	–	–	4.00	2379
62	$CrBr_2$	23.3	338.9	–	–	–	2.24	844	–	4.36	634.7
63	$CrBr_3$	53.1	426.8	–	–	–	2.57	1130	–	4.29	1356
64	$MnBr_2$	23.2	384.9	–	138.1	–	2.24	–	–	4.39	602.2
65	$FeBr_2$	24.4	249.8	238.1	140.6	80.2	2.2.4	684	–	4.64	657
66	$FeBr_3$	53.7	268.2	–	173.7	–	–	–	–	–	1365.4
67	$CoBr_2$	24.9	221	–	134	–	2.24	844	–	–	679.8
68	$NiBr_2$	25.5	212.1	–	136	–	2.24	963	–	5.10	700.6
69	$CuBr$	7.8	104.6	100.8	96.1	54.73	2.17	492	1345	4.98	260.5
70	$CuBr_2$	26.6	141.8	–	133.9	–	2.46	498	–	4.77	729.9
71	$ZnBr_2$	26.2	328.7	312.1	138.5	–	2.24	394	650	4.20	665.1
72	PBr_3	–	239.7	238.5	229.7	128	1.87	-46	91.3	2.64	1785
73	$AlBr_3$	54.3	527.2	504.4	180.2	101.7	2.21	97.5	262.3	2.64	1310.5
74	$GaBr_3$	56.5	386.9	–	180	–	2.28	121.5	278.8	3.69	1433
75	$SiBr_2$	–	–	–	–	–	–	–	–	–	676.5
76	$SiBr_4$	–	92.13	–	–	–	2.15	–	–	–	2492.2

Table 15A.5: Lewis acids

No	Descriptors									
	11	12	13	14	15	16	17	18	19	20
39	2878	113	–	–	–	–	–	9490	–	1.88
40	2845	112	–	–	–	–	–	2410	–	1.91
41	–	102	26.1	69.8	0.366	–	–	–	–	1.90
42	3046	91	–	–	–	–	–	1050	–	–
43	2930	99	27.2	63.3	0.452	-0.226	13.91	-38.2	–	–
44	–	154.3	71.5	86.7	0.504	-0.168	15.96	–	0	2.04
45	5924	141	44.5	100.4	0.729	-0.243	–	-13.4	–	1.61
46	6205	114	48.5	53.3	0.456	-0.152	–	–	–	1.81
47	–	141	–	–	–	-0.240	10.78	–	1.23	–
48	–	142.6	58.7	80.8	0.600	-0.150	15.19	–	0	1.90
49	–	112.5	59.8	51.2	0.416	-0.104	16.06	-50	0	2.01
50	2551	116	–	–	–	–	–	–	–	1.65
51	–	101	47.8	59.6	–	–	–	–	0	1.96
52	–	111.1	50.6	60.5	–	–	15.54	–	0	–
53	–	116.3	58.5	41.7	0.321	-0.107	12.30	–	2.59	2.18
54	5295	106	50.8	51.2	0.438	-0.146	–	-46	–	2.05
55	–	–	–	–	–	–	–	–	–	–
56	–	–	–	–	–	–	–	–	0	–
57	2360	–	–	–	–	–	–	640	–	1.54
58	5012	95	–	–	–	–	–	660	–	–
59	9288	89	–	–	–	–	10.55	–	0	–
60	5192	87	–	–	–	–	–	2896	–	–
61	–	90	–	–	–	–	–	–	–	–
62	2377	80	–	–	–	–	–	–	–	1.66
63	5355	72	–	–	–	–	–	–	–	–
64	2304	81	–	–	–	–	–	–	–	1.55
65	2464	83	–	–	–	–	–	13600	–	1.83
66	5269	71	–	–	–	–	–	–	–	–
67	2648	79	–	–	–	–	–	13000	–	1.88
68	2699	76	–	–	–	–	–	5600	–	1.91
69	879	80	33.6	36.5	0.235	-0.235	–	-49	–	1.90
70	2711	63	–	–	–	–	–	653.3	–	–
71	2632	66	35.7	30.3	0.266	-0.133	–	–	–	–
72	90	88	53.8	33.9	0.285	-0.095	–	–	–	2.04
73	5247	87	37.4	48.7	–	-0.170	10.91	–	–	1.61
74	4966	72.1	43	23	0.240	-0.08	–	–	–	1.81
75	–	86	–	–	–	–	12.00	–	–	–
76	–	78.8	–	–	0.360	-0.09	14.00	-128.6	0	1.90

Table 15A.5: Descriptors of Lewis acids

No	Lewis Acid	Descriptors									
		1	2	3	4	5	6	7	8	9	10
77	$GeBr_4$	101.7	330.8	–	396.9	–	2.31	26.1	186.5	3.13	2488.6
78	$SnBr_2$	22.1	264.8	248.9	146	–	2.55	215.5	620	5.12	581.9
79	$SnBr_4$	–	377.4	350.2	264.4	–	2.44	31	202	3.34	2227.7
80	PBr_5	–	269.9	–	–	–	2.10	100	106	–	4159.3
81	$AsBr_3$	54.7	130	159	363.9	76.16	2.33	32.8	221	3.54	1386.5
82	$SbBr_3$	48.4	–	–	–	–	2.51	96.6	280	4.15	1231
83	$SbBr_5$	–	–	–	–	–	–	–	–	–	3537
84	$AsBr_5$	–	–	–	–	–	–	–	–	–	3990
85	TiI_2	22.0	264	258.9	138.1	86.22	2.59	600	1000	4.99	585.6
86	TiI_3	48.9	322.2	318.5	192.5	116.8	–	–	727	–	1220.5
87	TiI_4	91.2	375.7	371.5	249.4	125.65	–	150	377.1	4.30	2220
88	VI_3	50.7	280.5	–	203.1	–	–	–	–	–	1296.5
89	VI_4	–	–	–	–	–	–	–	–	–	2374.9
90	CrI_2	22.4	158.3	–	–	–	2.43	868	800	5.20	634.7
91	CrI_3	51.9	205.1	–	199.6	–	–	600	350	4.92	1356
92	MnI_2	22.5	331	–	–	–	2.42	638	500	5.00	602.2
93	FeI_2	23.7	113	86.5	170	112.9	2.43	587	1093	5.32	657
94	FeI_3	52.8	71	–	–	–	–	–	–	–	1365.4
95	CoI_2	24.1	87.9	–	153.2	–	2.43	515	570	5.68	679.8
96	NiI_2	24.9	78.2	–	154	–	2.43	797	–	5.83	700.6
97	CuI	7.7	67.8	69.5	97.6	54.06	2.62	605	1290	5.62	260.5
98	CuI_2	25.9	7.1	–	–	–	2.43	–	–	–	729.9
99	ZnI_2	25.5	208.0	209.9	161.1	–	2.38	446	624	4.74	665.1
100	BI_3	–	71.1	20.7	349.2	70.79	2.10	–	–	–	1785
101	AlI_3	52.9	313.8	300.8	159	98.7	2.53	191	360	3.98	1310.5
102	GaI_3	55.4	239.4	–	203.9	–	2.44	212	345	4.15	1433
103	SiI_2	–	144	–	–	–	–	–	–	–	676.5
104	SiI_4	–	199	–	265.6	–	2.44	–	–	–	–
105	GeI_4	100.2	37.7	–	429.1	–	2.49	144	440	4.32	2488.6
106	SnI_2	21.3	143.5	–	167.8	–	2.73	320	714	5.29	581.9
107	SnI_4	–	143.9	143.9	168.6	84.9	2.67	144.2	364.5	4.47	2227.7
108	PI_5	–	–	–	–	–	–	–	–	–	4159.3
109	AsI_3	53.8	–	–	–	–	2.52	146	403	4.39	1386.5
110	SbI_3	47.3	–	–	–	–	2.67	170	401	4.92	1231
111	SbI_5	–	–	–	–	–	–	79	400.6	–	3537
112	AsI_5	–	–	–	–	–	–	76	–	3.93	3990
113	PCl_3	–	319.7	272.3	217.1	–	2.03	-112	75.5	1.58	–
114	PBr_3	–	185.4	175.7	240.2	–	2.18	-41.5	173.5	2.85	–

Table 15A.5: Lewis acids

No	Descriptors									
	11	12	13	14	15	16	17	18	19	20
77	–	67.2	46.4	17.1	0.188	-0.047	10.90	–	0	2.01
78	2211	80.5	–	–	–	–	6.84	–	–	1.65
79	7970	63.5	38.5	25.2	0.272	-0.068	11.10	-149	0	1.96
80	–	–	–	–	–	–	–	–	–	–
81	5497	61.2	45.2	10.8	0.114	-0.038	10.19	-106	1.7	2.18
82	4954	63.1	40.6	19.8	0.225	-0.075	9.77	-115	2.8	2.05
83	–	–	–	–	–	–	–	–	–	–
84	–	–	–	–	–	–	–	–	–	–
85	2259	–	–	–	–	–	–	1790	–	1.54
86	4845	80	–	–	–	–	–	160	–	–
87	9108	73	–	–	–	–	9.27	–	0	–
88	5058	72	–	–	–	–	–	–	–	–
89	–	–	–	–	–	–	–	–	–	–
90	2269	62	–	–	–	–	–	–	–	1.66
91	5201	54	–	–	–	–	–	–	–	–
92	2212	66	–	–	–	–	–	14400	–	1.55
93	2382	68	–	–	–	–	–	13600	–	1.83
94	5117	57	–	–	–	–	–	–	–	–
95	2569	63	–	–	–	–	–	10760	–	1.88
96	2607	62	–	–	–	–	–	3875	–	1.91
97	835	71	33	22.1	0.153	-0.153	–	-63	–	1.90
98	2640	48	–	–	–	–	–	–	–	–
99	2549	51	35	16.3	0.152	-0.076	–	-98	–	–
100	–	64.7	50	17	0.156	-0.052	–	–	–	2.04
101	5070	68	37	33.1	0.369	-0.123	9.66	–	–	1.61
102	4611	58.9	41.2	9.7	0.114	-0.038	9.40	-149	–	1.81
103	–	70	–	–	–	–	–	–	–	–
104	–	59.5	–	–	0.216	-0.054	–	–	0	1.90
105	–	51.4	44.1	4.3	0.052	-0.013	9.42	-174	0	2.10
106	2123	64	–	–	–	–	–	–	–	1.65
107	–	50	37.3	12.5	0.160	-0.04	–	–	0	1.96
108	–	–	–	–	–	–	–	–	–	–
109	4824	49	42.1	0.9	0.012	-0.004	9.00	-142	0.96	2.18
110	4867	48	38.5	7.8	0.096	-0.032	–	-147	1.58	2.05
111	–	–	–	–	–	–	–	–	–	–
112	–	–	–	–	–	–	–	–	–	–
113	–	78.5	–	–	–	–	9.91	-63.4	0.79	2.19
114	–	64.4	–	–	–	–	9.96	–	0.61	2.19

Table 15A.5: Descriptors of Lewis acids

No	Lewis Acid	Descriptors									
		1	2	3	4	5	6	7	8	9	10
115	PI_3	–	46.6	–	–	–	2.43	61.2	200	4.18	–
116	PF_3	–	918.8	897.5	273.2	58.7	1.55	-151.5	-101.8	3.90	–

[a] Descriptors: 1, Coordinate bond energy (eV); 2, negative of standard enthalpy of formation (kJ mol^{-1}); 3, negative of standard Gibbs energy of formation (kJ mol^{-1}); 4, standard entropy (J mol^{-1} K^{-1}); 5, heat capacity (J mol^{-1} K^{-1}); 6, mean bond lenght (Å); 7, melting point ($^\circ$ C); 8, boiling point ($^\circ$ C); 9, density (10^3 kg m^{-3}); 10, standard enthalpy of formation of M^{n+} species (kcal mol^{-1}); 11, lattice energy (exp. or calc.) (kJ mol^{-1}); 12, mean bond energy (kcal mol^{-1}); 13, covalent bond energy (kcal mol^{-1}); 14, ionic bond energy (kcal mol^{-1}); 15, partial charge on central atom (e); 16, partial charge on ligand atom (e); 17, ionization potential (gas phase) (eV); 18, magnetic susceptibility (measured at various temperatures) (10^{-6} cgs); 19, dipole moment (gas phase) (D); 20, atomic electronegativity of central atom in different oxidation states.

The descriptors are taken from Ref.[6]

Table 15A.5: Lewis acids

No	Descriptors									
	11	12	13	14	15	16	17	18	19	20
115	–	44	–	–	–	–	9.15	–	0	2.19
116	–	118.7	–	–	–	–	11.50	–	–	2.19

Table 15A.6: PC scores of the Lewis acids

No	Lewis acid	PC Scores t_1	t_2	No	Lewis acid	PC Scores t_1	t_2
1	$TiCl_2$	-1.89	2.63	39	CoF_2	-1.89	3.37
2	$TiCl_3$	0.71	0.94	40	NiF_2	-0.78	3.06
3	$TiCl_4$	4.36	-1.49	41	CuF	-3.23	3.53
4	VCl_3	1.61	2.06	42	CuF_2	-0.87	2.02
5	VCl_5	2.81	-1.16	43	ZnF_2	-1.06	3.76
6	$CrCl_2$	-2.17	1.47	44	BF_3	7.05	2.80
7	$CrCl_3$	-0.80	1.31	45	AlF_3	2.84	6.39
8	$MnCl_2$	-2.54	2.21	46	GaF_3	1.34	2.43
9	$FeCl_2$	-2.51	0.73	47	SiF_2	1.09	2.98
10	$FeCl_3$	-0.06	-0.49	48	SiF_4	7.76	2.90
11	$CoCl_2$	-1.73	-0.50	49	GeF_4	6.12	0.62
12	$NiCl_2$	-1.88	1.50	50	SnF_2	–	–
13	$CuCl$	-3.42	1.33	51	SnF_4	2.61	1.37
14	$CuCl_2$	-1.63	1.44	52	PF_5	7.94	1.19
15	$ZnCl$	-1.12	0.80	53	AsF_3	3.19	-0.21
16	BCl_3	2.92	-0.70	54	SbF_3	2.14	1.08
17	$AlCl_3$	0.97	1.84	55	SbF_5	6.17	0.46
18	$GaCl_3$	0.93	-0.40	56	AsF_5	–	–
19	$SiCl_2$	–	–	57	$TiBr_2$	-2.28	0.63
20	$SiCl_4$	3.18	-0.78	58	$TiBr_3$	0.28	-0.18
21	$GeCl_4$	3.92	-2.21	59	$TiBr_4$	3.50	-2.25
22	$SnCl_2$	-2.19	-0.36	60	VBr_3	0.40	0.20
23	$SnCl_4$	2.26	-2.22	61	VBr_4	–	–
24	PCl_5	3.36	-3.29	62	$CrBr_2$	-2.34	1.07
25	$AsCl_3$	0.62	-2.54	63	$CrBr_3$	-0.52	-0.45
26	$SbCl_3$	0.15	-1.67	64	$MnBr_2$	-2.29	1.12
27	$SbCl_5$	3.07	-3.18	65	$FeBr_2$	-2.48	0.38
28	$AsCl_5$	–	–	66	$FeBr_3$	0.26	-1.44
29	TiF_2	-0.82	2.68	67	$CoBr_2$	-2.48	0.50
30	TiF_3	1.56	5.69	68	$NiBr_2$	-2.11	0.16
31	TiF_4	5.92	3.60	69	$CuBr$	-3.44	0.79
32	VF_3	1.43	3.97	70	$CuBr_2$	-2.06	-1.38
33	VF_4	–	–	71	$ZnBr_2$	-1.94	-0.28
34	CrF_2	-1.37	3.71	72	PBr_3	0.51	-1.65
35	CrF_3	1.03	4.32	73	$AlBr_3$	0.35	0.11
36	MnF_2	-1.60	3.80	74	$GaBr_3$	0.29	-1.42
37	FeF_2	-1.29	3.26	75	$SiBr_2$	–	–
38	FeF_3	0.98	5.76	76	$SiBr_4$	1.71	-1.09

Table 15A.6: PC scores of the Lewis acids

No	Lewis acid	t_1	t_2	No	Lewis acid	t_1	t_2
77	$GeBr_4$	2.72	-3.43	97	CuI	-4.01	-0.47
78	$SnBr_2$	-2.70	-1.23	98	CuI_2	-2.75	-2.47
79	$SnBr_4$	1.38	-2.46	99	ZnI_2	-2.26	-1.77
80	PBr_5	4.30	-3.24	100	BI_3	-0.20	-2.99
81	$AsBr_3$	0.43	-3.59	101	AlI_3	-0.64	-0.92
82	$SbBr_3$	-0.23	-2.68	102	GaI_3	-0.45	-2.77
83	$SbBr_5$	–	–	103	SiI_2	–	–
84	$AsBr_5$	–	–	104	SiI_4	-1.13	-2.51
85	TiI_2	-2.74	-0.13	105	GeI_4	1.53	-4.78
86	TiI_3	0.14	-1.36	106	SnI_2	-2.80	-1.53
87	TiI_4	2.88	-2.90	107	SnI4	-1.44	-2.96
88	VI_3	–	–	108	PI5	–	–
89	VI_4	–	–	109	AsI_3	-0.35	-4.48
90	CrI_2	-2.98	-0.20	110	SbI_3	-0.65	-3.85
91	CrI_3	0.03	-1.78	111	SbI_5	–	–
92	MnI_2	-3.02	0.29	112	AsI_5	–	–
93	FeI_2	-2.89	-1.01	113	PCl_3	1.32	-2.70
94	FeI_3	–	–	114	PBr_3	0.25	-3.25
95	CoI_2	-2.42	-1.21	115	PI_3	-1.06	-4.24
96	NiI_2	-2.33	-1.17	116	PF_3	5.27	0.13

References to Appendix 15A

1. C. Reichardt
   Solvent and Solvent Effects in Organic Chemistry. Second Revised and Enlarged Edition, Verlag Chemie, Weinheim 1988.

2. C. Hansch and T. Fujita
   J. Am. Chem. Soc. 86 (1964) 1616.

3. R. Carlson, T. Lundstedt and C. Albano
   Acta Chem. Scand. B 39 (1985) 79.

4. R. Carlson, M.P. Prochazka and T. Lundstedt
   Acta Chem. Scand. B 42 (1988) 145.

5. R. Carlson, M.P. Prochazka and T. Lundstedt
   Acta Chem. Scand. B 42 (1988) 157.

6. R. Carlson, T. Lundstedt, Å. Nordahl and M.P. Prochazka
   Acta Chem. Scand. B 40 (1986) 522.

Appendix 15B: On factorization of matrices

Target rotation

Principal Components Analysis and Factor Analysis transform the original space spanned by the descriptor variables to another space spanned by the eigenvectors of the correlation matrix. The eigenvectors are linear tranformations of the vectors which span the descriptor space. The factors and the principal components are mathematical constructs which describe directions of variation. It is, however, not always possible to interpret the eigenvectors in terms of physical models. Sometimes, it is of interest to make such interpretations. One method to achieve this is to make a rotation of the eigenvector system so that the rotated vectors represent contributions of the descriptors which can be linked to some physical model believed to give a more honest representation of a fundamental property of the system. This technique is known as *Target Transformation Factor Analysis* (TTFA). It is accomplished by the following procedure.

Let X be the mean centred (and, if necessary, scaled) ($N \times K$) data matrix. It was seen that if the variance-covariance matrix (correlation matrix) $X'X$ has K distinct eigenvalues it is possible to factorize X as the product of a matrix of object scores, T, and the transpose matrix, P', of the eigenvectors.

$$X = TP'$$

A rotation can be described by an orthogonal rotation matrix R. Orthogonality implies that $R' = R^{-1}$, i.e. $R'R = RR' = I$ (unit matrix). Insertion of RR' in the factor model gives

$$X = TRR'P'$$

which can be written

$$X = TR(PR)'$$

This expression describes a rotation of both the score vectors and the loading vectors.

For more details of this technique, see Malinowski an Hovey[1] and Jöreskog et al.[2].

Singular value decomposition

It can be shown[3,4] that any matrix X can be factorized into two orthonormal matrices, V and U, and a diagonal matrix Σ.

$$X = V\Sigma U'$$

The diagonal elements in Σ are called the singular valuees of X and are actually the square roots of the eigenvalues to $X'X$. This factorization is called *singular value decomposition, SVD*. An efficient algorithm for this has been given by Golub and Van Loan[4]. SVD is the most efficient method available for factor analysis if all eigenvalues are desired.

We have previously seen that it is possible to factorize a matrix **X** by determining the eigenvectors to **X'X**. This is not, however, a very efficient method from a computational point of view, especially not with large matrices.

From the relations below, it is seen that **V** corresponds to the orthonormalized score matrix, and **U'** is the transpose of the orthonormalized loading matrix.

Factorization of **X** by principal components analysis gives

X = TP'

We can therefore write **X'X** as

X'X = (TP')'(TP') = P(T'T)P'

The matrix **T'T** is symmetric and actually a diagonal matrix since the score vectors are orthogonal. Any real symmetric matrix can be written as the product of the transposed orthonormal eigenvector matrix, **V'**, a diagonal matrix of eigenvalues, **G** , and **V**.

X'X = P(V'GV)P'

The diagonal matrix **G** can be written as the product $G^{\frac{1}{2}}G^{\frac{1}{2}}$, in which the matrix $G^{\frac{1}{2}}$ is also a diagonal matrix with the element equal to the square roots of the elements in **G**.

$$X'X = PV'G^{\frac{1}{2}}G^{\frac{1}{2}}VP'$$

Since $G^{\frac{1}{2}}$ is a diagonal matrix it does not matter if it is premultiplied or postmultiplied by another matrix, and as $G^{\frac{1}{2}} = G^{\frac{1}{2}\prime}$ we can write

$$X'X = (PG^{\frac{1}{2}\prime}V)(VG^{\frac{1}{2}}P) = (VG^{\frac{1}{2}}P)'(VG^{\frac{1}{2}}P)$$

From this expression it is seen that $G^{\frac{1}{2}}$ corresponds to Σ in the singular value decomposition.

References

1. F. Malinowski and D. Howery
   Factor Analysis in Chemistry
   Wiley, New York 1980.

2. K. Jöreskog, J. Klovan and R. Reyment
   Geological Factor Analysis
   Elsevier, Amsterdam 1976.

3. G. Strang
   Linear Algebra and Its Applications, 3rd Ed.
   Hacourt Brace Jovanovich, San Diego 1988.

4. G. Golub and C. Van Loan
   Matrix Computations
   The Johns Hopkins University Press, Oxford 1983.

Appendix 15C: The NIPALS algorithm[1]

Although a matrix, X, can be factorized by singular value decomposition, or by computing the eigenvectors from $X'X$, this is not necessary when a few principal components are desired. In such cases it is more advantageous to use the NIPALS algorithm by which the principal components can be determined one at a time in decreasing order according to how much variance they describe. This algorithm is a variant of the power method.[2] It is iterative and generally converges after few (< 20) iterations. It can be summarized as follows:

Let X be the centred (and scaled) data matrix.

(1)   Use any column vector in X as a starting score vector $t_{(0)}$.

(2)   Determine a first loading vector as $p'_{(0)} = t_{(0)}'X/t_{(0)}'t_{(0)}$.
This affords a least squares estimation of the elements in p as the slopes of the linear regression of $t_{(0)}$ on the columns x_i in X.

(3)   Normalize $p_{(0)}$ to unit length by multiplication with
$1/\|p_{(0)}\| = (p_{(0)}'p_{(0)})^{-1/2}$.

(4)   Compute a new score vector $t_{(1)} = Xp_{(0)}$.
This gives an estimate of the scores as the regression coefficients of p on the corresponding row vector in X.

(5)   Check the convergence by comparing $t_{(1)}$ to $t_{(0)}$. This can be made by computing the norm $\|t_{(1)} - t_{(0)}\|$ and comparing this value to a criterion of convergence, say $\|t_{(1)} - t_{(0)}\| < \epsilon = 10^{-10}$. The value of ϵ can be set according to the precision in the calculations by the computer.
If the score vectors have converged go to (6), else return to (2).

(6)   Remove the variation described by the first component and form the matrix of residuals $E = X - t_1p'_1$. Use E as the starting matrix for the next principal component dimension, and return to (1).

References

1.   H. Wold
      Non linear estimation by iterative least squares procedures
      in F. David (Ed.)
      Research Papers in Statistics
      Wiley, New York 1966, pp. 411–444.

2.   G. Strang
      Linear Algebra and its Applications, 3rd Ed.
      Hacourt Brace Jovanovich, San Diego 1988, pp. 370–372.

Chapter 16

Strategies for the selection of test systems

In the preceding chapter it was shown how the axes of the reaction space can be quantified by the principal properties. The variation of the principal properties over the set of possible test candidates is visualized by the score plots. Compounds with similar properties will be close to each other in these plots, while compounds whose properties are radically different will be projected far from each other. These projections therefore offer a means of ensuring that in the set of compounds actually selected for experimental work, a sufficient range of variation has been achieved in all properties initially considered.

The points of the selected test compounds in the score plots should have a sufficient spread. In this chapter it is discussed how such selections can be made to cope with some common problems.

To illustrate the general principles, we shall use selections of solvents in the examples below.

Remember that statistical methods are but tools which can guide the chemist towards good chemistry. Statistics can never substitute chemical knowledge. No statistical analysis will make sense if the experiment does not make sense. It would, for instance, be rather senseless to use glacial acetic acid as a solvent in a base catalyzed reaction, or to use cyclohexane as a solvent in a ionic reaction in which a solid inorganic salt is used to furnish the reagent ion without also adding a phase-transfer catalyst. Any selection base on the principal properties must therefore be carefully analyzed with regard to what is already known. If that requirement is met, the principal properties offer efficient methods for systematical exploration of the reaction space.

1. Selection of solvents by their principal properties

When a new reaction is studied, previous experience is scarce and some common questions in this situations are:

* Do the properties of the solvent have any influence at all?

* What type of solvent is the most useful?

* Is it possible to improve the result by using another solvent?

* Will the properties of the solvent have a systematic influence on yield and selectivity?

* Will a change of solvent modify the reactivity of substrate and reagent so that more dimensions of the reaction space must be considered?

These questions are different and have been arranged in approximately increasing order of complexity. The appropriate experimental designs for finding the answers to these questions will be different.

In the sections below, the following principles for selection are discussed:

* Selections which afford a maximum spread in the principal properties.

* Selection of typical items from subgroups of similar compounds.

* Selections which afford a uniform spread in the principal properties.

* Simplex search of a suitable solvent or reagent.

* Selections by a factorial or fractional factorial design to explore several dimensions of the reaction space.

It is possible to adopt a stepwise approach to these experiments so that additional information can be gained by running a complementary set of experiments. The tools for an efficient analysis will be more elaborate the more complex the questions are which are posed to the experimental system. Some additional examples on the use of designs based on principal properties will therefore be discussed in the next chapter, after the PLS method[1] has been discussed.

1.1. Maximum spread in the principal properties

A selection based on this principle will answer the question "Do the properties of the solvent have any influence at all?" The selected test solvents should have a maximum spread in all their properties. Such a selection is accomplished by chosing

---

[1] PLS (Projections to Latent Structures) is a method by which it is possible to obtain quantitative relations between a matrix of descriptors (independent variables), called the **X** block, and a matrix of one or more response variables, called the **Y** block. The method is based on projections, similar to principal components analysis, and the quantitative relation between the two blocks is obtained by a correlation between the components of the respective block.

maximum spread in all their properties. Such a selection is accomplished by chosing solvents projected on the periphery of the score plot and as far as possible from each other, see Fig. 16.1.

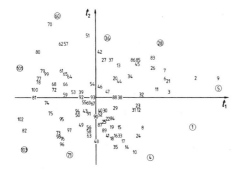

Fig.16.1: Selection of solvents which gives a maximum spread in the principal properties.

Some cautions should be exercised when items are selected by this principle: this method can be safely used in screening experiments aimed at exploring the role of the solvent, but in the selection of substrates or reagents it should be remembered that objects projected on the periphery of the score plot are in several respects extreme objects. Some of them will probably be borderline outliers. Conclusions as to the scope and limitations of a reaction by a "maximum spread" design can therefore be dangerous. A better design in such cases would be to use a "good spread" design, i.e. to select test objects projected far from each other, but *not* on the extreme periphery of the plot.

1.2. Example: Solvent selection in the reduction of an enamine

The reduction of the morpholine enamine from camphor by the reaction with formic acid was discussed in Chapter 5. In the example below, this reaction was applied to the reduction of the corresponding pyrrolidine enamine. The reaction was conducted without any solvent present, i.e. by adding formic acid dropwise to the hot enamine, and afforded an almost quantitative conversion to the corresponding bornyl pyrrolidine. However, the reaction was not stereoselective and a mixture of *endo* and *exo* isomers was formed. The proportions were: *endo* isomer (85 %), *exo* isomer (15 %). This was regarded as a promising result.

432

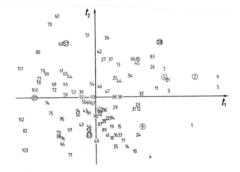

endo   exo

The reaction showed promising stereoselectivity when it was run without any solvent. As the reaction might involve charged species[2], it was quite natural to examine whether or not the selectivity could be increased in the presence of a solvent. The solvents used to investigate this were selected according to a "maximum spread" design in their principal properties, see Fig. 16.2.

However, no improved stereoselectivity was found in any of the solvents examined. The proportion of the *endo* isomer was 85−87 %, and of the *exo* isomer 13−15 %. This was the same result as was obtained without any solvent. The tested solvents span a large range in properties. It can, of course, not be excluded that some solvent might have unique properties for giving an improved selectivity in this reaction. In view of the results obtained, the probability of finding such a solvent must, however, be considered extremely small. The attempts to find a suitable solvent was therefore abandoned, and a method to separate the isomers was developed instead.[1]

Fig.16.2:   Solvents in the study on enamine reduction. Selected test items: Formamide (2), sulfolane (28), 1,2-dichlorobenzene (57), cyclohexane (81), tetrahydrofuran (63), methoxyethanol (8), 2-methyl-2-butanol (40), diethylene glycol (6).

---

[2] By experiment with deuterated formic acid, it was later found that the reaction proceeds in two steps, The first step is a rapid and reversibe protonation to yield an iminium formate, followed by a slow hydride transfer from the formate ion.[1]

1.3. Selection of typical representatives from subgroups of similar test items

Subgroups defined by different classes of solvents are depicted in Fig. 16.3 below. If the problem is to determine in which *type* of solvent the reaction can be run, an evident principle is to select representative test solvents from each class. In this respect, representative members of a class would be items which are not at the extreme ends of the subgroups in the score plot. For example, it would be better to choose isopropanol (17) as a *typical* alcohol rather than methanol (4).

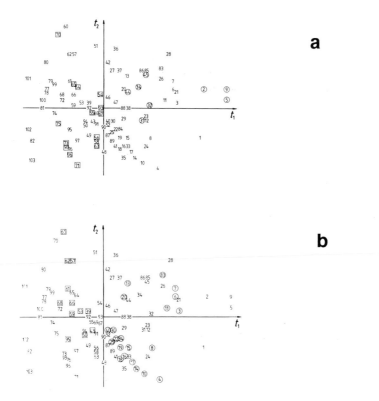

Fig.16.3: Subgroups of solvents: *(a)* Squares show monofunctional ethers, circles show amides; *(b)* Squares show halocarbons, circles show alcohols.

1.4.   Example: Scope of the Willgerodt-Kindler reaction with regard to amine variation[2]

This is not an example of solvent selection. It is given to show a selection of co-substrates from subgroups in the determination of the scope of a reaction, viz. amine variation in the Willgerodt-Kindler reaction. The general features of this reaction were discussed in Chapter 6.

It has for long been assumed that morpholine is the preferred amine in this reaction, and that other amines generally give inferior results. With the aim of examining the scope of the reaction with regard to amine variation, the reaction was run with a series of amines selected by their principal properties. Acetophenone was used as the ketone substrate in these reactions, and quinoline was used as a solvent. The reason for using quinoline was that it permits a large span of the reaction temperature (b.p 237 ° C).

The selection of amines was based on a preliminary study of the principal properties determined from a set of 29 primary and secondary amines characterized by seven descriptors.[2] The corresponding score plot is given in Fig 16.4a which also shows the subgroups formed by primary amines (P), open secondary amines (S), and cyclic secondary amines (C). The selection was made so that three members from each subgroup were included in the test set. The selected items are encircled in Fig. 16.4a. The distribution of these amines in the score plot obtained from the much larger data set of amines given in the preceeding chapter is shown in Fig. 16.4b. The amines selected were: *Primary amines*: Isopropylamine (112), *sec*-butylamine (83), octylamine (22). *Open secondary amines*: Diethylamine (94), dipropyamine (106), dipentylamine (105). *Cyclic secondary amines*: 4-Methylpiperidine (32), pyrrolidine (120), morpholine (115). The identification numbers corresponding to Fig.14.4b are given within parentheses.

For each amine, the experimental conditions were adjusted to achieve a maximum yield. A response surface model was determined for each amine. The optimum conditions and the maximum yields obtained are summarized in Table 16.1.

The results in Table 16.1 show that the selected amines afforded good to excellent yields of the thioamide. Morpholine yielded the best result, but it is seen that it is by no means unique. Another result, is that the optimum conditions are changed when the amines are changed. The yield is sensitive to variations in the experimental conditions, and these results are yet another illustration of the danger of using "standardized" experimental conditions in determining the scope of reactions.

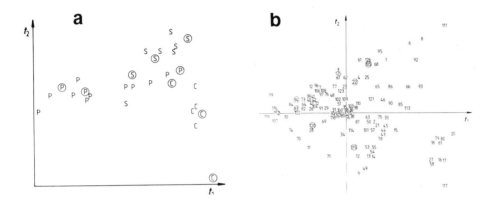

Fig.16.4: Selection of amines in the Willgerodt-Kindler reaction: *(a)* Selection of three items from each subgroup, *(b)* Overview of the selected items in the score plot from the enlarged study of the principal properties of amines.

1.5. Uniform spread in the principal properties

This principle is applicable when

*   a screening experiment is aimed at finding a solvent which is sufficiently promising to motivate future development.

*   one wishes to find out if there is a gradual and systematic change in the performance of the reaction which could be related to the properties of the solvent. A systematic variation can be analyzed by PLS. The PLS model can then be used to predict an expected result in new and yet untested solvents.

*   it is of interest to determine the scope of a reaction with respect to solvent variation. This is similar to the preceeding problem. The difference is that a *yes/no* answer to the question whether or not a solvent is useful can be sufficient.

Table 16.1: Optimum conditions and optimum yield in the Willgerodt-Kindler rection with different amines

Amine	Optimum conditions[a]			Yield (%)	
	u_1	u_2	u_3	y_{GLC}[b]	y_{Isol}[c]
Isopropylamine	10.3	4.8	133	89	87
sec-Butylamine	9.5	5.5	132	88	85
Octylamine	12.5	11.7	137	65	62
Diethylamine	8.4	3.2	135	86	83
Dipropylamine	12.5	6.6	142	81	80
Dipentylamine	16.5	10.4	138	76	71
4-Methylpiperidine	8.0	7.5	148	80	78
Pyrrolidine	10.0	4.0	135	86	83
Morpholine	7.5	10.3	123	94	90

[a] u_1, The amount of sulfur/ketone (mol/mol); u_2, the amount of amine/ketone (mol/mol); u_3, the reaction temperature ($^\circ$ C). [b] Yields determined by gas chromatography directly on the reaction mixture (internal standard technique). [c] Yield of isolated product after chromatography on silica gel.

Selection of solvents that will produce a uniform spread in all properties considered is accomplished by choosing solvents so that their projected points are uniformly spread and form a fairly regular lattice in the score plot.

It is seen that selection of test sets by any of the previously described principles can be augmented to give a test set in which the properties are uniformly spread. Sometimes, a point in the lattice corresponds to a solvent which is *known* to be incompatible with the reaction. If this should happen, take a nearby solvent instead.

The first example of a "uniform spread" design in the study of solvent varation was applied to the Willgerodt-Kindler reaction.[3] It is not reproduced here, since it is very similar to the above example of amine variation. A very clear application of the principle of "uniform spread" is given in an example on the Fischer indole synthesis in the next chapter.

1.6.   Exploration of the vicinity of a promising candidate to find the best choice

It is often known in which solvent the reaction can be run, but for some reason this solvent is considered not to be the preferred one. An example is *benzene* which has been extensively used as a solvent in the past, but today its toxic properties are fully realized and it should be replaced by a less toxic solvent. Another example is when a reaction is scaled up. In lab-scale applications of a synthetic method, the price of the solvent is not too important, whereas the price becomes a factor of tremendous importance in production scale applications.

A similar problem is encountered when a *promising* solvent has been found in a screening experiment, e.g. by a "uniform spread" design. It is reasonable to assume that the preferred solvent has properties which are similar to those of a promising candidate. The next step is then to explore the solvents projected in the vicinity of the promising candidate in the score plot. This can be accomplished by a small "uniform spread" selection around the winning candidate, or by a simplex search described below.

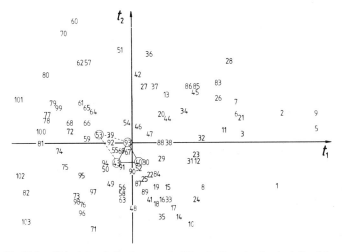

Fig.16.5:   Principle of simplex search: The starting simplex is defined by Dichloromethane (43), 1,4-dioxane (67), and 2-methyl-2-butanol (40). If (40) affords the poorest result, it should be replaced by chloroform (53).

Simplex search

It is possible to adopt a simplex strategy to explore the neighbourhood of a promising solvent. The score values are not continuous and it is therefore not possible to make reflections of the worst vertex in a strict geometrical sense. It is, however, possible to make a simplex search in an approximate way. In the exploration of the solvent space, there are two principal properties to consider. The simplex is therefore a triangle and will be defined by three solvent points in the score plot. Let one vertex correspond to the promising candidate, or to a hitherto known "useful" solvent. The other vertices are chosen not too far from the first one. Run the reaction in the three solvents selected and determine in which experiment the oucome is least favourable. Discard this point and run a new experiment in a solvent chosen so that the points of the two remaining solvents and the new one form a new simplex in the score plot. The new simplex will be oriented away from the poorest solvent of the first simplex. This is repeated until a suitable solvent has been found. The principles are illustrated in Fig. 16.5. The simplex method makes it possible to systematically explore the variation spanned by the principal properties. I have personally found this principle very useful for finding solvents for recrystallization.

1.7. Selection according to the principles of a factorial or fractional factorial design in the principal properties

The principles hitherto discussed are useful for "one-dimensional" studies of the reaction space, i.e. when variations along one axis only are considered. It was emphasized in Chapter 14 that interaction effects are to be expected when the solvent is varied with a simultaneous variation of substrate and/or reagent. To take such interaction effects into account it is necessary to use a multivariate design for selecting the test systems, so that the principal properties are jointly varied over the set of test systems. Examples of situations when this problem is encountered are:

* A screening experiment aimed at determining a suitable combination of reagent and solvent for a given transformation. Hopefully, this may furnish a good starting point for optimizing the procedure.

* It is of interest to determine the general scope of a new reaction. This will necessitate a study of possible variations with regard to the structure of the substrate as well as the nature of both the reagent(s) and the solvent. If the reaction can be run in a variety of solvents, it is possible to use the reaction in conjunction with other reactions without having the necessity to isolate intermediary products. This will offer good opportunities to adopt efficient one-pot procedures.

* It is desirable to know *which* properties of the reaction system are responsible for the result, and hence critical to control with a view to obtaining an optimum result.

In the section below, three examples are given of how the principles of factorial and fractional factorial designs can be applied in the selection of test systems. In the next chapter, an example is given of how a multi-level factorial design in the principal properties was used in conjunction with PLS modelling to analyze which properties of the reaction system are responsible for controlling the selectivity in the Fischer indole reaction.

2. Examples of the use of experimental designs in principal properties

2.1. A screening experiment for determining a suitable combination of reagent and solvent[4]

The synthesis of 2-trimethylsilyloxy-1,3-butadiene by treatment of methyl vinyl ketone with chlorotrimethylsilane, lithium bromide and triethylamine in tetrahydrofuran was discussed in section 12.5.6. It was discussed how the stoichiometry of the reaction was determined by canonical analysis of the response surface model, and how this analysis made it possible to establish experimental conditions which afforded a quantitative conversion. However, before the response surface model could be established it was necessary to find a reaction system worth optimizing.

Previously known procedures for the synthesis of silyl enol ethers indicated that chlorotrimethylsilane in the presence of a Lewis acid and an amine base would be worth trying. Triethylamine was considered a convenient amine base, and the base was not varied in the experiments. The problem was thereby reduced to finding a suitable combination of Lewis acid and solvent. As zinc chloride has been used by Danishefsky for the synthesis of 1-methoxy-3-trimethylsilyloxy-1,3-butadiene[5], and as it is known that certain oxophilic Lewis acids are prone to cleave silyl enol ethers[6], it was assumed that a suitable Lewis acid would be a soft type Lewis acid. These acids are projected in the lower left quadrant of the principal property score plot shown in the preceding chapter. It has also been found that the presence of sodium iodide in the reaction mixture increases the rate of formation of silyl enol ethers[7] and it has been suggested that the reason for the increased rate is the formation *in situ* of a reactive iodotrimethylsilane. It can also be that sodium ion is an electrophilic catalyst. If the reaction is facilitated by electrophilic catalysts, a lithium salt might also be useful as a co-reagent. Based on these considerations, six electrophilic reagents were selected for testing: Sodium iodide, lithium bromide, and four Lewis acids from the lower left part of the score plot, zinc dichloride (15), copper(I) iodide (97), zinc diiodide (99), and iron trichloride, see Fig.16.6a. Four solvents were selected from the solvent score plot: Acetonitrile (24), dimethyl sulfoxide (DMSO) (26), tetrahydrofuran (63), and benzene (72), see Fig.16.6b.

The reaction was tested in a series of experiments in which all electrophiles were used in all solvents. The results are summarized in Table 16.2. In all experiments in which iron trichloride was used, a rapid polymerization of the starting ketone occurred. No diene was observed in these experiments. Dimethyl sulfoxide was deoxygenated under the reaction conditions and was therefore not a useful solvent

440

for this reason. For these reasons, the experiments with FeCl$_3$ and DMSO are omitted from Table 16.2.

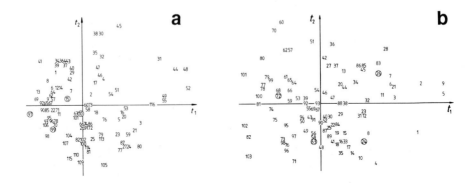

Fig.16.6: Selection of test systems in the synthesis of 2-trimethylsilyloxy-1,3-butadiene: (a) Lewis acid selection; (b) Solvent selection.

Table 16.2: Yield (%) of 2-trimethylsilyloxy-1,3-butadiene

Electrophile:	Solvent:		
	MeCN	THF	Benzene
CuI	6	4	5
ZnCl$_2$	8	6	6
ZnI$_2$	15	10	11
LiBr	0	62	6
NaI	5	8	0

The result in Table 16.2 is self-explanatory. The most promising system for future development is to use lithium bromide in tetrahydrofuran.

2.2. Combinations of reagent and substrates in the synthesis of bezamides[8]

There are a number of methods available for the synthesis of carboxamides from carboxylic acids and amines. In most methods, the carboxylic acid is first converted into a more reactive intermediate, e.g. an acid chloride, which is then reacted with the amine. For practical reasons, it is preferable if the reactive intermediate can be generated *in situ*.

One reaction of this type is to treat the carboxylic acid and a primary or secondary amine with a Lewis acid in the presence of a tertiary amine base.[9] Which type of Lewis acid should be used could depend on the nature of the amine. To investigate this assumption a study was undertaken in which both the Lewis acid and the amine were varied. The reaction between benzoic acid and benzylamine in the presence of boron trifluoride etherate had been described in the literature.[9a] and to allow a comparison with this method, the same carboxylic acid and the same solvent (toluene) were used in the present study. Benzylamine was also used as one of the amines.

The following test systems were selected: *Lewis acids*: Boron trifluoride etherate, titanium tetrachloride, aluminum trichloride, zinc dichloride, and zinc diiodide; *Amines*: Benzylamine, butylamine, morpholine, and dipropylamine. The distribution of these items in the corresponding score plots are shown in Fig.16.7.

A series of experiments was run in which all combinations of Lewis acids and amines were tested. These experiments are summarized in Table 16.3.

Table 16.3: Yields of carboxamides (%)

Amine	Lewis acid				
	$BF_3 \cdot OEt_3$	$TiCl_4$	$AlCl_3$	$ZnCl_2$	ZnI_2[a]
Benzylamine	86	15	63	0	0
Butylamine	81	19	68	0	trace
Morpholine	91	81	72	0	0
Dipropylamine	45	44	8	0	0

[a] Zinc iodide was included in the study when it was found that zinc chloride was not useful.

Table 16.4: Isolated yields of benzamides in preparative scale runs

Amine	Lewis acid	Reaction time (h)	Yield (%)
Benzylamine	$BF_3 \cdot OEt_2$	96	89
Butylamine	$BF_3 \cdot OEt_2$	96	83
Morpholine	$TiCl_4$	10	86
Dipropylamine	$TiCl_4$	10	66

The results in Table 16.3 show that the hard oxophilic Lewis acids afford the carboxamide, but that the soft zinc halides cannot be used. An interesting result is that poor yields are obtained when titanium tetrachloride is used with primary amines. The reason for this is that the primary amines are partially decomposed by this rather aggressive Lewis acid. For the synthesis of carboxamides from the secondary amines it is, however, the preferred reagent.

Yields obtained in preparative scale runs are summarized in Table 16.4. The formation of carboxamides with boron trifluoride etherate is quite slow, whereas the formation is fairly rapid with titanium tetrachloride.

The results obtained in this study show that it is necessary to consider possible interaction effects between reagents and substrates. The initial assumptions were thereby confirmed. The only way to detect such interaction effects is to run experiments in which both reagents and substrates are jointly varied.

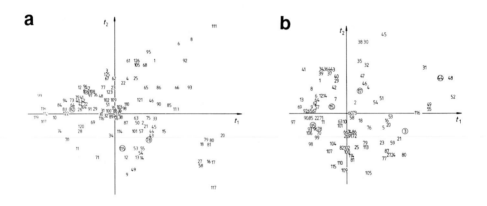

Fig.16.7:   Score plots: *(a)* Lewis acids; *(b)* Amines. The selected test items are:
(a) BF$_3$ · OEt$_2$ (44), TiCl$_4$ (3), AlCl$_3$ (17), ZcCl$_2$ (15), ZnI$_2$ (95);
(b) Benzylamine (78), butylamine (82), morpholine (115), dipropylamine (106).

2.3.   Scope and limitations of a reaction

To assess the preparative utility of an organic reaction, a knowledge of its scope and limitations is essential. In a narrow sense, this means a knowledge of how the reaction performs with different substrates. This usually entails studies on the influence of steric and electronic effects and the presence of functional groups which might interfere with the desired reaction. In a broader sense, the scope of a reaction will encompass also knowledge of the influence of variations in the structure of the attacking reagent(s) as well as of the nature of the solvent, which means that the entire reaction space must be taken into account.

It is impossible to evaluate all possible combinations of substrates, reagents, and solvents by experiments. It is quite cumbersome, even to run a complete factorial design with selected substrates, reagents and solvents, as was described in the examples above. To achieve a more manageable number of test systems, it is possible to use the principles of fractional factorial designs to select test systems by their principal properties[3]. To illustrate this, we shall once more make use of the Willgerodt-Kindler reaction.

Let us pretend that the Willgerodt-Kindler reaction with acetophenone has been recently discovered. Let us also assume also that it has been shown that it can be applied also to substituted acetophenones, provided that they do not carry strong withdrawing substituents on the aromatic ring.

To assess the general scope of the reaction it would therefore be necessary to consider the structure of the substrate, the structure of the amine, and the nature of the solvent.

If it is assumed that the structure of the substrates can be described by the nature of the substituents, it would be possible to use common substituent parameters as descriptors of the substrates. Several studies on substituent parameters by principal components analyses have now been published.[11] When the experiment described here were carried out, one such study was available.[11a] This study described the variation of the properties of aromatic substituents by a two-component model, and this model was used for the selection of the substrates in the present study. The principal properties of amines and solvents are also described by two-component models. Each axis in the reaction space is therfore two-dimensional, and the whole reaction space is six-dimensional. To span the whole reaction space by selecting test systems from each "corner" would require $2^6 = 64$ different test systems. It would be

---

[3] Fractional factorial designs in principal properties have also been used to design aminoacid sequences in peptides in studies on quantitative structure activity relations.[10]

444

quite an effort to carry out the corresponding experiments. By the principles of a fractional factorial design it is possible to select subsets of the 64 different systems so that these subsets span as much as possible of the entire space.

Each constituent of the reaction system is described by two principal properties. The score vectors are mutually orthogonal and independent. It is therefore possible to use a two-level six-variable fractional factorial design to define uncorrelated settings which correspond to different combinations of the principal properties. The smallest fractional factorial design which can accomodate six variables is a 2^{6-3} design with eight runs. If we let the principal property variables, $z_1 - z_6$, in pairs define items from the corresponding quadrants of the principal property score plots, a selection of eight systems can be accomplished as shown in Fig.16.8. Such a selection will ensure that a large span in the principal properties is covered by the selected systems.

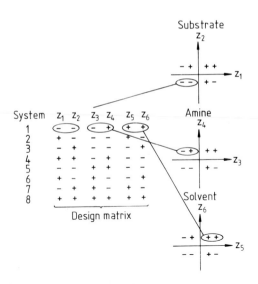

Fig.16.8: Selection of test systems by a fractional factorial design.

As the points in the score plots are often unevenly distributed, it will usually not be possible to select items which define perfect settings in each quadrant so that an orthogonal design *as expressed in the score values* is obtained. This does not really matter, since the objective of the design is to ensure a sufficient spread of the properties of selected test systems, and *not* to compute "effects" of the principal properties. All attempts to derive quantitative relations between the properties of the reaction space and observed results should be made by PLS modelling. A critical discussion on this is given in Chapter 17.

Another way to comprehend the above design is to regard the couples (z_1, z_2), (z_3, z_4), (z_5, z_6) as dummy variables, and use the combinations of their settings, $[(-),(-)]$, $[(-),(+)]$, $[(+),(-)]$, and $[(+),(+)]$ to define four different test items selected to span a sufficient variation in their principal properties. In that respect, the variations in substrates, reagents, and solvents will be orthogonal to each other.

Fig.16.9 shows the score plots of substituents, amines, and solvents. Four items were selected from each plot, and these are encircled in the plots. They correspond to the assignments given in Table 16.5.

Table 16.5: Assignments of design settings to selected test items

Assignment z_i	z_j	Substituent	Amine	Solvent
–	–	Cl-	Isopropylamine	Benzene
+	–	H-	Morpholine	Ethanol
–	+	MeO-	Diethylamine	Quinoline
+	+	PhO-	Dipentylamine	Triethylene glyclol (TEG)

If we limit the study in a first run to include only *para* substituted acetophenones, a series of eight systems as defined by the design in Fig 16.8 would be sufficient to afford a first check of the scope and limitations of the reaction. This would correspond to the systems summarized in Table 16.6. These systems were tested experimentally.[12]

To make fair comparisons, the experimental conditions were optimized to afford a maximum yield for each system. For this response surface technique was used. The optimum conditions and the yields obtained under these conditions are also given in Table 16.6.

446

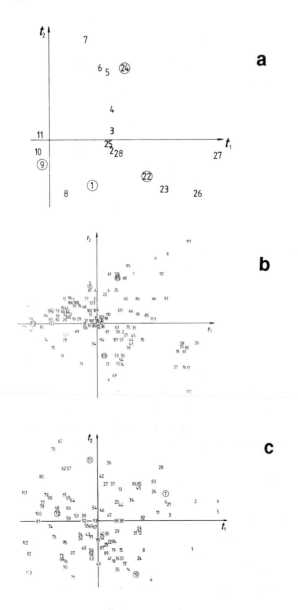

Fig.16.9:   Score plots: *(a)* Substituents; *(b)* Amines; *(c)* Solvents The selected items were:
Substituents: Chloro (9), hydrogen (1), methoxy (22), phenoxy (24);
Amines: Isopropylamine (112), morholine (115), diethylamine (94),
dipentylamine (105); *Solvents*: Benzene (72), Ethanol (10), quinoline (51),
triethylene glycol (7).

Table 16.6: Selected test systems and optimized experimental conditions

No	Test system			Optimum conditions[a]			Yield[b]
	Subst.	Amine	Solvent	u_1	u_2	u_3	(%)
1	Cl-	Et_2NH	TEG	8.4	5.3	123	89
2	H-	i-$PrNH_2$	Quinoline	10.3	4.8	133	89
3	MeO-	i-$PrNH_2$	EtOH	3.8	6.6	80[c]	91
4	PhO-	Et_2NH	Benzene	9.6	5.8	80[c]	85
5	Cl-	Pe_2NH	Benzene	13.6	8.5	80[c]	68
6	H-	Morholine	EtOH	3.7	13.4	80[c]	86
7	MeO-	Morholine	Quinoline	9.3	8.9	130	100
8	PhO-	Pe_2NH	TEG	13.0	8.3	118	73

[a] u_1, amount of sulfur/ketone (mol/mol); u_2, amount of amine/ketone (mol/mol); u_3, reaction temperature (° C).
[b] Determined by gas chromatography (internal standard technique)
[c] The reaction was carried out under reflux.

All test systems afforded moderate to excellent yield, and a first conclusion is therefore that the reaction seems to have a broad scope. It is, of course, not possible to establish the general scope of any reaction from such a small set of test systems. However, the principle is important, and when a fractional factorial design is employed to select a first set of test system, it will be easy to augment the study by selecting complementary fractions so that a better coverage of the reaction space is obtained. One important point is, that even from a small initial subset of test systems, like this one, it is possible to establish a PLS model which can relate the properties of the reaction systems to the observed experimental result. By the PLS model it is then possible to make predictions of the expected results with new systems. These predictions can then be validated by experimental runs. For every new experiment which becomes available, the PLS model can be updated. This can be repeated until the PLS model gives good predictions. This makes it possible to use a step-wise approach to the problem of the scope and limitations. To establish a PLS model it is essential that the calibration set (i.e. the initial design) spans the range of interest of the principal properties.

The limitations of a synthetic method is defined by those combinations of substrates, reagents, and solvents which fail to give the desired reaction. If a selected test system does not give the expected result, it might be an indication that an unfavourable combination of the principal properties is involved. To confirm any such assumptions that the principal properties of the system impose limitations on the scope, it is advisable to select a new series of experiments from test systems selected in the vicinity of the failure point in the reaction space.

3. Conclusions

In experimental sciences any experiment is carried out within a framework of known facts and mere assumptions. An inherent danger when experimental results are interpreted is that assumptions which later on turn out to be false might interfere with the conclusions drawn. This problem is imminent when a newly discovered chemical reaction, or an *idea* to a new reaction are subjected to experimental studies. In both cases, detailed knowledge at the mechanistic level is scarce. If mere speculations on reactions mechanisms are the sole basis for the selection of test systems in such situations, the choice may be too narrow, and there is a risk that a useful new reaction runs is overlooked.

By analyzing the background information and by scrutinizing the constituents of the reaction system, it is possible to select pertinent descriptors of the system. This allows all aspects on the reaction to be taken into account prior to any experiment. Principal components analysis of the variation of the descriptors over the whole set of possible constituents of the system will reveal the principal properties. Sometimes a descriptor variable which does not contribute at all to any systematic variation over the set of compounds has been included among the descriptors. In such cases, it can reasonably be assumed that the descriptor has little relevance to the problem under study. Irrelevant descriptors are easily detected by principal components analysis.

A design based on the principal properties will make it possible to select test items in such a way that variations of all pertinent properties of the reaction system are covered by the set of selected test systems.

A proper design will keep the redundancy small and this will ensure that each experiment will furnish new, and in some respect unique, information on the roles played by the properties of the reaction system. Strategies based on principal properties make experimentation efficient, and this will, hopefully, shorten the time spent on elaborating an idea of a new reaction into a synthetic method.

References

1.   R. Carlson and Å. Nilsson
      Acta Chem. Scand. B 39 (1985) 181.

2.   T. Lundstedt, R. Carlson and R. Shabana
      Acta Chem. Scand. B 41 (1987) 157.

3.   R. Carlson, T. Lundstedt and R. Shabana
      Acta Chem. Scand. B 40 (1986) 694.

4. L. Hansson and R. Carlson
    Acta Chem. Scand. 43 (1989) 304.

5. S. Danischefsky and T. Kitahara
    J. Am. Chem. Soc. 96 (1974) 7807.

6. T. Mukaiyama, K. Banno and K. Narasaka
    J. Am. Chem. Soc. 96 (1974) 7503.

7. *(a)* P. Cazeau, F. Duboudin, F. Molines, O. Babot and J. Dunoque
    Tetrahedron Lett. 28 (1987) 2089;

    (b) C. Rochin, O. Babot, J. Dunogue and F. Duboudin
    Synthesis (1986) 667.

8. Å. Nordahl and R. Carlson
    Acta Chem. Scand. B 42 (1988) 145.

9. *(a)* J. Tani, T. Oine and I. Inoue
    Synthesis (1975) 714.

    (b) J.D. Wilson and H. Weingarten
    Can. J. Chem. 48 (1970) 983.

10. S. Hellberg, M. Sjösröm, B. Skagerberg, C. Wikström and S. Wold
    Acta Pharm Jugosl. 37 (1987) 53.

11. *(a)* S. Alunni, S, Clemeti, U. Edlund, D. Johnels, S. Hellberg and S. Wold
    Acta Chem. Scand. B 37 (1983) 47;

    (b) D.Johnels, U. Edlund, H. Grahn, S. Hellberg, M. Sjöström, S Wold, S. Clementi,
    W.J. Dunn III
    J. Chem. Soc. Perkin Trans. II (1983) 863;

    (c) B. Skagerberg, D. Bonelli, S. Clementi, G. Cruciani and C. Ebert
    Quant. Struct-Act. Relat. 8 (1989) 32.

12. R. Carlson and T. Lundstedt
    Acta Chem. Scand. B 41 (1987) 164.

Chapter 17

Quantitative relations between observed responses and experimental variations

1. The problem

When a synthetic reaction is studied, the number of possible variations of the reaction is very large. We have introduced the concepts *reaction space* and *experimental space* to describe these possible variations. We have seen that these spaces are interrelated and that a variation in the reaction space will induce an altered influence of the experimental variables. For instance, the optimum experimental conditions will be different for different substrates.

There is often more than one response of interest, e.g. yields of main product and of byproducts, selectivity, rates, and costs. This means that also the response variables can define a multidimensional space, see Fig.17.1

In this chapter it is discussed how these different "spaces" can be linked together by quantitative models. By means of such models it is possible to analyze the interdependencies of possible variations. Analyses based on such models will be helpful to clarify a number of the *Which?* and the *How?* questions. Examples of such questions are:

* How are the yield and the selectivity influenced by the properties of the solvent? Can these responses be further improved by adjusting the experimental conditions, and how are these adjustments dependent on the solvent properties?

* How should the experimental conditions be adjusted so that optimum yield is obtained when the substrate is changed?

* Which properties of the reaction system must be controlled to achieve a selective transformation?

* How is the optimum yield influenced by variation in the quality of the starting materials?

The appropriate tools to cope with these questions are PLS models. The PLS method is described in detail in this chapter and several examples of its application to organic synthesis are given. But before entering into the details of this method, a few other methods will be discussed, with the aims of pointing out when they can be applied, respectively *should not* be applied to the type of problems mentioned above.

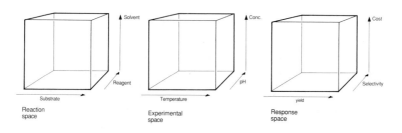

Fig.17.1: Multidimensional variations of a synthetic reaction.

2. Multiple regression cannot always be used

2.1. There are not always cause-effect relations

A large part of this book deals with methods which can be used to explore the experimental space. These comprise methods for designing and analyzing screening experiments, methods for locating an optimum experimental domain, and methods for analyzing the optimum domain by means of response surface models. It has also been discussed how more than one response can be analyzed simultaneously through overlay of the correspondning response surface models. Behind all these methods lies, however, an assumption that there is a *cause-effect* relation between the actual settings of the independent experimental variables and the observed response. Some other assumptions are also involved: it is assumed that the independent variables are totally relevant to the problem, and their settings in the experiment are perfectly known (without errors). It is also assumed that the experimetal errors are independent and have a fairly constant variance in the experimental domain. These assumptions are reasonable, provided that *(a)* the experimeter has control over the experiment and can reproduce the results in repeated runs, an *(b)* the experimental domain is not too vast. Under these circumstances it is possible to establish a sufficiently good description of the variation of the response by fitting an approximate model to known experimental results. A good experimental design ensures that the estimated model parameters afford adequate measures of the influence of the experimental variables. When a polynomial model is used to describe the functional relations between the experimental variables and the response, least squares multiple linear regression is an appropriate method to fit the models.

However, when the reaction space is explored, the assumptions underlying multiple regression modelling are no longer valid. It is not possible to assume cause-effect relations between *measures* of variation (descriptors of reagents and solvents) and the observed response. It would be quite foolish to assume that, for instance, the melting point of a compound is *responsible*, in a philosophical sense, for the reactivity of the compound in a chemical reaction. Another cirumstance is that it is not possible to induce independent variations of the molecular descriptors over a set of compounds. An example will illustrate this: A change of a methyl substituent for a chlorine will alter a number of molecular properties, e.g. dipole moment, lipophilicity, refractive index, and ionization potential. As a consequence any attempts to derive a model by regression directly to measured properties will be made from ill-conditioned experimental designs and any interpretation becomes dubious, since the estimated model parameters will be highly biassed.

Nevertheless, there *are* relations between the properties of the reaction system and its chemical behaviour, as can be seen from the following example:

The structure of a substrate is changed. If this change influences the electron density in a bond being broken in the reaction, a highly probable consequence is that this will lower the activation energy of the reaction. This in turn, will change the temperature dependence of the reaction rate and a likely consequence of this is that the optimum conditions with regard to the reaction temperature is changed.

The fundamental molecular property underlying such an observed change of the behaviour, is the electronic distribution over the substrate. This will determine its interaction with the other constituents of the reaction system. However, this fundamental intrinsic property of the substrate cannot be measured directly.

The principal components are orthogonal to each other. The measures of the principal properties will therefore be independent and uncorrelated. As uncorrelated independent variables are necessary for obtaining stable models by multiple linear regression, one way to derive quantitative relations between the reaction space and observed results would be to use the principal properties as variables in multiple regression analysis. This technique is known as *Principal components regression*[1] and it has been used to derive regression models when the original settings of the variables are ill-conditioned. This is a better method than using the original descriptors in multiple regression, but it is not good enough for our purposes. The problem is that this method too assumes a cause-effect relation between the descriptors and the response. However, when the descriptors for determining the principal properties are compiled, we use our prior assumptions as to their relevance to the problem to decide which descriptors should be included in the data set. We can, however, not be *sure* that the ultimately responsible intrinsic property for the problem under study really has been picked up by the principal properties actually determined.[2]

It is less drastic to assume that there are *relations* between the properties of the reaction system and the observed repsonse(s), but that these relations are not necessarily cause-effect relations. The properties of reaction space and the responses are related because they *depend* on the same intrinsic properties.

This view is closely associated with the way chemists comprehend phenomena in organic chemistry. The outcome of an experiment is often "explained" in terms of various "effects" (inductive effects, mesomeric effects, steric effects etc.) which in the present context fairly closely correspond to the "principal properties".

Quantitative relations between the reaction space and the observed responses can be described as follows: The data are divided into two blocks X and Y; The X block contains the descriptors of the reaction system, and the Y block contains the responses observed with these systems. We are thus looking for a relation between the data structure of these blocks.

One way to do this is to compute principal components models for each block and then determine the correlation between the components from each block. This method is usually called *Canonical correlation*.[3] From a philosophical point of view, canonical correlation is appealing since it does not assume anything about the *direction* of the interdependencies of the two blocks. From a practical point of view, it is, however, not very useful as a tool for problem solving. It is not certain that variables which describe a large variation in the X block are related to large variations in the Y block. It may be that only a small part of the variation in the X block is strongly related to a large and systematic variation of the responses in the Y block.

When we make experiments with a view to studying a synthetic reaction, our interest is to learn how the *results* depend on different perturbations of the experimental system. This means that we wish to obtain a quantitative model which describes as much as possible of the variation in *the response variables* in relation to the properties of the experimental system. This is exactly what PLS will do. PLS models can be established with cross validation to ensure that they have good properties for predictions. We shall, however, not immediately turn to the PLS technique. We shall first discuss an example of a screening experiment when there are many response variables to consider to show that a combination of principal components analysis and multiple regressions can be used to study relations between the experimental space and the response space. It should be mentioned that the same results would have been obtained by a PLS model, but the results, in this case, are more easily interpretated from the regression model.

2.2. Principal components analysis combined with multiple regression. Screening of significant experimental variables when there are several response variables[5]

It is rare that a synthetic reaction proceeds cleanly and affords the desired compound as the sole product. In practice, there is therefore often more than one product in the reaction mixture. An obvious response for the optimization is, of course, the yield of the desired product. Often an initial screening experiment is run to determine the important variables for the optimization step. When there are several products formed, their distribution in the different experiments may furnish extra information on the roles played by the experimental variables. This may give clues to a better understanding of the basic mechanism of the reaction. One important question to answer is therefore: Is there a systematic variation in the product distribution which can be related to the experimental settings? In Chapters 4–6 it was discussed how screening experiments can be laid out by fractional factorial designs, Plackett-Burman designs, and D-optimal designs. A common feature of these designs is that the variables are uncorrelated (or almost uncorrelated with D-optimal designs) over the set of experiments. Such designs therefore have excellent properties for multiple linear regressions. It is also reasonable to assume a cause-effect relation between the variables and the response. The significant experimental variables can be identified from the regression models after a least squares fit to linear or second order interaction models.

When there are several response variables to consider, it is possible to evaluate each response separately. However, by this technique the response variables must be evaluated one at a time. This is manageable when there are only a few response variables, but it becomes cumbersome when there are many responses.

A common problem when series of experiments are run, is that for some runs there are missing data. This occurs rather frequently when seveal responses are measured. Orthogonal designs are balanced designs and each experiment is necessary, and equally important, for the overall picture. Missing data can therefore ruin the analysis and give highly biassed estimates of the effects of the variables.

Another disadvantage of evaluating each response separately is that such an analysis ignores the fact that the different responses have been obtained from the *same experiment*. Correlations among the responses over the set of experiments can be assumed to reflect a similar dependency of the experimental variables. A problem is that colinearities among the responses are not easily detected when they are evaluated separately. A joint evaluation of several responses can be accomplished by overlaying the contour plots of their response surfaces. This is a useful technique when there are few responses to consider. With many responses this can be very difficult.

Colinearities

Colinearities among the responses are sometimes introduced unintentionally in the course of raw data treatment. It is common to normalize chromatograms to give a sum of peak areas equal to 100 percent. If such data are used to analyze the product profile in a synthetic experiment there will be a purely artificial correlation between the relative amounts of the products in each experiment, i.e. if the amounts of one product increase, the amounts of the remaining products will decrease. Such data can lead to conclusions which are totally erroneous. To avoid such pitfalls, it is advisable to determine product distribution from chromatograms by using internal standard technique.

Another common way of introducing colinearities is to use the measured responses to *derive* other responses, e.g. to obtain a measure of selectivity. It is often seen that ratios of sums of responses are used this way. This causes another problem: The error distribution of such *derived* responses will rarely be approximated by the normal distribution. Hence, any attemps to evaluate the significance of the estimated effects by the t distribution or by the F distribution must be considered highly dubious.

It is strongly recommended that such derived responses should be avoided in the evaluation of experiments. Sums and differences of responses are acceptable as long as they are not "normalized", but do not use ratios or products of responses. If possible, use the raw data directly as they are measured.

A protective procedure against erroneous conclusions due to colinearities among the responses has been suggested in a prize-winning paper by Box et al.[4] Box recommends that multiple response matrices, Y, should always be analyzed by eigenvector decomposition of the matrix $Y'Y$. Eigenvalues equal to zero or of the same magnitude as the experimental error variance indicate the presence of linear dependencies among the response variables. Box advises that such dependencies should be removed prior to any model building. Modelling should always be done with linearly independent original responses or by using orthogonal (and hence, independent) linear combinations of the original responses. Such orthogonal linear combinations of the responses can be obtained by projecting the original responses on the eigenvectors corresponding to non-zero eigenvalues. This is equivalent to using the score vectors from the principal components analysis of **Y** as response variables.

Principal components analysis of the response matrix reveals the *systematic* variation of the responses over the set of experiments, and this is exactly what is desired in a screening experiment. As there will almost always be some missing data in the response matrix, the NIPALS algorithm is the method of choice for computing the principal components. The cross validation criterion ensures that only significant components are included in the model. The errors of the original responses are generally assumed to be approximately normally distributed. As the scores are linear combinations of the original responses, they will also have an approximate normal error distribution. This offers a simple graphic method of discerning significant effects of the variables from error noise variation in screening experiments also when the scores are used as responses, viz. plotting the estimated effects on normal probability paper.

In the example below, it is shown how this principle can be applied to the evaluation of significant variables in factorial and fractional factorial experiment.[5]

2.3. Example: Identification of significant experimental variables to control the product distribution in the synthesis of substituted pyridines from pyrylium salts

Treatment of 2-methyl-2-pentene with acetyl chloride in the presence of a Lewis acid affords a mixture of substituted pyrylium salts.[6] Further treatment of this mixture with aqueous ammonia yields a mixture of substituted pyridines, see formula scheme on next page.

To gain an insight into which experimental variables influence the course of the reaction, the authors varied four variables in a two-level factorial design. The variables are specified in Table 17.1, and the design is given in Table 17.2. The responses recorded were the yields of the various pyridines. The responses are summarized in Table 17.3.

Table 17.1: Variables and experimental domain in the synthesis of pyridines

Variables		Domain	
		−	+
x_1:	Amount of acetyl chloride/Lewis acid (mol/mol)	0.8	1.7
x_2:	Amount of acetyl chloride/olefin (mol/mol)	2.0	3.0
x_3:	Reaction temperature (°C)	0	25
x_4:	Type of Lewis acid	$FeCl_3$	$AlCl_3$

In the original work[6] the authors analyzed five derived responses, calculated from the originally recorded yields. The derived responses were analyzed one at a time. We shall see that the important variables can be identified from the principal component score calculated from the original yields.

Principal components analysis

Response y_{11} was deleted from the response matrix prior to analyis. It does not vary over the set of experiments. The remaining responses afforded one significant component according to cross validation. This component described 81.7 % of the total variance in the **Y** matrix. All responses were measured in the same units, percent yield, and were therefore not scaled prior to computing the principal components model. The data were, however, centred by removing the averages. The calculated scores are given in Table 17.2. The loadings showing the contribution of the responses to the systematic variation are given in the bottom line of Table 17.3.

A least squares fit of a second order interaction model (without the constant term, as the data had been centred prior to computing the PC model) to the score values afforded the estimated coefficients shown in Table 17.4.

A normal probability plot of the estimated coefficients is shown in Fig.17.2. It is seen that variables x_1 and x_4 are significant, and that they also show a significant interaction effect. These are the same conclusions as were reached in the original report, where five derived responses were anaylzed one at a time.

Additional examples of the use of this technique for the evaluation of screening experiments when there are many responses are given in Ref.[5]

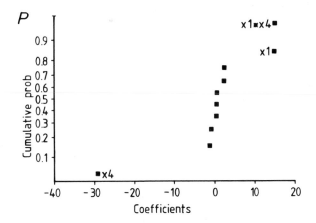

Fig.17.2: Normal probability plot of the estimated coefficients of the model determined from the score values

Table 17.2: Experimental design and score of the significant principal component

Exp no	Experimental variables				PC Score
	x_1	x_2	x_3	x_4	t_1
1	−	−	−	−	31.79
2	+	−	−	−	23.23
3	−	+	−	−	22.32
4	+	+	−	−	28.52
5	−	−	+	−	34.41
6	+	−	+	−	35.73
7	−	+	+	−	30.50
8	+	+	+	−	22.59
9	−	−	−	+	−51.40
10	+	−	−	+	−4.25
11	−	+	−	+	−67.01
12	+	+	−	+	2.66
13	−	−	+	+	−58.09
14	+	−	+	+	2.56
15	−	+	+	+	−57.06
16	+	+	+	+	3.54

Table 17.3: Yield of substituted pyridines

Exp no	Responses[a]										
	y_1	y_2	y_3	y_4	y_5	y_6	y_7	y_8	y_9	y_{10}	y_{11}
1	1.0	1.0	0	72.7	9.0	0.3	4.4	0	6.0	0	0
2	0	0	0	61.3	11.0	0.5	0.2	7.6	0.7	12	0
3	0	0	0	59.4	10.8	0.5	0.5	10.7	0.3	15.3	0
4	0	0	0	70.5	12.9	0.7	0.1	4.5	0.4	4.9	0
5	0.9	0.7	0	75.5	8.8	1.3	1.3	0.3	4.1	0.4	0
6	0	1.0	0	76.3	8.3	1.7	0	5.2	0	4.8	0
7	0.1	0.7	0	69.4	8.5	1.0	0.5	7.3	0.7	5.5	0
8	0.5	0.8	0	68.9	8.3	1.0	1.0	0.3	6.6	0.6	0
9	2.8	2.1	23.1	0	47.2	0	0.1	0.1	0.1	0	0
10	1.0	3.1	0	42.4	30.7	5.5	6.1	0.9	0.1	0	0
11	7.5	0	9.8	0	77.3	0	0	0	0	0	0
12	0	0.7	0	29.0	7.4	2.8	1.5	23.5	2.5	28.7	0
13	16.8	0.4	18.5	0	57.0	0.4	0.2	0.1	0.2	0.1	0
14	0	0.6	0	33.1	9.7	5.5	2.0	9.7	2.1	14.8	0
15	19.0	0.3	16.1	0	55.7	0	0	0.8	0	0	0
16	0.3	0.8	0	34.7	11.0	6.2	2.1	14.4	1.6	17.7	0
p^b	0.128	-0.001	-0.187	0.775	-0.581	-0.010	0.008	0.067	0.013	0.075	*

[a] Response variables y_i are the yield (%) of the pyridine "i" in the formula scheme above.
[b] Loadings of the first principal component.

Table 17.4: Estimated coefficients from the score values

Coefficient[a]	Estimated value[b]
b_1	14.32
b_2	-1.75
b_3	1.67
b_4	-28.63
b_{12}	1.75
b_{13}	0.02
b_{14}	15.44
b_{23}	-0.14
b_{24}	0.91
b_{34}	-0.40

[a] b_i is the coefficient of the linear terms of x_i, b_{ij} is the coeffiecient of the cross product terms $x_i x_j$.
[b] Significant coefficients have been shadowed. The estimated 95 % confidence limits is ± 3.47

3. PLS: Partial Least Squares Projections to Latent Structures

The PLS method was developed by Wold.[7] It has now been applied to problem solving in so many different areas of science and technology that it is not possible to give even a brief comment to them all. We shall see that PLS is a very useful tool, also in organic synthesis.

PLS is a modelling and computational method, by which quantitative relations can be established between blocks of variables, e.g. a block of descriptor data for a series of reaction systems (X block) and a block of response data measured on these systems (Y block). By the quantitative relation between the blocks, it is possible to enter data for a new system to the X block and make predictions of the expected responses. For example, if a reaction has been run in a series of solvents, we can use a PLS model to relate the properties of the solvents to the observed optimum conditions in these solvents. By subsequently entering the property descriptors of a new solvent to the PLS model it is possible to predict the optimum conditions of the reaction in the new solvent. For this to be efficient, it is necessary that the solvents used to determine the PLS model have a sufficient spread in their properties. To ensure this, a design in the principal properties is most useful.

The model can also, conversely, be used to find which conditions **X** give a certain desired **Y**-profile.

One great advantage of the method is that the results can be evaluated graphically by different plots. In most cases, visual interpretations of the plot are sufficient to obtain a good understanding of different relations between the variables.

3.1. Geometrical description of the PLS method

The principles behind PLS are simple and easily understood from a geometrical illustration. The method is based upon projections, similar to principal components analysis. The two blocks of variables are given by the matrices **X** and **Y**. The following notations will be used:

The number of independent variables (e.g. descriptors) in the X block is given by K, and the number of dependent variables in the Y block is given by M. The number of objects (e.g. experiments) both in the X block and the Y block is N. The X block is a $(N \times K)$ matrix and the Y block is a $(N \times M)$ matrix.

It is a common misunderstanding that in a statistical analysis the number of objects must exceed the number of variables. This is true of regression analysis but it is not true of PLS. Since PLS is based upon projections, we can easily have more variables than objects. Any object included in the analysis can be characterized by a large number of descriptors, and the outcome of the experiment can be characterized by several response variables. K and M do not have to exceed N. To analyze the interdependencies between the variables in the respective block, it is therefore possible to start with only a few objects. To obtain any PLS model, the number of objects should be at least five. It is evident that a model obtained from such a small number of objects will not, however, be very precise. The precision will increase with an increasing number of objects. The model can be updated and refined by including more experimental data as they become available. It is therefore possible to adopt a stepwise approach and start with relatively few objects in a first calibration set. Predictions of the expected results with new objects can be validated by experimental runs, and these experimental results can then be used to refine the model. This process is carried on until a satisfactory description has been obtained and satisfactory answers have been obtained to all questions posed to the experimental system.

As was discussed in Chapter 15, any matrix can be represented geometrically by a swarm of points in a multidimensional space. The K variables in the X space will thus define a K-dimensional X space, and the M response variables an M-dimensional Y space. The N objects define a swarm of N points in each space.

In PLS modelling as much as possible of the variation in the Y space should be modelled with a simultaneous modelling of the variations in the X space in such a way that these variations can be related to each other.

The modelling process constitutes a projection of the object points in each space down to PLS components. The projections are made so that the variations in the swarm of points are well described by the PLS components with the constraint that for each PLS dimension, j, the PLS scores of the Y block (denoted u_j) should have a maximum correlation to the scores of the X block (denoted t_j). The principles are illustrated by a one-component model in Fig.17.3.

Predictions by the PLS model, as illustrated in Fig.17.3, can be explained as follows: For an object "i", the corresponding x variables in the **X** matrix define a point in the X space. This point is projected on the first PLS(X) component to afford the score t_{1i}. From the correlation, called the *inner relation*, the corresponding score, u_{1i}, along the PLS(Y) component of the Y block is determined. This score corresponds to a point in the Y space, and this point in turn corresponds to the predicted $\hat{y}_i$ values of each response variable. These can then be compared to the observed responses in the **Y** matrix.

464

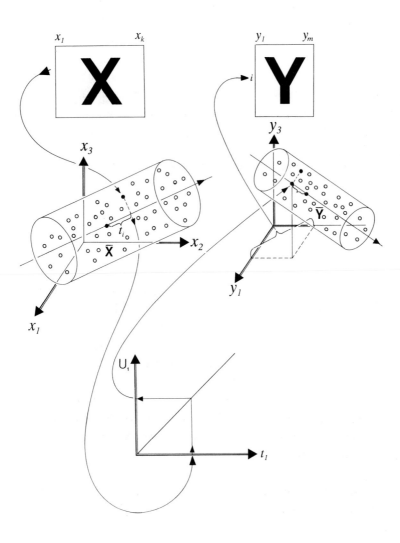

Fig.17.3: The PLS projections.

After fitting the first PLS component, there may still be systematic variation left in the Y space which can be described by a second PLS component. As in principal components analysis, the PLS components can be peeled off one dimension at a time, until the systematic variation in the Y space has been described. The model can be determined by cross validation to ensure valid predictions. The number of significant PLS dimensions is usually denoted by A.

The residual variance after fitting the model can be used to compute tolerance limits around the model. In Fig.17.3, these tolerance limits are illustrated by cylinders. To obtain reliable predictions for a new object, the corresponding object point should be within the tolerance limits of the model in the X space.

3.2. Mathematical description of the PLS model

In this section, the matrices **X** and **Y** denote the scaled and centred matrices of the variables in the X space and the Y space, respectively. Scaling to unit variance is usually employed.

The PLS model is defined by the following matrix relations, which involve factorization of the **X** and **Y** matrices into scores and loading matrices.

$$\mathbf{X} = \mathbf{TP'} + \mathbf{E} \qquad \text{(X block)}$$
$$\mathbf{Y} = \mathbf{UC'} + \mathbf{F} \qquad \text{(Y block)}$$
$$\mathbf{U} = \mathbf{DT} + \mathbf{H} \qquad \text{(inner relation)}$$

The matrices **T** and **U** are matrices of the score vectors, **P'** and **C'** are the transposed matrices of the loading vectors, and **E** and **F** are the matrices of the residuals. The inner relation which describes the correlation between the Y space and the X space is given by the **D** which is a diagonal matrix. The diagonal elements, $\{d_{ii}\}$, are the correlation coefficients for the linear relations between u_i and t_i. The matrix **H** is the residuals from the correlation fit.

In principal components analysis, the score vectors t_i are eigenvectors to **XX'**, whereas in PLS the first score vector of the X block is an eigenvector to **XX'YY'**. The first score vector of the Y block is an eigenvector to **YY'XX'**. Fot the second PLS dimension, the score vectors t_2 and u_2 are eigenvectors to $(\mathbf{X} - t_1 p_1')'(\mathbf{X} - t_1 p_1')$ and to $(\mathbf{Y} - u_1 c_1')'(\mathbf{Y} - u_1 c_1')$, respectively. If more PLS dimensions are desired, the corresponding score vectors are defined analogously as the preceding ones. It is, however, not necessary to compute the eigenvalues and the eigenvectors by diagonalization of matrices as the PLS alorithm, briefly described below, is based upon the NIPALS algorithm and computes the scores and loadings of both blocks simultaneously for one PLS dimension at a time.

There is one important matrix, **W**, which does not appear in the above matrix description of the model. This matrix is created during the computations by the PLS algorithm and is used to compute the **T**, **P**, **U**, and **C** matrices. The column vectors $(w_1, w_2, ..., w_A)$ in **W** are called the weight vectors and contain the PLS weights which show the contribution of each x variable to the description of the systematic variation in the Y space. For instance, $\mathbf{w}_1 = [w_{11} \, w_{12} \, ... \, w_{1K}]'$ shows the contribution of $x_1 - x_K$ to the first PLS component. The weight vectors are orthogonal to each other and normalized to unit length. The role of the weight vector is seen in the algorithm below.

PLS algorithm

A brief description of the algorithm is given here to show that there is nothing mysterious involved in the modelling process, for a detailed account, see Ref.[8]

Start with a guessed first score vector, $u_{1(0)}$, of the Y block, e.g. the column in **Y** with the largest variation.

(1) Compute a weight vector by

$$w_{1(0)}' = u_{1(0)}'\mathbf{X} \qquad \text{(normalize)}$$

(2) Compute the score vector of the X block:

$$t_{1(0)} = \mathbf{X}w_{1(0)}$$

(3) Use this score vector to compute the loading vector of the Y block:

$$c_{1(0)}' = t_{1(0)}'\mathbf{Y} \qquad \text{(normalize}^1\text{)}$$

(4) Use this loading vector to compute an updated score vector, $u_{1(1)}$, of the Y block:

$$u_{1(1)} = \mathbf{Y}c_{1(0)}$$

Check the convergence by computing:

$$\|u_{1(0)} - u_{1(1)}\|/u_{1(1)}'u_{1(1)} < \epsilon \text{ (e.g. } 10^{-10})$$

If the criterion is fulfilled, go to (5), else go back to (1) and use $u_{1(1)}$ as the guessed vector.

---

[1] If the computation of the loading vector is made as $t\mathbf{Y}/t't = c$ and c is left unnormalized, the correlation coefficient between **t** and **u** vectors will be one (the matrix **D** is then the unit matrix).

(5) After convergence, compute the PLS loadings of the X block:

$$p_1' = t_1'X \qquad \text{(normalize)}$$

Compute the first correlation coefficient d_{11} of the inner relation, diag$\{d_{ii}\}$:

$$d_{11} = t_1'u_1/(t_1't_1)$$

(6) Remove the variation described by the first PLS component and determine the matrices E and F of the residuals:

$$E = X - t_1p_1'$$
$$F = Y - d_1t_1c_1'$$

Use E and F as starting matrices for the next PLS dimension, and go to (1).

The above algorithm has been modified in different ways, and there are several variants developed for other types of modelling problems. For details on this, see the dissertation by Martens in the reference list.

4. Plots from the PLS model

4.1. Score plots

A plot of the score vectors against each other, e.g. t_2 against t_1, shows the positions of the projected object points in the plane spanned by the PLS vectors in ths X space, and a plot of u_2 against u_1 shows the same for the Y space. These plots are therefore similar to the score plots from principal components analysis.

A plot of u_1 aginst t_1 shows the linear relation between the PLS components. The slope is the coefficient of the inner relation.

4.2. Loading plots and weight plots

Loading plots are obtained by plotting the loading vectors against each other, similar to loading plots in principal components analysis. In PLS there is one more trick available. It is possible to plot, for instance, (p_2 and c_2) against (p_1 and c_1) in the same plot. Such combined loading plots are very useful for the understanding of how the variables in the Y space and the X space are related to each other. Variables which are strongly related will have their loadings plotted either close to each other or opposite to each other with respect to the origin. It is therefore possible to use the combined loading plots to determine how the independent (x) variables should be adjusted to afford a desired change in the dependent (y) variables. An example on the use of loading plots in the optimization of an enamine synthesis is given below.

Weight plots are obtained by plotting the weight vectors, w_i, against each other. As the weight vectors are mutually orthogonal, these plots show the independent contribution of the x variables to the description of the systematic variation in the Y space. Weight plots are therefore useful when we wish to understand the roles of the independent variables in the reaction under study. A plot of w_2 against w_1 will be similar, but not identical to a plot of p_2 against p_1. The w_i vectors are orthonormal, while the p_i vectors are not. An example on the use of weight plots in a study of the Fischer indole synthesis is given below.

5. Examples on the use of PLS modelling in organic synthesis

PLS is a flexible tool which can be used to describe interdependencies of variables in many situations encountered when synthetic reactions are explored. In the examples given below, it is shown how PLS can be used for linking together *the reaction space, the experimental space* and *the response space*. The following examples are discussed:

* *The experimental space linked to the response space*, illustrated by a study on the suppression of a by-product in the synthesis of an enamine.

* *The reaction space linked to the experimental space and the response space*, illustrated by prediction of optimum conditions for new substrates in the Willgerodt-Kindler reaction.

* *The reaction space linked to the response space*, illustrated by a study on the Fischer indole synthesis. The objective of the study was to determine which properties of the reaction system are necessary to control with a view to obtaining regioselectivity in the reaction.

5.1. Suppression of a by-product in the synthesis of an enamine

The same example as was discussed in section 12.8.4 will be used here to show that also PLS can be used to identify which variables must be adjusted to maximize the yield of the desired product with a simultaneous suppression of the formation of a by-product.

The reaction studied was the synthesis of the morpholine enamine from 3,3-dimethyl-2-butanone. The undesired by-product was formed by self-condensation of the ketone.

The experimental design matrix and the yields of the enamine and the by-product were used to define the X block and the Y block, respectively. These data are recapitulated in Table 17.5.

In the X block for the PLS model were also included the squared variables x_{ii}^2 and the cross-product variables $x_i x_j$. The Y block contained the two response variables.

Prior to computing the PLS model, the blocks were centred by subtracting the averages, and the variables were scaled to unit variance over the set of experiments. For the scaled and mean centred x variables, the following notation is used: The scaled variable from x_i is denoted $\tilde{x}_i$; the scaled variable from the squared variable x_i^2 is denoted $\tilde{x}_{ii}$; the scaled variable from the cross-product $x_i x_j$ is denoted $\tilde{x}_{ij}$. For the scaled response variable, y_i, the symbol $\tilde{y}_i$ is used.

A two-component PLS model was sigificant (cross validation) and accounted for 96.2 % of the variance in the Y block. The first PLS component described 47.4 % of the variance, and the second component 48.8 %.

Evaluation from the mathematical expression of the model

We shall first go through the mathematical description of the PLS model, just to show that there is nothing mysterious involved. Such a detailed treatment is by no means necessary for the evaluation of the result.

The PLS scores are defined by the following relations:

$t_1 = 0.1814\,\tilde{x}_1 - 0.9711\,\tilde{x}_2 - 0.1317\,\tilde{x}_3 - 0.0622\,\tilde{x}_{11} + 0.0315\,\tilde{x}_{22} -$
$\quad 0.2267\,\tilde{x}_{33} + 0.0326\,\tilde{x}_{12} - 0.0149\,\tilde{x}_{13} + 0.0057\,\tilde{x}_{23}$

$t_1 = 0.7863\,\tilde{x}_1 + 0.1249\,\tilde{x}_2 + 0.4986\,\tilde{x}_3 - 0.1648\,\tilde{x}_{11} - 0.1241\,\tilde{x}_{22} - 0.0728\,\tilde{x}_{33}$
$\quad + 0.2199\,\tilde{x}_{12} + 0.0951\,\tilde{x}_{13} + 0.1111\,\tilde{x}_{23}$

$u_1 = -0.6637\,\tilde{y}_1 - 0.7480\,\tilde{y}_2$

$u_2 = 0.7600\,\tilde{y}_1 - 0.6499\,\tilde{y}_2$

Table 17.5: Experimental design and responses in the PLS study on by-product formation

Exp no	Variables[a]			Responses[b]	
	x_1	x_2	x_3	y_1	y_2
1	−1	−1	−1	41.6	14.6
2	1	−1	−1	45.1	6.7
3	−1	1	−1	51.7	26.2
4	1	1	−1	64.7	17.7
5	−1	−1	1	47.8	11.9
6	1	−1	1	57.1	7.5
7	−1	1	1	63.0	26.1
8	−1	1	1	77.8	11.0
9	1.414	0	0	66.7	8.1
10	−1.414	0	0	49.5	22.2
11	0	1.414	0	70.4	18.9
12	0	−1.414	0	43.9	8.0
13	0	0	1.414	66.4	9.8
14	0	0	−1.414	52.4	17.3
15	0	0	0	56.5	13.8
16	0	0	0	60.0	12.3
17	0	0	0	58.6	12.6
18	0	0	0	57.2	13.6

[a] The experimental domain is shown in Table 12.12, p. 307, and is not reproduced here. The variables are: x_1, the amount of morpholine; x_2, the amount of titanium tetrachloride; x_3, the reaction temperature.
[b] The responses are the yields (%) of: y_1, the enamine; y_2, the by-product.

The coefficients in the above relations are the elements of the loading vectors, p_1 and p_2 for the X block, and c_1 and c_2 for the Y block. These loading matrices are therefore:

$$
\begin{array}{cc}
& \begin{array}{cc} p_1 & p_2 \end{array} \\
P = & \begin{bmatrix}
0.1814 & 0.7863 \\
-0.9711 & 0.1249 \\
-0.1317 & 0.4986 \\
-0.0622 & -0.1648 \\
0.0315 & -0.1241 \\
-0.2267 & -0.0728 \\
0.0326 & 0.2199 \\
-0.0149 & 0.0951 \\
0.0057 & 0.1111
\end{bmatrix}
\end{array}
\qquad
\begin{array}{cc}
& \begin{array}{cc} c_1 & c_2 \end{array} \\
C = & \begin{bmatrix}
-0.6637 & 0.7600 \\
-0.7480 & -0.6499
\end{bmatrix}
\end{array}
$$

The inner relation between the score vectors are

$u_1 = 0.9751\, t_1 + e$

$u_2 = 0.8802\, t_2 + e$

The diagonal matrix which defines the inner relation between the score vectors is therefore:

$$
D = \begin{bmatrix}
0.9751 & 0 \\
0 & 0.8802
\end{bmatrix}
$$

Remember that the PLS model in matrix notation is written

$X = TP' + E$

$Y = UC' + F$

$U = DT + H$

The plots of the score vectors u_i aginst t_i are shown in Fig.17.4.

From the loadings of the Y block, the following approximate relations between the responses and the Y block scores can be obtained. The matrix C is not exactly orthogonal, but nearly so.

$\tilde{y}_1 = -0.5499\, u_1 + 0.7600\, u_2$

$\tilde{y}_2 = -0.7480\, u_1 - 0.6637\, u_2$

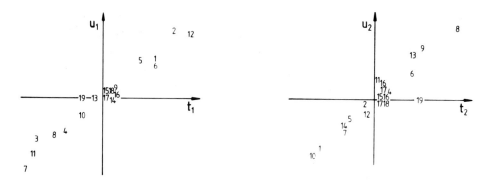

Fig.17.4: Plots of PLS scores.

It is seen that in order to increase the yield of enamine, $\tilde{y}_1$, the experimental conditions should be adjusted so that u_1 moves in the negative direction of the first PLS(Y) component, while u_2 moves in the positive direction on the second PLS(Y) component. A decrease in the yield, $\tilde{y}_2$, of the by-product is expected if both u_1 and u_2 are moved in the positive direction. For u_1, the conclusions are conflicting. Let us continue the analysis and see what can be done to solve this conflict.

As the u and t vectors are parallel, we shall adjust the t vectors accordingly. From the loadings of the X block it is seen that t_1 is mainly related (negative loading) to $\tilde{x}_2$ (the amount of titanium tetrachloride). A decrease in t_1 with a view to improving $\tilde{y}_1$ (but with the risk of aggravating $\tilde{y}_2$) can therefore be accomplished by increasing the amount of titanium terachloride. For the second component, two variables of the X block are largely responsible for the variation of t_2, viz. $\tilde{x}_1$ (the amount of morpholine), and $\tilde{x}_3$ (the reaction temperature). Both these variables should be increased in order to increase t_2.

The conclusion of the above analysis is, that an improved result can be expected by increasing the amount of morpholine, and by increasing the temperature. The amount of titanium tetrachloride to use should be further analyzed to determine conditions which give an acceptable level of the by-product.

This is exactly the same conclusions as were reached by analysis of the response surface models, as was discussed in chapter 12. The difference between the two approaches to the problem is that PLS takes both responses into account in the *same* model.

Evaluation by plots

Analysis by the algebraic expression of the PLS model is not easy. Fortunately, it is rarely necessary to carry out such an analysis as the same conclusions can be reached by analysis of plots. Such analyses are much more easy to make. The conclusions above follow immediately from an analysis of the plot of the combined loadings against each other. Fig. 17.5 shows the plot of (p_2 and c_2) against (p_1 and c_1).

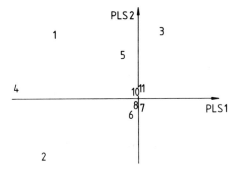

Fig.17.5: Combined loading plot of the PLS components.The numbers represent the following variables: 1 ($\bar{y}_1$), 2 ($\bar{y}_2$), 3 ($\bar{x}_1$), 4 ($\bar{x}_2$), 5 ($\bar{x}_3$), 6 ($\bar{x}_{11}$), 7 ($\bar{x}_{22}$), 8 ($\bar{x}_{33}$), 9 ($\bar{x}_{12}$), 10 ($\bar{x}_{13}$) and 11 ($\bar{x}_{23}$)

The loading plots are projections of the *variables* in the X space and the Y space. Variables which contribute to the first PLS dimension (PLS1) are displayed along the PLS1 axis. Variables with minor influence are close to the origin. It is seen in Fig.17.5 that PLS1 contributes to the description of both $\bar{y}_1$ and $\bar{y}_2$ and that the variation in this dimension is largely described by $\bar{x}_2$. Here comes the interesting use of the laoding plots: If we increase $\bar{x}_2$ we can say that this corresponds to a movement in the positive direction (to the right) along the PLS1 axis. This will also correspond to a movement to the right of the associated response variables, i.e. both $\bar{y}_1$ and $\bar{y}_2$ will increase. In the second PLS dimension, it is seen that $\bar{x}_1$, $\bar{x}_3$, and $\bar{y}_1$ are projected at the positive end of the PLS2 axis, while $\bar{y}_2$ is on the opposite side of the origin and at the negative end. An increase in $\bar{x}_1$ and $\bar{x}_3$ should therefore result in an increase in $\bar{y}_1$ *and* a decrease in $\bar{y}_2$. The remaining variables are close to the origin and do not contribute much to the description of the systematic variation.

These are the same conclusions as were reached above. It is seen that a graphic evaluation is much simpler than a detailed analysis of the algebraic expression of the model.

Experimental design

It is desirable that the PLS model covers the X space as well as possible. This implies that the experimental design is very important (as always!). In this example, a composite design was used which made it possible also to consider quadratic variations.

If a PLS model is determined with as many PLS dimensions as there are independent variables in the X block, this simply means that the original coordinate system of the X space is rotated and translated into the system spanned by the PLS components. It will then be possible to rotate the PLS description of the dependencies between the variables into the description given by multiple linear regression. This means that the PLS method includes the method of multiple linear regression as a limiting case.

Hence, it is evident why a proper design is necessary for discerning the roles played by the variables. All arguments presented in the initial chapters of this book, as to the importance of a good design in multiple regression, therefore also apply to PLS.

5.2 Prediction of optimum experimental conditions for new substrates from a series of test substrates

Yet another example on the Willgerodt-Kindler reaction is shown.[9] It also illustrates that a stepwise approach to the study of discrete variations is possible by PLS.

A preliminary study had shown that the reaction failed with strong electron withdrawing substituents (e.g. NO_2, CN), and the study was limited to include *donor*, *alkyl*, and *halogen* substituents.[2]

---

[2] Analysis of NMR spectra of substituted aromatics by principal components analysis had shown that the substituents could be grouped into four stable classes: Donors, acceptors, alkyls, and halogens. It was also shown that these groups were different to each other in their influence on the spectra.[10] As the NMR spectra reflect the electronic properties of the molecules, it was assumed that these groups would also represent different properties with regard to their chemical behaviour in chemical reactions.

In the initial set of test compounds five *para* substituted acetophenones were selected (Y = H, Me, Cl, MeO, Me$_2$N) to span a variation along the "substrate" axis in the reaction space.

The experimental conditions for optimum yield were determined by response surface technique for each of these substrates. The optimum conditions are summarized in Table 17.6, Entry 1−5.

To characterize the compounds, common substituent parameters, summarized in Table 17.7, were used.

A first PLS model was determined from Entry 1-5 in Table 17.6. The X block contained the corresponding substituent parameters from Table 17.7. The Y block contained the optimum settings of the natural variables, z_1 - z_3, and the experimental optimum yield, y_{Obs}. Three components accounted for 80.4 % of the variance in the Y block. This shows that there *is* systematic variation, but nothing more with that few objects. The model was then used to predict the optimum conditions for the fluoro substituted acetophenone, Entry 6*. These predictions were checked by determining the response surface model for the fluoro compound. The predicted optimum by the response surface model is given under Entry 6, together with the experimental yield. The method for computing the predictions is described below.

The X block was augmented by the descriptors for the fluoro substituent, and the Y block was augmented by the predicted optimum conditions from the response surface model, and the corresponding experimental yield. The model was recalculated and then used to predict the result with the bromo compound, Entry 7*. Validation of the predictions by response surface modelling and experimental confirmation of optimum predicted by the response surface model is shown in Entry 7. Augmenting the X block and the Y block and recalculation of the PLS model afforded the predictions for *p*-methylthioacetophenone. An experimental yield of 95 % was obtained under the optimum conditions predicted by the PLS model. The predicted yield by the PLS model was 97.7 %. It was assumed that the experimental yield was reasonably close to the maximum yield, and that the predictions were reliable.

Table 17.6: Optimum conditions in the Willgerodt-Kindler reaction. Predictions of optimum conditions by RSM and PLS, and the experimental yields obtained under predicted optimum conditions

Entry	Subst.	Predicted optimum[a]			Method	Yields (%)	
		z_1	z_2	z_3		y_{Pred}	y_{Obs}[b]
1	Me	9.6	8.4	133	RSM	96.0	96
2	MeO	9.3	8.9	130	RSM	98.2	100
3	H	7.5	10.3	123	RSM	90.8	94
4	Cl	9.7	9.9	119	RSM	95.0	100
5	Me$_2$N	8.8	8.3	122	RSM	89.0	89
6*	F	7.8	11.0	112	PLS	89.3	---
6	F	8.3	10.6	116	RSM	94.0	93
7*	Br	10.4	9.3	123	PLS	98.4	---
7	Br	10.2	9.5	121	RSM	95.0	95
8	MeS	10.4	8.4	124	PLS	97.0	95

[a] To avoid confusion with score vectors, the natural variables are denoted by z_i: z_1, The amount of sulfur/ketone (mol/mol); z_2, the amount of morpholine/ketone (mol/mol); z_3, the reaction temperature ($^\circ$ C)
[b] The experimental yield was determined by gas chromatography (internal standard technique) directly on the reaction mixture.

Table 17.7: Descriptors of the substituents

Subs	Descriptors													
	1	2	3	4	5	6	7	8	9	10	11	12	13	14
Me	−0.01	−0.17	−0.04	−0.13	−1.24	0	0.52	3.00	1.52	1.90	1.90	2.04	5.65	0.56
MeO	0.3	−0.27	0.26	−0.51	−0.55	−0.23	0.36	3.98	1.35	1.90	1.90	2.78	7.87	−0.02
H	0	0	0	0	0	0.32	0	2.06	1.00	1.00	1.00	1.00	1.03	0
Cl	0.47	0.23	0.41	−0.15	−0.97	−0.56	0.55	3.52	1.80	1.80	1.80	1.80	6.03	0.71
Me$_2$N	---	−0.83	0.10	−0.92	---	---	0.43	3.53	1.50	2.56	2.80	2.80	15.55	0.18
F	0.54	0.06	0.43	−0.34	−0.46	−0.14	0.27	2.65	1.35	1.35	1.35	1.35	0.92	0.14
Br	0.47	0.23	0.44	−0.17	−1.16	−0.84	0.65	3.83	1.95	1.95	1.95	1.95	8.88	0.86
MeS	0.3	0	0.2	−0.18	−1.07	−0.75	0.64	4.3	1.70	1.90	1.90	3.26	13.82	0.61

Descriptors: 1 = σ_I (Taft inductive parameter); 2 = σ_p (Hammett parameter for *para* substituent); 3 = *F* and 4 = *R* (Swain-Lupton dual substituent); 5 = E_S and 6 = $E_S c$ (Taft steric parameters); 7 = v (van der Waals radius); 8 = *L*, 9 = B_1, 10 = B_2, 11 = B_3, 12 = B_4 (Verloop sterimol parameters); 13 = *MR* (molar refractivity); 14 = π (Hansch lipophilicity parameter).

Predictions

When the PLS model is used to predict responses for a new object, "*i*", its scores in the X space, $t_{1i} - t_{Ai}$, must first be determined by projection down to the PLS(X) components. This is accomplished by the weight matrix **W**. Multiplication of **W** by the row vector of the descriptors afford a column vector of the corresponding scores:

$$[x_{1i}\, x_{2i}\, \dots\, x_{2i}]\ \mathbf{W} = \begin{bmatrix} t_{1i} \\ t_{2i} \\ \cdot \\ \cdot \\ t_{Ai} \end{bmatrix}$$

The t_{ji} score values are then entered into the inner relation to predict the corresponding u_{ji} scores. These are then used for predictions of the responses.

To obtain reliable predictions, the object point "i" should be within the tolerance limits of the model in the X space.

Experimental design

To achieve good predictions with the PLS model, it is much safer to make these by interpolation than by extrapolation. It is therefore important that the calibration set used to establish the PLS model should cover the range of the variation of interest in the X space. To ensure this, a design in the principal properties is helpful. When a stepwise approach to a problem is desired and a small initial set of test system is selected, the design used to select the calibration set should therefore be a "maximum spread" design. Validation experiments can then be selected from the interior of the reaction space. In this way it is possible to augment the "maximum spread" design into a "uniform spread" design, if this is of interest. A PLS model established by a "uniform spread" design can be used to analyze the relations to the properties of the reaction space in more detail. It is never known beforehand whether a new synthetic reaction is interesting enough to motivate detailed studies. A stepwise strategy as indicated above may then be useful.

5.3. Properties of the reaction system related to the selectivity of a reaction. Regioselectivity in the Fischer indole synthesis.[11]

Background

This example describes how a PLS model can be used to identify those properties of a reaction system which determine the selectivity of a reaction. This technique has been applied to the Fischer indole synthesis. A short description of the reaction is:

A carbonyl compound, aldehyde or ketone, is condensed with phenylhydrazine to yield a phenylhydrazone. In the presence of an acid, the phenylhydrazone tautomerizes into an enehydrazine which undergoes a sigmatropic rearrangment followed by ring closure and elimination of ammonia to yield an indole.

The reaction was discovered by Fischer in 1884 [12] and has been extensively used for the preparation of indoles. The reaction is of wide scope and a recent monograph[13] on its applications contains more than 2700 references.

The reaction is catalyzed by acids, and over the years a number of both proton acids and Lewis acids has been suggested as preferred catalysts.

A problem when the reaction is applied to dissymmetric ketone substrates with both α and α' methylene groups is that two isomeric indoles are possible.

Test systems

It has been suggested that the problem of selectivity can be overcome, so that the reaction can be controlled to yield only one isomer by a proper choice of acid catalyst and solvent. However, the literature is bewildering and contradictory on this point. With a view to clarifying this problem, the following study was undertaken: A series of test systems was selected to span a large variation in the reaction space. The objective was to determine suitable combinations of catalysts and solvent which would yield give a reliable selectivity.

The number of combinations of possible carbonyl substrates, substituted phenylhydrazines, acid catalysts, and solvents is overwhelmingly large. The present study was limited to include dissymmetric ketone substrates with α and α' methylene groups, phenylhydrazine, Lewis acids, and common solvents. The selection of test systems was based on the principal properties of the ketone, the Lewis acids, and the solvents. The selection was made to achieve approximately uniform distributions of the selected items in the score plots, see Fig.17.6.

Five ketones, twelve Lewis acids and ten solvents were selected. These give a total of 600 different combinations. From these, 296 systems were selected and examined by experimental studies. The Fischer indole reaction was obtained with 162 of these systems. The PLS study was based on the 162 successful systems, in the hope that they would reveal a systematic pattern. A schematic illustration of the distribution of these systems in the reaction space is shown in Fig.17.7.

Analysis

To characterize the reaction system for the PLS analysis, the principal property score values were used as descriptors. For the ketones, two additional descriptors were used to describe the steric environment of the carbonyl group. The v parameter given by Charton[14] was used to describe the size of the ketone side chain. This parameter is a measure of the van der Waals radius of the substituent, and can be regarded as a measure of how large the side chain appears to be when

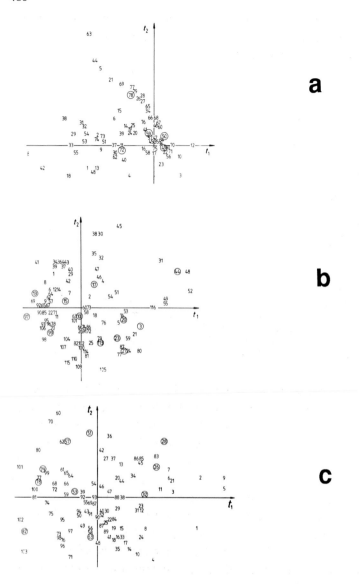

Fig.17.6: Selection of test items in the Fischer indole synthesis: *(a)*
Ketones: 3-Hexanone (50), 2-hexanone (49), 3-undecanone (78),
1-phenyl-2-butanone (72), 5-methyl-3-heptanone (59);
(b) Lewis acids: BF_3 (44), CuCl (13), ZnI_2 (99), $TiCl_4$ (3), $ZnCl_2$ (15),
$SbCl_5$ (27), PCl_3 (113), CuI (97), $FeCl_3$ (10), $SiCl_4$ (20), $AlCl_3$ (17), $SnCl_4$ (23);
Solvents: Sulfolane (28), carbon disulfide (78), *N,N*-dimethylacetamine (32),
quinoline (51), 1,2-dichlorobenzene (57), dimethyl sulfoxide (26),
carbon tetrachloride (79), chloroform (53), tetrahydrofuran (63), hexane (82).

viewed from the position of the carbonyl group. Attempts at using the van der Waals volumes of the side chains afforded poorer PLS models. The descriptors of the reaction systems are summarized in Table 17.8.

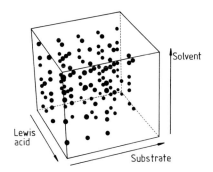

Fig.17.7: Distribution of the successful reaction systems in the reaction space. The axes have been graduated by ordinal scales: Five ketones were selected, and the ketones are represented by five equally spaced divisions of the "substrate" axis. Similarly, the Lewis acids and the solvents correspond to twelve, respectively ten equally spaced divisions of the "reagent" and the "solvent" axes.

The reactions were allowed to proceed for 48 h and the course of the reaction was monitored by gas chromatography. The relative distribution of indole isomers was not changed over time. It was therefore concluded that the isomer distribution was dependent on the difference in the rates of formation of the isomers. The isomer distribution was determined from the integrated peak areas of the isomers in the gas chromatograms. The response used for the PLS analysis was the *regioisomeric excess, RE* = Amount (%) of major isomer − amount (%) of minor isomer. An account of the identities of the indole isomers is given in Ref.[11], but we will not enter into details of this here, since it is not needed for the PLS analysis.

The descriptors given in Table 17.8 were used to define the X block.[3] To take non-linear effects and possible interaction effects into account, also the squares and

---

[3] A first PLS model was established from 124 reaction systems. To ensure that this set of reaction systems was not selected in such a way that the descriptor variables were correlated, a principal component analysis was made of the variation of the eight descriptors over the set. This analysis afforded eight significant principal components according to cross validation. This showed that the variance-covariance matrix of the descriptors was a full rank matrix and that there were no severe colinearities among the descriptors.

the cross-products of the descriptors were included as variables. This gives a total of 44 variables in the X block. These variables are numbered as follows: Variables *1, 2* are the principal property descriptors t_1 and t_2 of the Lewis acids; *3, 4* are the principal properties of the solvents; *5, 6* are the principal properties of the ketones; *7, 8* are the steric parameters of the ketones; *9 - 16* are the squared variables from *1 - 8*; *17 - 44* are the cross-product *1*2 - 7*8*. These numbers are used to identify the variables in the discussions below.

The analysis was carried out in two steps: The first step was a calibration of an initial PLS model from a set of 124 different reaction systems. The set was defined by the first 124 systems given in *Appendix 17A*. The initial model was then validated by predictions for the remaining 38 systems. The correlation between y^{Pred} and $y^{Observed}$ of these systems afforded a correlation coefficient of 0.93 . This was considered a satifactory result. We will proceed directly to the final analysis, and will not enter into details on this correlation.

The final analyis was made from the set of all 162 systems. A three-component PLS model was significant according to cross validation. The model accounted for 86.9 % (64.1 + 16.3 + 6.4 %) of the total variance of the response.

To determine *which* variables in the X block contributed to the description of the systematic variation of the regioselectivity, the plots of the PLS weights were used. These plots are shown in Fig.17.8. Variables with a high degree of contribution are projected at the periphery of the plots. Variables with no or minor influence are projected close to the origin.

There is another criterion, *the modelling power*, which can be used to assess the contribution of a variable to the model. The criterion is based on the residuals between the object point and its projected point. The residuals in the X space describe to what extent the variation in the independent variables is not described by the PLS(X) components. The residual sum of squares can be partitioned over the different variables, to give a residual variance, $s^2(x_i)$ for each variable x_i. Variables associated with large residual variances will therefore contribute only marginally to the description of the systematic variation. Let $s_0^2(x_i)$ be the variance of the variable before the model was fitted. The modelling power, MPOW, is defined

$$MPOW(x_i) = 1 - [s^2(x_i) / s_0^2(x_i)]^{1/2}$$

The modelling power is zero for variables which do not at all contribute to the model. It is equal to one for variables which are fully used to describe the variation. The modelling power of the variables in the PLS model of the Fischer indole systems are summarized in Table 17.9. To simplify the table, only variables with MPOW > 0.1000 are given.

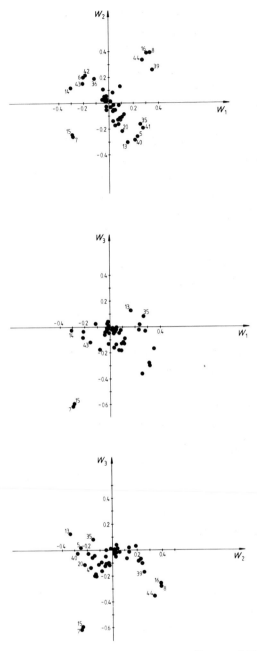

Fig.17.8: Plots of the PLS weights. Significant variables are projected at the periphery of the plots, while insignificant variables are projected close to the origin.

Table 17.8: Descriptors used to characterize the reaction systems in the study of the Fischer indole reaction.

System constituent	Principal properties		Steric parameter, v	
	t_1	t_2	R^1CH_2	R^2CH_2
Ketones:				
3-Hexanone	2.34	−0.16	0.68	0.56
2-Hexanone	2.46	−0.15	0.52	0.68
3-Undecanone	−0.52	2.16	0.68	0.56
1-Phenyl-2-butanone	−0.74	−0.42	0.56	0.70
5-Methyl-3-heptanone	1.17	0.74	0.56	1.00
Sovents:				
Sulfolane	3.29	2.21		
Carbon disulfide	−3.27	0.98		
N,N-Dimethylacetamide	1.81	0.15		
Quinoline	−0.29	3.13		
1,2-Dichlorobenzene	−0.99	2.27		
Dimethyl sulfoxide	2.54	0.95		
Carbon tetrachloride	−2.49	0.38		
Chloroform	−1.56	−0.59		
Tetrahydrofuran	−1.04	−1.57		
Hexane	−3.00	−1.20		
Lewis acids:				
BF_3	7.05	2.80		
CuCl	−3.42	1.33		
ZnI_2	−2.25	−1.77		
$TiCl_4$	4.35	−1.48		
$ZnCl_2$	−1.11	0.79		
$SbCl_5$	3.07	−3.18		
PCl_3	1.32	−2.70		
CuI	−4.00	−0.47		
$FeCl_3$	−0.06	−0.48		
$SiCl_4$	3.18	−0.78		
$AlCl_3$	0.97	1.83		
$SnCl_4$	2.20	−2.22		

Table 17.9: Modelling power of the variables in the PLS model of the Fischer indole reaction systems

Variable no	Identification	Variable in model	MPOW	Variable no	Variable in model	MPOW
3	Solvent, t_1	3	0.1661	33	3*7	0.1686
4	Solvent, t_2	4	0.1031	34	3*8	0.1763
5	Ketone, t_1	5	0.4472	35	4*5	0.3644
6	Ketone, t_2	6	0.3226	36	4*6	0.1557
7	Ketone, v_1	7	0.4643	37	4*7	0.1082
8	Ketone, v_2	8	0.7892	38	4*8	0.1214
12		4^2	0.1082	39	5*6	0.4361
13		5^2	0.5993	40	5*7	0.4308
14		6^2	0.4320	41	5*8	0.4250
15		7^2	0.4643	42	6*7	0.3205
16		8^2	0.7825	43	6*8	0.3241
20		1*5	0.1077	44	7*8	0.7501

From the plots in Fig.17.8 and from Table 17.9 it is seen that by far the most important factor for controlling the regiochemistry is, not unexpectedly, the steric properties of the ketone. Also the principal properties of the ketones and of the solvents participate in the model. The most surprising result is, that the properties of the Lewis acids do not contribute *at all* to the model. The conclusions must therefore be, that the properties of the Lewis acid catalyst do not have any influence on the regiochemistry of the Fischer indole reaction. This result shows that it is not possible to achieve general control of the regioselectivity of the Fischer indole synthesis *in solution* by the choice of the Lewis acid catalyst and the solvent. Previous claims in the literature in this direction are therefore incorrect. For certain substrates the choice of catalyst and solvent may well give a selective reaction, but this is not valid as a general principle of the method.

The results obtained in the above analysis point in another direction. As the steric effects play a dominant role for the selectivity, one possibility to achieve regiocontrol would be to amplify the effects of steric congestion and force the reaction to occur close to, or on, the surface of a solid acid catalyst. In this context, zeolites were considered as potentially useful catalysts. Zeolites have very specific surface geometries and clearly defined cavities and channels. The assumptions that selective transformations in the Fischer indole reaction might be achieved by zeolite catalysis could be fully confirmed in a subsequent study. It was found that the mode of ring closure was highly dependent on the type of zeolite used, and that the choice of catalyst offered a means of influencing and to some extent also controlling the regioselectivity of the reaction.[15]

Conclusions from the example on Fischer indole synthesis

The above example shows a selection of test systems by a design affording a uniform spread in the principal properties. The objective was to establish whether there was a gradual change in the performance of the reaction which could be related to the properties of the reaction system. The aim was to determine those properties which have an influence on the selectivity so that these properties could be controlled. Both these objectives were attained. The results would have been very difficult to achieve without the PLS method and without using a multivariate design for selecting test items.

It is true that the above example is quite extreme with respect to the number of investigated systems. The study was made with a view to clarifying a controversial question, and to do so a sufficiently large number of degrees of freedom in the statistical analysis was desired. This was provided by the above example.

6.   Conclusions

When a chemical reaction is studied with the objective to develop a synthetic method, a number of questions must be answered. At the beginning of this process, very little is known. Gradually, a better knowledge is aquired. In this learning process, experiments play a vital role. Inferences from experimental observation will, hopefully, lead to a new a useful synthetic method. Many questions arise due to the multidimensionality of the problems encountered when a synthetic reaction is evaluated by experimental studies.

In this chapter it has been shown how quantitative relations can be established between the observed responses and the variations induced by the experiments. It has also been shown how such relations can be evaluated to provide the answers to many different types of questions.

The most flexible tool in this context is the PLS method, although there are other methods which can be used in certain situations. In one example it was shown how a combination of multiple regression and principal components analysis can be used in screening experiments with many responses.

In the general case, the PLS method is the preferred choice. It can be used to link the variations in the experimental space, the reaction space, and the response space. As the PLS method does not involve any assumptions as to cause-effect relations, it is in that respect a method for unprejudiced analysis.

Analysis of the PLS model will reveal regularities in the experimental observations. This is, however, possible only if the experiments have been properly designed. The appropriate experimental design is dependent on the problem. For explorations of the reaction space, such designs can be accomplished by the principal properties.

As PLS models can be updated and refined when more experimental observations become available, it is possible to adopt a stepwise approach to solve the problems.

References

1. H. Hotelling
   J. Educat. Psycol. 24 (1933) 417, 489.

2. I.T. Jolliffe
   Appl. Statist. 31 (1982) 300.

3. F.H.C. Mariott
   The Interpretation of Multiple Observation
   Academic Press, London 1986.

4. G.E.P. Box, W.G. Hunter, J.F. McGregor and J. Erjavec
   Technometrics 15 (1973) 33.

5. R. Carlson, Å. Nordahl, T. Barth and R. Myklebust
   Intell. Lab. Syst. (submitted).

6. A. Diallo, H. Hischmueller, M. Arnaud, C. Roussel, D. Mathieu and R. Phan-Tan-Luu
   Nouv. J. Chim. 7 (1983) 433.

7. H. Wold
   Chapter 1: *Soft Modelling. The Basic Design and Some Extensions*
   in K.-G. Jöreskog and H. Wold (Eds.)   '
   Systems Under Indirect Observation, Vols I, II
   North Holland, Amsterdam 1982.

8. S. Wold, A. Ruhe, H. Wold and W.J. Dunn III
   SIAM J. Scient. Statist. Comput. 5 (1984) 735

9. R. Carlson, T. Lundstedt and R. Shabana
   Acta Chem. Scand. B 40 (1986) 534.

10. S. Alunni, S. Clementi, U. Edlund, D. Johnels. S. Hellberg, M. Sjöström and S. Wold
    Acta Chem. Scand. B 37 (1983) 47.

11. M.P. Prochazka and R. Carlson
    Acta Chem. Scand. 43 (1989) 651.

12. E. Fischer and V.F. Hess
    Ber. dtsch. Chem. Ges. 17 (1884) 559.

13. B. Robinson
    The Fischer Indole Synthesis
    Wiley, Chichester 1982.

14. M. Charton
    Topics Curr. Chem. 114 (1983) 57.

15. *(a)* M.P. Prochazka, L. Eklund and R. Carlson
    Acta Chem. Scand. 44 (1990) 610;

(b) M.P. Prochazka and R. Carlson
Acta Chem. Scand. 44 (1990) 614.

Suggestions to further reading

P. Geladi and B.R. Kowalski
Partial Least Squares Regression (PLS): a tutorial
Anal. Chim. Acta 185 (1986) 1–17.

A. Höskuldson
PLS Regression Methods
J. Chemometrics 2 (1988) 211–228.

S. Wold, C. Albano, W. Dunn III, U. Edlund, K. Esbensen, P. Geladi, S. Hellberg, E. Johansson,
W. Lindberg and M.Sjöström
Multivariate Data Analysis in Chemistry
in B.R. Kowalski (Ed.)
Chemometrics: Mathemathics and Statistics in Chemistry
Reidel, Dordrecht 1984, pp. 17–95.

S. Wold, C. Albano, W. Dunn III, K. Esbensen, P. Geladi, S. Hellberg, E. Johansson, W. Lindberg,
M. Sjöström, B. Skagerberg, C. Wikström and J. Öhman
Multivariate Data Analysis. Converting Chemical Data Tables to Plots
in J. Brandt and I. Ugi (Eds.)
Computer Applications in Chemical Research and Education
Dr Alfred Hühtig Verlag, Heidelberg 1989.

H. Martens
Multivariate Calibration
Diss., Technical University of Norway, Trondheim 1985.

Some available programs for principal components analysis and PLS modelling.

The programs use the NIPALS algorithm and can treat data matrices even if there are some data missing.

SIMCA-R (V. 4.3)

Umetri AB
P.O.Box 1456
S-901 24 Umeå, Sweden

MDS, Inc.
371 Highland Ave.
Winchester, MA 01890, USA

SIRIUS

Chemical Institute
Att. O.M. Kvalheim
University of Bergen
N-3007 Bergen, Norway

UNSCRAMBLER II

Camo AS
Jarleveien 4
N-7041 Trondheim, Norway

Appendix 17A: Reaction systems in the PLS study of the Fischer indole synthesis

The table below specifies the reaction systems used in the study of the Fischer indole reaction. Reaction systems 1–124 were used to establish an initial PLS model. This model was validated by predictions of the remaining systems 125–162. The final model was then accomplished from all 162 reaction systems. The response given, RE, is the regioisomeric excess.

Table 17A.1: Reaction systems and observed regioisomeric excess

No	Reaction system Ketone	Lewis acid	Solvent	RE
1	3-Hexanone	ZnI_2	Sulfolane	57.4
2	3-Hexanone	$TiCl_4$	Sulfolane	48.8
3	3-Hexanone	$ZnCl_2$	Sulfolane	52.4
4	3-Hexanone	PCl_3	Sulfolane	51.2
5	3-Hexanone	$FeCl_3$	Sulfolane	58.8
6	3-Hexanone	$AlCl_3$	Sulfolane	63.2
7	3-Hexanone	BF_3	Carbon disulfide	50.6
8	3-Hexanone	ZnI_2	Carbon disulfide	43.6
9	3-Hexanone	$TiCl_4$	Carbon disulfide	54.6
10	3-Hexanone	$ZnCl_2$	Carbon disulfide	44.0
11	3-Hexanone	$SbCl_5$	Carbon disulfide	54.0
12	3-Hexanone	$FeCl_3$	Carbon disulfide	61.0
13	3-Hexanone	$AlCl_3$	Carbon disulfide	60.4
14	3-Hexanone	BF_3	1,2-Dichlorobenzene	36.2
15	3-Hexanone	ZnI_2	1,2-Dichlorobenzene	42.0
16	3-Hexanone	$TiCl_4$	1,2-Dichlorobenzene	22.0
17	3-Hexanone	$ZnCl_2$	1,2-Dichlorobenzene	39.0
18	3-Hexanone	PCl_3	1,2-Dichlorobenzene	52.4
19	3-Hexanone	$FeCl_3$	1,2-Dichlorobenzene	54.8
20	3-Hexanone	$AlCl_3$	1,2-Dichlorobenzene	42.0
21	3-Hexanone	$FeCl_3$	Dimethylsulfoxide	58.0
22	3-Hexanone	BF_3	Tetrahydrofuran	57.4
23	3-Hexanone	$TiCl_4$	Tetrahydrofuran	58.6
24	3-Hexanone	$FeCl_3$	Tetrahydrofuran	58.0
25	3-Hexanone	$AlCl_3$	Tetrahydrofuran	58.0
26	2-Hexanone	BF_3	Sulfolane	100.0
27	2-Hexanone	ZnI_2	Sulfolane	100.0
28	2-Hexanone	$TiCl_4$	Sulfolane	100.0
29	2-Hexanone	$ZnCl_2$	Sulfolane	100.0
30	2-Hexanone	PCl_3	Sulfolane	100.0
31	2-Hexanone	$FeCl_3$	Sulfolane	100.0
32	2-Hexanone	$AlCl_3$	Sulfolane	100.0
33	2-Hexanone	BF_3	Carbon disulfide	100.0
34	2-Hexanone	ZnI_2	Carbon disulfide	100.0
35	2-Hexanone	$ZnCl_2$	Carbon disulfide	100.0
36	2-Hexanone	BF_3	1,2-Dichlorobenzene	98.0

Table 17A.1 (Continued)

No	Reaction system Ketone	Lewis acid	Solvent	RE
37	2-Hexanone	ZnI_2	1,2-Dichlorobenzene	100.0
38	2-Hexanone	$TiCl_4$	1,2-Dichlorobenzene	100.0
39	2-Hexanone	$ZnCl_2$	1,2-Dichlorobenzene	98.0
40	2-Hexanone	PCl_3	1,2-Dichlorobenzene	100.0
41	2-Hexanone	$FeCl_3$	1,2-Dichlorobenzene	100.0
42	2-Hexanone	$AlCl_3$	1,2-Dichlorobenzene	100.0
43	2-Hexanone	BF_3	Tetrahydrofuran	90.0
44	2-Hexanone	$TiCl_4$	Tetrahydrofuran	84.0
45	2-Hexanone	PCl_3	Tetrahydrofuran	88.0
46	2-Hexanone	$FeCl_3$	Tetrahydrofuran	92.0
47	2-Hexanone	$AlCl_3$	Tetrahydrofuran	98.0
48	3-Undecanone	BF_3	Sulfolane	23.6
49	3-Undecanone	ZnI_2	Sulfolane	32.0
50	3-Undecanone	$TiCl_4$	Sulfolane	30.0
51	3-Undecanone	$ZnCl_2$	Sulfolane	30.0
52	3-Undecanone	PCl_3	Sulfolane	31.0
53	3-Undecanone	$FeCl_3$	Sulfolane	36.0
54	3-Undecanone	$AlCl_3$	Sulfolane	36.0
55	3-Undecanone	BF_3	Carbon disulfide	21.0
56	3-Undecanone	ZnI_2	Carbon disulfide	34.2
57	3-Undecanone	PCl_3	Carbon disulfide	33.0
58	3-Undecanone	$AlCl_3$	Carbon disulfide	30.0
59	3-Undecanone	BF_3	1,2-Dichlorobenzene	23.0
60	3-Undecanone	ZnI_2	1,2-Dichlorobenzene	30.4
61	3-Undecanone	$ZnCl_2$	1,2-Dichlorobenzene	29.0
62	3-Undecanone	PCl_3	1,2-Dichlorobenzene	31.2
63	3-Undecanone	$FeCl_3$	1,2-Dichlorobenzene	34.0
64	3-Undecanone	$AlCl_3$	1,2-Dichlorobenzene	28.0
65	3-Undecanone	BF_3	Dimethylsulfoxide	32.2
66	3-Undecanone	ZnI_2	Dimethylsulfoxide	2.6
67	3-Undecanone	$ZnCl_2$	Dimethylsulfoxide	14.0
68	3-Undecanone	$FeCl_3$	Dimethylsulfoxide	22.0
69	3-Undecanone	BF_3	Terahydrofuran	35.0
70	3-Undecanone	ZnI_2	Tetrahydrofuran	12.0
71	3-Undecanone	$TiCl_4$	Tetrahydrofuran	40.0
72	3-Undecanone	PCl_3	Tetrahydrofuran	33.0
73	3-Undecanone	$FeCl_3$	Tetrahydrofuran	34.0
74	3-Undecanone	$AlCl_3$	Tetrahydrofuran	38.0
75	1-Phenyl-2-butanone	BF_3	Sulfolane	68.0
76	1-Phenyl-2-butanone	ZnI_2	Sulfolane	44.0
77	1-Phenyl-2-butanone	$TiCl_4$	Sulfolane	56.0
78	1-Phenyl-2-butanone	$ZnCl_2$	Sulfolane	38.0
79	1-Phenyl-2-butanone	PCl_3	Sulfolane	26.0
80	1-Phenyl-2-butanone	$FeCl_3$	Sulfolane	46.0
81	1-Phenyl-2-butanone	$AlCl_3$	Sulfolane	28.0
82	1-Phenyl-2-butanone	BF_3	Carbon disulfide	88.0
83	1-Phenyl-2-butanone	ZnI_2	Carbon disulfide	58.0
84	1-Phenyl-2-butanone	$TiCl_4$	Carbon disulfide	30.0

491

Table 17A.1 (Continued)

No	Reaction system			RE
	Ketone	Lewis acid	Solvent	
85	1-Phenyl-2-butanone	$ZnCl_2$	Carbon disulfide	70.0
86	1-Phenyl-2-butanone	PCl_3	Carbon disulfide	66.0
87	1-Phenyl-2-butanone	$FeCl_3$	Carbon disulfide	28.0
88	1-Phenyl-2-butanone	$AlCl_3$	Carbon disulfide	88.0
89	1-Phenyl-2-butanone	BF_3	1,2-Dichlorobenzene	62.0
90	1-Phenyl-2-butanone	ZnI_2	1,2-Dichlorobenzene	54.0
91	1-Phenyl-2-butanone	$TiCl_4$	1,2-Dichlorobenzene	62.0
92	1-Phenyl-2-butanone	$ZnCl_2$	1,2-Dichlorobenzene	46.0
93	1-Phenyl-2-butanone	PCl_3	1,2-Dichlorobenzene	52.0
94	1-Phenyl-2-butanone	$FeCl_3$	1,2-Dichlorobenzene	90.0
95	1-Phenyl-2-butanone	$AlCl_3$	1,2-Dichlorobenzene	42.0
96	1-Phenyl-2-butanone	BF_3	Dimethylsulfoxide	50.0
97	1-Phenyl-2-butanone	$FeCl_3$	Dimethylsulfoxide	66.0
98	1-Phenyl-2-butanone	BF_3	Tetrahydrofuran	60.0
99	1-Phenyl-2-butanone	ZnI_2	Tetrahydrofuran	88.0
100	1-Phenyl-2-butanone	$TiCl_4$	Tetrahydrofuran	76.0
101	1-Phenyl-2-butanone	PCl_3	Tetrahydrofuran	38.0
102	1-Phenyl-2-butanone	$FeCl_3$	Tetrahydrofuran	56.0
103	1-Phenyl-2-butanone	$AlCl_3$	Tetrahydrofuran	50.0
104	5-Methyl-3-heptanone	BF_3	Sulfolane	100.0
105	5-Methyl-3-heptanone	ZnI_2	Sulfolane	100.0
106	5-Methyl-3-heptanone	$TiCl_4$	Sulfolane	100.0
107	5-Methyl-3-heptanone	$ZnCl_2$	Sulfolane	100.0
108	5-Methyl-3-heptanone	PCl_3	Sulfolane	100.0
109	5-Methyl-3-heptanone	$FeCl_3$	Sulfolane	100.0
110	5-Methyl-3-heptanone	$AlCl_3$	Sulfolane	100.0
111	5-Methyl-3-heptanone	BF_3	Carbon disulfide	100.0
112	5-Methyl-3-heptanone	ZnI_2	Carbon disulfide	100.0
113	5-Methyl-3-heptanone	BF_3	1,2-Dichlorobenzene	100.0
114	5-Methyl-3-heptanone	ZnI_2	1,2-Dichlorobenzene	100.0
115	5-Methyl-3-heptanone	$TiCl_4$	1,2-Dichlorobenzene	100.0
116	5-Methyl-3-heptanone	$ZnCl_2$	1,2-Dichlorobenzene	100.0
117	5-Methyl-3-heptanone	PCl_3	1,2-Dichlorobenzene	100.0
118	5-Methyl-3-heptanone	$FeCl_3$	1,2-Dichlorobenzene	100.0
119	5-Methyl-3-heptanone	BF_3	Dimethylsulfoxide	100.0
120	5-Methyl-3-heptanone	$FeCl_3$	Dimethylsulfoxide	100.0
121	5-Methyl-3-heptanone	BF_3	Tetrahydrofuran	100.0
122	5-Methyl-3-heptanone	$TiCl_4$	Tetrahydrofuran	100.0
123	5-Methyl-3-heptanone	PCl_3	Tetrahydrofuran	100.0
124	5-Methyl-3-heptanone	$AlCl_3$	Tetrahydrofuran	100.0
125	3-Hexanone	$TiCl_4$	Quinoline	34.0
126	3-Hexanone	$SiCl_4$	Quinoline	56.4
127	3-Hexanone	BF_3	Carbon tetrachloride	31.8
128	3-Hexanone	CuCl	Carbon tetrachloride	43.2
129	3-Hexanone	ZnI_2	Carbon tetrachloride	48.2
130	3-Hexanone	$ZnCl_2$	Carbon tetrachloride	44.2
131	3-Hexanone	PCl_3	Carbon tetrachloride	60.0
132	3-Hexanone	$SiCl_4$	Carbon tetrachloride	56.8

Table 17A.1 (Continued)

No	Reaction system			RE
	Ketone	Lewis acid	Solvent	
133	3-Hexanone	BF_3	Chloroform	43.2
134	3-Hexanone	ZnI_2	Chloroform	36.0
135	3-Hexanone	$TiCl_4$	Chloroform	36.0
136	3-Hexanone	$ZnCl_2$	Chloroform	44.2
137	3-Hexanone	PCl_3	Chloroform	57.2
138	3-Hexanone	$FeCl_3$	Chloroform	53.4
139	3-Hexanone	$SiCl_4$	Chloroform	42.8
140	3-Hexanone	$AlCl_3$	Chloroform	45.8
141	3-Hexanone	$SnCl_4$	Chloroform	33.0
142	3-Hexanone	$SbCl_5$	Tetrahydrofuran	66.0
143	3-Hexanone	$SnCl_4$	Tetrahydrofuran	62.0
144	2-hexanone	$SbCl_5$	Tetrahydrofuran	100.0
145	2-hexanone	$SnCl_4$	Tetrahydrofuran	88.0
146	2-hexanone	$SiCl_4$	Sulfolane	100.0
147	2-hexanone	$SnCl_4$	Sulfolane	100.0
148	3-Undecanone	BF_3	Hexane	22.0
149	3-Undecanone	ZnI_2	Hexane	28.0
150	3-Undecanone	$ZnCl_2$	Hexane	14.0
151	3-Undecanone	PCl_3	Hexane	32.0
152	3-Undecanone	$FeCl_3$	Hexane	16.0
153	1-Phenyl-2-butanone	BF_3	Hexane	70.0
154	1-Phenyl-2-butanone	ZnI_2	Hexane	66.0
155	1-Phenyl-2-butanone	$TiCl_4$	Hexane	56.0
156	1-Phenyl-2-butanone	PCl_3	Hexane	48.0
157	1-Phenyl-2-butanone	$FeCl_3$	Hexane	66.0
158	5-Methyl-3-heptanone	$SiCl_4$	Sulfolane	100.0
159	5-Methyl-3-heptanone	$SnCl_4$	Sulfolane	100.0
160	5-Methyl-3-heptanone	$SiCl_4$	1,2-Dichlorobenzene	100.0
161	5-Methyl-3-heptanone	$SiCl_4$	Tetrahydrofuran	100.0
162	5-Methyl-3-heptanone	$SnCl_4$	Tetrahydrofuran	100.0

Chapter 18

A method for determining a suitable order of introducing reagents in "one-pot" procedures

1.1. The problem

To achieve efficient multistep synthetic procedures, it is often convenient to carry out consecutive transformations by one-pot procedures. In such procedures, the various reagents are introduced in sequence to the reaction mixture. Sometimes it is evident in which order this must be done. For instance, a deprotonation by adding a base to yield an enolate must precede further transformation of the enolate. In such evident cases, an optimization of the procedure must be made with a given order of introducing the reagents as a constraint. The problem is then reduced to determining the optimum settings of the experimental variables, e.g. rates of adding the reagents, hold-times before the next addition and temperature profile during the reaction. An optimization in such cases can therefore be achieved by the methods described in the first part of this book.

However, sometimes the order of introducing the reagents is not at all evident. Many established procedures belonging to this category smack of "witchcraft chemistry", where some innovator presented a procedure which is then followed by practioners in the field.

When a new reaction is studied, the reaction mechanisms remain to be established. The method must then be investigated through experimental studies. As most reactions involve several chemical species which can interact, the order of their introduction to the reaction mixture may sometimes be very important for the result. In our own activities, we have come across several cases where the order of introduction was critical. For instance, the yield of enamine by the titanium tetrachloride method was increased from 70 % overnight to 95 % after a few minutes by reversing the order of introduction of amine and titanium tetrachloride.[1] Other examples are the synthesis of silyloxydienes [2], and the synthesis of carboxamides using Lewis acids.[3]

494

These observations from the author's own experience indicate that the problem must be a very common one when new methods are developed. It is therefore astonishing that this general problem does not seem to have been subject to any systematic studies. Papers on new methods rarely mention that the order of introducing reagents has been studied.[1] It can therefore not be excluded that promising new ideas have been abandoned too early, i.e. if an initial experiment failed to give the expected result, due to an inadvert order of introduction.

New ideas for synthetic reactions should never be abandoned until the consequences of altering the order of intoduction has been considered. A method for such studies is briefly discussed below.

Scope of the problem

Assume that there are N different constituents which should be introduced into the reaction flask. The number of possible orders of introduction will then be $N!$. With five constituents, for instance, there are 120 different orders of introduction; often there are more than five constituents to mix. If also the detailed experimental procedure (rates of addition, hold-times, temperatures) are to be taken into account, the number of experimental runs would be absurd as a proper experimental design for the exploration of the importance of the experimental settings would have to be drawn up for each one of these orders of introduction. It is evident, that some strategy for a selection is necessary. An attempt in this direction is outlined below.

1.2. Strategy

There may be several reasons why the order of introducing reagents is critical. One obvious reason is that two constituents must interact to form a *necessary* intermediate which is the active reagent for the transformation. Another reason is that in the absence of the necessary counterpart, a constituent may react with other reagents or with the substrate to form an undesired complex or a product in a parasite reaction. If events like these are occurring in the reaction mixture, the following principles offer a good chance to detect important differences between altered orders of mixing the constituents.

---

[1] A search in *Chemical Abstracts* (up to 1990) using the key-words: "Order", "introduction" sequence", "organic synthesis", "reagent(s)", "one-pot", afforded only *two*(!) references.[4]

With N constituents *(A, B, C,...M, N)* which should be introduced into the reaction mixture, run N experiments in such a way that the order of introduction is a cyclic permutation of the sequence *ABC...MN*;

i.e.　　　　　*ABC...KLMN*

　　　　　　　NABC...KLM

　　　　　　　MNABC...KL　etc.

In this subset of N items out of $N!$ possible combinations, the items represent orders in which each constituent is added as the *last* components to the reaction mixture.

Should it be found that the order of introduction is important, this is an indication that two constituents interact, either to produce a necessary intermediate, or to induce undesired side reactions. In the set of experiments defined by the cyclic permutation above, there will be at least one experiment in which such interaction cannot occur until the last addition. A likely consequence is that the result will differ between the experiments. The next steps to be taken, depend on the results in the N experiments.

(a)　If the results in the N experiments are *similar*, this is an indication that the order of introduction is not important. Further studies on this will probably not improve the result.

(b)　A clear difference is observed between the N runs and one combination gives a *satisfactory* result. The next step would be to adjust the experimental conditions to an optimum performance by a proper design for this specific order of introduction .

(c)　A clear difference is observed, and *one* of the experiments afforded a *promising* result. The next step would then be to keep the last added constituent as such, and use the remaining $(N-1)$ constituents in a series of $(N-1)$ cyclic permutations to establish whether this will result in further improvements. This process can be repeated until a suitable order of introduction has been established.

(d)　The situation will be more complicated if there are more than one promising order of introduction. To draw any safe conclusions as to which order will be the optimum choice, it would be necessary to examine each of them according to *(c)*. From a practical point of view, it would be easiest to start with the most promising one, or the most convenient with regard to the experimental procedure, and

proceed with this according to *(c)* and stop when a satisfactory result has been obtained.

Experiments which yield poor results by the above principles may still help to elucidate the reaction mechanisms. They represent introduction orders which involve a limiting factor: For instance, a critical intermediate is not formed rapidly enough; an intermediate is formed which undergoes a side reaction; a higher-order complex between three or more constituents is necessary and the formation of such complexes is dependent on the order of introducing the reagents.

The above principles are not fool-proof. It is evident that running N experiments out of $N!$ possible combinations involves a risk of missing some details. Nevertheless, such a selection offers a good chance of detecting important effects of altered orders of introduction. The number of experiments is not too frightening, and they may prevent that a good idea from being prematurely discarded.

As the detailed experimental procedure is not subjected to variations by the above strategy, it is advised that the chemist should use his/her experience and chemical intuition to determine the concentrations of the reagent(s) to be added, rates of addition, hold times etc. This is an exception to what has been claimed throughout this book. It is, of course, possible to lay out fractional factorial designs to accomodate also such variations but this would lead to an unnecessary proliferation of the number of experimental runs. If it is suspected that also the detailed experimental procedure may interact with the order of introduction, the experimental procedure as a whole, of course, must be analyzed. If this is not the case, the detailed analysis of the experimental conditions can wait until a suitable introduction order has been found.

1.3.   Example: Self-condensation of 3,3-dimethyl-2-butanone[5]

In Chapters 12 and 17 it was discussed how an undesired side reaction could be suppressed by adjusting the experimental conditions, in the synthesis of the enamine from 3,3-dimethyl-2-butanone. The undesired product was formed by self-condensation of the ketone. By using a tertiary amine, e.g. triethylamine, in combination with titanium tetrachloride the self-condensation reaction can be made to be the dominant reaction.

Initial experiments had shown that the presence of both titanium tetrachloride and amine was necessary to obtain the condensation product.

If we exclude the solvent (hexane), there are three constituents to mix: *A*, the ketone; *B*, triethylamine; *C*, titanium tetrachloride.

This gives a total of 3! = 6 different orders of introduction. Three of them represent a cyclic permutation subset. The results obtained in the experiment of this subset are summarized in Table 18.1.

Table 18.1: Yields of condensation product after cyclic permutations of the order of reactant introduction

Order of introduction	Yield (%)
ABC	43.5
CAB	1.5
BCA	66.4

The results are clearly different and the order of mixing the constituents is critical. To elaborate the procedure, the formation of TiCl₄–amine complex should be further analyzed.

1.4. A note on "ad hoc" explanations

It is quite common in organic chemistry to give *ad hoc* "explanations" of why an experiment has given an observed result. Sometimes, such proposals may give valuable suggestions as to further studies to probe mechanistic details. Sometimes, however, such "explanations" may be totally misleading when it is not realized that they are nothing else than speculations.

In the reaction above, at least three hypothetical mechanistic "explanations" can be furnished, to fit any of the three different orders of introduction.

(1)   The essential feature is the formation of a TiCl₄–ketone complex which polarizes the carbonyl group. Addition of triethylamine yields a titanium enolate by proton abstraction. The enolate attacks a

498

titantium coordinated unenolized ketone, followed by base-induced elimination of the elements of a titanium oxide.

(2) Titanium tetrachloride and amine yield an equilibrium mixture of Lewis acid-base complexes. At least one of these complexes has an increased reactivity towards the carbonyl oxygen and forms a ternary ketone—amine—$TiCl_x$ complex [1] which in the presence of excess amine yields a titanium enolate. The reaction then proceeds as in *(1)*.

(3) The ketone yields an enolate in the presence of an amine base, but due to steric congestion the enolate has a low reactivity towards the carbonyl of the unenolized ketone. Addition of $TiCl_4$ to the mixture, quenches the enolate, forms a titanium enolate and polarize the unenolized ketone. The reaction then proceeds as in *(1)*.

The driving force to overcome the steric congention is the formation of the thermodynamically stable titanium dioxide.

If the reaction would have been studied utilizing only one of the orders of introduction, and depending on which one was actually used, one of the above "mechanisms" could have been appended to "explain" the result. It is not likely that all of them are true. In view of the result of changing the order of introduction, *(1)* and *(3)* are the least probable "explanations".

The moral of this is, that great care should be exercised in giving *ad hoc* explanations, especially if the experimental background is meager.

References

1. Carlson, R., Nilsson, Å. and Strömquist, M.
    Acta Chem. Scand. B 37 (1983) 7.

2. Hansson, L. and Carlson, R.
    Acta Chem. Scand. 43 (1989) 1888.

3. Nordahl, Å. and Carlson, R.
    Acta Chem. Scand. B 42 (1988) 28.

4. *(a)* Fukuda, T., Matsuura, Y. and Kusumoto, S.
    Nippon Shokuin Gakkaishi 28 (1981) 606, [*Chem. Abstr. 96* (1982) 85000],

    (b) Soai, K. and Ookawa, A.
    J. Chem. Soc. Perkin Trans.I (1986) 759.

5. Carlson, R., Nordahl, Å. and Kraus, W.
    Acta Chem. Scand. 45 (1991) 46.

Chapter 19

Concluding remarks

1.   Comments on statistics and chemistry

The basis on which all strategies in organic synthesis are founded are the individual synthetic reactions. An overall plan for the construction of a complex target molecule may be ingenious and elegant, but it is sufficient that one single step fails for the whole plan to fail. To develop new reactions or to develop known reactions into reliable and efficient synthetic *methods* is therefore crucial to the art and science of organic synthesis.

This book deals with problems involved when a synthetic reaction is elaborated, and it shows how various statistical tools may be helpful in this context. More specifically it describes, how such methods may help the experimenter to obtain experimental data which permit reasonably safe conclusions to be drawn, and how these conclusions can be linked to a certain degree of probability.

I am fully aware that some of my fellow colleagues will raise objections to the whole idea of using statistics in the domain of organic synthesis, a field that has long been regarded as an "art in the midst of science". It has even been claimed that statistical priciples are unnecessary, since an experienced and knowledgeable organic chemist will intuitively do the right things. It is true that well-experienced chemists with a profound knowledge often succeed in making good experiments and in obtaining excellent results. Unfortunately, the contrary is also true. The literature shows that a great deal of mediocre chemistry yielding poor results is published by less experienced chemists, who draw far-reaching conclusions from poorly designed experiments.

Chemical intuition is always a result of experience and detailed knowledge. An intuitive solution to a problem is often associated with a strong feeling that the solution is right, although it is sometimes not possible to state clearly why it is right. I believe that intuition is a result of lateral thinking where several different aspects are considered simultaneously. When these aspects are processed at a subconcious level, they may sometimes fit into a clear pattern which gives the intuitive solution.

Remember the Kekulé revelation on the structure of benzene. In that respect, an intuitive solution to a problem is the result of multidimensional thinking.

The outlines of methods found in this book describe the use of multidimensional models in synthetic chemistry, and they can therefore be seen as methods for quantification of chemical intuition. Models derived from experiments where several factors have been jointly considered, furnish quantitative relations and discernable patterns. They must, however, be analyzed by statistical methods to ensure that they represent real phenomena and not merely spurious correlations due to random events and experimental noise.

The use of statistics in chemistry, and especially model fitting to experimental results, is sometimes met with scepticism and sometimes even with scorn by "orthodox" chemists who regard such matters as "pure empiricism" which has nothing in common with "real chemistry" and chemical theory. In my opinion, such attitudes are wrong and, hopefully, based on a misunderstanding of when and where statistics should be applied to chemistry. Statistics can never substitute chemical knowledge. Chemical theory should be used in contexts where it is powerful. This involves all aspects when a problem is formulated so that its connection with known facts and previous knowledge is fully realized. It is also of great importance for determining the initial experimental setup, and for the choice a proper analytical procedure. However, chemical theory should *not* be immediately used for analysis of the recorded experimental data. Due to the omnipresent random experimental error, all experimentally determined data must be evaluated by statistical methods so that systematic variation can be distinguished from noise. Statistically significant observed relations should then be scrutinized in the light of chemical theory to determine whether they also are chemically significant. Chemical theory and statistical analysis are therefore complementary and should always be used in combination whenever chemical *experiments* are used to solve a chemical problem.

Such problems are legion when a synthetic reaction is elaborated. Often these problems can be translated into questions which can be answered by experiments. Strategies for developing synthetic reactions will therefore largely be strategies for designing and analyzing synthetic experiments. The chapters of this book have highlighted different aspects of these topics. Which tool will be appropriate will depend on the complexity of the questions posed to the chemical system, and there is always an interplay between the chemical problem which must be solved, the comprehension of the problem, the design of the experiments and the analysis of the experimental results. In this context, the benefit of a stepwise approach has been emphasized throughout the book. This is a consequence of acknowledging that experiments can be regarded as a means of acquiring *new* knowledge. With increasing experience we will gain a better understanding of problems, and the more we know, the more detailed questions can be posed to the chemical system.

The methods described in this book can now be summarized in an outline of an interactive, stepwise, strategy for the development of synthetic methods.

2.   Strategies for analyzing synthetic reactions

The problems in synthesis are truly multidimensional. We have introduced the concepts of the *experimental space*, and the *reaction space* to describe different types of variation of the conditions of a synthetic reaction. As there are often more than one response to consider, we also have introduced a *response space*. The tools for exploring these spaces will be different and dependent on the questions posed to the chemical system.

2.1.   The experimental space

The variables in the experimental space are continuous.[1] The relations between the settings of the experimental variables and the observed response can reasonably be assumed to be cause-effect relations. The appropriate method for establishing quantitative relation is to use multiple linear regression for fitting response surface models to observed data. For this purpose, an experimental design with good statistical properties is essential.

At the outset of the study of a new chemical reaction, little is known with certainty.

A *screening experiment* will reveal which experimental variables have a real influence. These variables can then be adjusted in subsequent experiments with a view to *optimizing* the reaction conditions. During this process the tools will be different, depending on the questions posed to the system.

*   Screening experiments with one response variable can be accomplished by a two-level design from which a linear or a second order interaction model can be obtained by least squares fit to the experimental results. Significant variables are identified by comparison to the experimental error variance through statistical tests, t test, F test, or by plotting the estimated effects on normal probability paper. These methods were outlined in Chapters 4—8.

---

[1] Discrete variations are sometimes also included in two-level designs to allow a choice between two alternatives. Such experiments are based on an assumption that the chemical systems are *similar* even when the conditions involve discrete variations.

502

* When there is more than one response variable to consider in a two-level screening experiment, it was shown in Chapter 17 that principal components analysis of the response matrix offers a means of identifying those experimental variables which are responsible for the systematic joint variation of the observed responses. This technique is useful when the reaction gives rise to several products and when the distribution of these product is of importance in the identification of suitable conditions for selectivity.

* The initial experimental domain is often chosen by intuition. A screening experiment may suggest that a better domain is likely to be found outside the explored domain. By the method of Steepest ascent (Chapter 10), or by a simplex search (Chapter 11) an near-optimum domain can be reached by a limited number of experiments.

* A quantitative model which describes the variation of the response in the optimum experimental domain can be accomplished by a quadratic response surface model. From such models a more precise location of the optimum conditions can be achieved. Canonical analysis of the model may reveal underlying interdependencies of the experimental variables, and this may give clues to an understanding of the basic mechanism of the reaction. When there are quite few responses to consider, a joint analysis of the response surface models from each response can afford a simultaneous optimization of all responses.

* Another technique for evaluating several responses is to determine quantitative relations between the experimental space and the response space by means of a PLS model. This technique is especially helpful when there are *many* response variables to consider and when one wants to know how the variations in the experimental space is coupled to the variation of *all* responses.

* A peculiar but important problem is encountered when several reagents are to be introduced in sequence to the reaction mixture. It is sometimes found that the order of introduction is critical. To detect whether or not this is the case, a method for analyzing the order of introducing reagents was suggested in Chapter 18.

2.2   The reaction space

The variations in the reaction space are discrete and comprise all possible combinations of substrates, reagents, and solvents which can be used for a given reaction. Interactions between the constituents are always to be expected. These interactions depend on the molecular properties of the constituents, and their interdependencies as well as their relations to the observed response(s) are most probably very complictated from a theoretical point of view. It is not possible to assume any cause-effect relations between the *observable* macroscopic properties of the constituents of the reaction system and their chemical behaviour. The chemical behaviour is determined by intrinsic properties at the molecular level. Such properties are not accessible through direct observations. In Capter 15 it was discussed how principal components models can be used to determine the *Principal properties*. These properties can be assumed to reflect intrinsic molecular properties, and the principal components scores afford *measures* of how the properties vary over a set of possible reaction systems. The principal properties therefore offers a means of establishing experimental designs by which test systems can be selected so that the set of selected items have a sufficient and desired spread with respect to their intrinsic molecular properties.

Designs based upon principal properties can be used in many contexts when it is necessary to consider discrete variations of the reaction systems. Some examples of their applications are:

*   In assessing the general scope of a reaction from a limited number of test systems.

*   In screening experiments when the aim is to identify a suitable reagent or solvent to accomplish a given transformation.

*   In optimization experiments when the objective is to determine an optimum reagent or solvent, or an optimum combination of both in order to accomplish a given transformation.

*   In experiments for analyzing *which* properties of the reaction system are involved to determine the experimental result.

The danger of using "standardized" experimental conditions has been pointed out several times in this book. To be able to make fair comparisons of the performance of different reaction systems, it will in most cases be necessary to adjust the experimental conditions. The PLS method, described in Chapter 17, is an excellent tool in this context which makes it possible to use a stepwise approach to this problem.

* Select an initial set of test systems to span the range of interest of the variation in the principal properties. Run the experiments, and adjust the experimental conditions for each of these systems towards an optimum performance. Fit an initial PLS model which relates the properties of the reaction systems to the optimum experimental conditions. Use the model to predict the optimum conditions for *new* reaction systems. Validate the predictions by experiments, and update the model by including the validation experiment. Continue the process until a sufficiently good mapping of the reaction space is obtained which permit reliable predictions. At this point, the questions posed to the reaction systems can very likely be adequately answered.

The PLS method is also the appropriate tool for determining *which* properties of the reaction system have an influence on the experimental results. An extensive study of the Fischer indole synthesis was given as an example. For analysis of this type of problem, an experimental design which affords a uniform spread in the principal properties should be used.

The PLS method can accomodate any number of response variables in the Y block and any number of molecular descriptors and/or independent experimental variables in the X block. It is thus a very flexible method which provides the tools by which all the methods described in this book can be integrated into an overall multivariate strategy for describing and analyzing systematic interrelations between the experiments and the observed results.

2.3. Conclusions

Proper experimental designs to span the variations of interest, both in the experimental space and the reaction space will ensure that each individual experiment provides a maximum of information on the problem under study. This will ensure efficient experiments which can provide the desired answers to the *Which?* and *How?* questions. Often this is enough to achieve an optimum synthetic method.

A good experimental design will also ensure that the observed data are *consistent*. The access to consistent data, and the answers already obtained to the *Which?* and *How?* questions will afford an excellent opportunity to proceed with other experiments which may provide answer also to the *Why?* questions. Any explanations in this direction must be compatible with the significant results already obtained.

Remember! Statistics is always secondary to chemistry in the domain of organic synthesis. It does not matter how statistically significant an analysis may turn out to be, if the chemistry does not afford the desired result. Therefore, any conclusions from a model *must be confirmed by an experiment.*

Statistics provides methods by which good chemists will be able to do even better chemistry and the strategies proposed above will, hopefully, make the confirmatory experiment also a *satisfactory* experiment.

Synthetic chemistry is fun!

Epilogue

I have tried to write the kind of textbook I myself would have liked to read when I was a young student of organic chemistry, and was asked by my professor to investigate a reaction by the method of "trial and error". In this book I have also tried to express my personal conviction that it is necessary to link organic synthetic experimental chemistry with statistics. In this respect, my intentions have been to explain the underlying principles in sufficient detail to show that there are no mysteries involved in statistical analysis. Whether or not the text is clear enough on these points is for the reader to decide. I will be grateful for comments and suggestions from readers of the book.

Rolf Carlson
Department of Organic Chemistry
University of Umeå
S-901 87 Umeå
Sweden.

Appendix 1: Matrix calculus

This appendix has been included in the book to give a brief recapitulation of some details of matrix algebra and matrix calculus.

Definitions

A matrix is a rectangular array of numbers, a table.

Examples:

$$\begin{bmatrix} 1 & 2 \\ 3 & 4 \end{bmatrix} \qquad \begin{bmatrix} 1 & 2 & 3 \\ 4 & 5 & 6 \end{bmatrix}$$

The ways these number are arranged in the matrix are called *rows* and *columns*

$$\begin{bmatrix} r\ o\ w & & c \\ & & o \\ & & l \\ & & u \\ & & m \\ & & n \end{bmatrix}$$

Dimension

The number of rows i by the number of columns j is called the *dimension* of the matrix and is usually denoted $(i \times j)$.

$$\begin{bmatrix} 1 & 2 \\ 3 & 4 \\ 5 & 6 \end{bmatrix} \quad \text{is a (3 x 2) matrix}$$

$$\begin{bmatrix} 1 & 2 & 3 & 4 \\ 5 & 6 & 7 & 8 \end{bmatrix} \quad \text{is a (2 x 4) matrix}$$

It is common to use capital letters (bold face) to denote matrices.

Example

$$\mathbf{A} = \begin{bmatrix} a_{11} & a_{12} \\ a_{21} & a_{22} \end{bmatrix}$$

in which a_{ij} is the *matrix element* in row i and column j

Row vector, column vector

A *row matrix* has only one row and is called a *row vector*. It is a $(1 \times i)$ matrix.
A *column matrix* has only one column and is called a *column vector*. It is a $(j \times 1)$ matrix.
When the vord *vector* is used, it usually denotes a column vector.

The matrix $\mathbf{B}$ below is a (2×3) matrix. It has two row vectors and three column vectors.

$$\mathbf{B} = \begin{bmatrix} b_{11} & b_{12} & b_{13} \\ b_{21} & b_{22} & b_{23} \end{bmatrix} \begin{matrix} \rightarrow \text{row vectors} \\ \rightarrow \end{matrix}$$

$$\downarrow \quad \downarrow \quad \downarrow$$
column vectors

Null matrix, null vector

A matrix (vector) in which all elements are zero is called a *null matrix*, **O**, (*null vector, o*).

Diagonal matrix

A matrix in which all elements $a_{ij} = 0$, for $i \neq j$ is called a *diagonal matrix*.

Example:

$$\begin{bmatrix} 1 & 0 & 0 & 0 \\ 0 & 2 & 0 & 0 \\ 0 & 0 & 3 & 0 \\ 0 & 0 & 0 & 4 \end{bmatrix}$$

Symmetric matrix

In a symmetric matrix $a_{ij} = a_{ji}$, for $i \neq j$

Example:

$$\begin{bmatrix} 1 & 5 & 6 & 7 \\ 5 & 2 & 0 & 0 \\ 6 & 0 & 3 & 8 \\ 7 & 0 & 8 & 4 \end{bmatrix}$$

Unit matrix, identity matrix

A diagonal matrix in which all diagonal elements are equal to one, $a_{ii} = 1$ for all i, is called a *unity matrix* or an *identity matrix*. It is denoted by **I**. Sometimes the dimension is specified by an index.

Example:

$$\begin{bmatrix} 1 & 0 & 0 & 0 \\ 0 & 1 & 0 & 0 \\ 0 & 0 & 1 & 0 \\ 0 & 0 & 0 & 1 \end{bmatrix} = \mathbf{I}_4$$

Transpose matrix

When a matrix is transposed the rows and columns change places in such a way that the first row becomes the first column, the second becomes the scond column etc. There are several notations for the transpose of a matrix **A**, e.g. $\mathbf{A}^T$, $^t\mathbf{A}$, $\mathbf{A}'$. The prime, ', notation has been used in this book.

Example:

$$\mathbf{A} = \begin{bmatrix} a & b \\ c & d \end{bmatrix} \qquad \mathbf{A'} = \begin{bmatrix} a & c \\ b & d \end{bmatrix}$$

$$y = \begin{bmatrix} y_1 \\ y_2 \\ y_3 \\ y_4 \end{bmatrix} \qquad y' = [y_1 \ y_2 \ y_3 \ y_4]$$

A $(i \times j)$ matrix which is transposed becomes a $(j \times i)$ matrix.

Matrix calculus

Matrix addidtion

The matrices must have the same dimension.

Example:

$$\mathbf{A} = \begin{bmatrix} a_{11} & a_{12} \\ a_{11} & a_{12} \end{bmatrix} \qquad \mathbf{B} = \begin{bmatrix} b_{11} & b_{12} \\ b_{11} & b_{12} \end{bmatrix}$$

$$\mathbf{A} + \mathbf{B} = \begin{bmatrix} a_{11} + b_{11} & a_{12} + b_{12} \\ a_{11} + b_{21} & a_{12} + b_{22} \end{bmatrix}$$

Example:

$$\begin{bmatrix} 1 & 2 \\ 3 & 4 \end{bmatrix} + \begin{bmatrix} 5 & 6 \\ 7 & 8 \end{bmatrix} = \begin{bmatrix} 6 & 8 \\ 10 & 12 \end{bmatrix}$$

The following rules apply for matrix addition:

$\mathbf{A} + \mathbf{B} = \mathbf{B} + \mathbf{A}$

$(\mathbf{A} + \mathbf{B}) + \mathbf{C} = \mathbf{A} + (\mathbf{B} + \mathbf{C})$

O null matrix $a_{ij} = 0$ for all i, j

$\mathbf{O} + \mathbf{A} = \mathbf{A}$

Multipication by a scalar

$$\mathbf{A} = \begin{bmatrix} a_{11} & a_{12} \\ a_{21} & a_{22} \end{bmatrix}$$

$$\text{ß} \mathbf{A} = \text{ß} \cdot \begin{bmatrix} a_{11} & a_{12} \\ a_{11} & a_{12} \end{bmatrix} = \begin{bmatrix} \text{ß}a_{11} & \text{ß}a_{12} \\ \text{ß}a_{21} & \text{ß}a_{22} \end{bmatrix}$$

where ß is any number

Example:

$$2 \cdot \begin{bmatrix} 4 & 2 \\ 3 & 1 \end{bmatrix} = \begin{bmatrix} 8 & 4 \\ 6 & 2 \end{bmatrix}$$

The following applies

$$1 \cdot \mathbf{A} = \mathbf{A}$$

$$0 \cdot \mathbf{A} = \mathbf{O}$$

$$(\alpha + \beta) \, \mathbf{A} = \alpha \cdot \mathbf{A} + \beta \cdot \mathbf{B}$$

$$\alpha \, (\mathbf{A} + \mathbf{B}) = \alpha \cdot \mathbf{A} + \alpha \cdot \mathbf{B}$$

Matrix multiplication

$$\mathbf{A} = \begin{bmatrix} a_{11} & a_{12} \\ a_{21} & a_{22} \end{bmatrix} \qquad \mathbf{B} = \begin{bmatrix} b_{11} & b_{12} \\ b_{21} & b_{22} \end{bmatrix}$$

The multiplication **AB** is defined according to the example below.

$$\mathbf{AB} = \begin{bmatrix} a_{11}b_{11} + a_{12}b_{21} & a_{11}b_{12} + a_{12}b_{22} \\ a_{21}b_{11} + a_{22}b_{21} & a_{21}b_{12} + a_{22}b_{22} \end{bmatrix}$$

Example

$$\begin{bmatrix} 1 & 0 \\ 2 & 3 \end{bmatrix} \begin{bmatrix} 0 & 2 \\ 1 & 3 \end{bmatrix} = \begin{bmatrix} (1 \cdot 0 + 0 \cdot 1) & (1 \cdot 2 + 0 \cdot 3) \\ (2 \cdot 0 + 3 \cdot 1) & (2 \cdot 2 + 3 \cdot 3) \end{bmatrix} = \begin{bmatrix} 0 & 2 \\ 3 & 13 \end{bmatrix}$$

The multiplication is defined only if the number of columns in **A** is equal to the number of rows in **B**. If **A** is a $(k \times n)$ matrix, and **B** is a $(n \times p)$ matrix, the product matrix will be a $(k \times p)$ matrix.

Examples

$$\begin{bmatrix} 1 & 2 \\ 2 & 3 \end{bmatrix} \begin{bmatrix} a \\ b \end{bmatrix} = \begin{bmatrix} a + 2b \\ 3a + 4b \end{bmatrix}$$

$$\begin{bmatrix} x & y \end{bmatrix} \begin{bmatrix} x \\ y \end{bmatrix} = x^2 + y^2$$

$$\begin{bmatrix} x \\ y \end{bmatrix} \begin{bmatrix} x & y \end{bmatrix} = \begin{bmatrix} x^2 & xy \\ yx & y^2 \end{bmatrix}$$

Matrix multiplication is *not* commutative

AB ≠ BA

Example:

$$\mathbf{A} = \begin{bmatrix} 10 & 4 \\ 2 & 1 \end{bmatrix} \qquad \mathbf{B} = \begin{bmatrix} 0 & 1 \\ 2 & 3 \end{bmatrix}$$

$$\mathbf{AB} = \begin{bmatrix} 8 & 22 \\ 2 & 5 \end{bmatrix} \qquad \mathbf{BA} = \begin{bmatrix} 2 & 1 \\ 26 & 11 \end{bmatrix}$$

Matrix multiplication is

associative   (AB)C = A(BC)
distributive   (A + B)C = AC + BC

Multiplication by a unit matrix

The unit matrix **I** fulfils

A I = A and
I A = A

Transpose of a product matrix

The transpose of a product matrix is product in reversed order of the transposed matrices factors, i.e.

(AB)' = B' A'

Scalar product

The scalar product, $v \cdot u$, of two vectors v and u is a number which describes the degree to which these vectors points in the same direction. Let v and u be the vectors below

$$v = \begin{bmatrix} v_1 \\ \cdot \\ \cdot \\ v_n \end{bmatrix} \qquad u = \begin{bmatrix} u_1 \\ \cdot \\ \cdot \\ u_n \end{bmatrix}$$

The scalar product is defined as

$$v \cdot u = \|v\| \cdot \|u\| \cdot \cos\Theta$$

where $\|v\|$ and $\|u\|$ are the norms (length) of the vectors and are computed as

$$\|v\| = (v_1^2 + \dots + v_n^2)^{1/2}$$

$$\|u\| = (u_1^2 + \dots + u_n^2)^{1/2}$$

and $\cos\Theta$, is the cosine of the angle between the vectors.

When the vectors are defined in an orthonormal vector space, as is the case in the examples in this book, the scalar product is computed as

$$v \cdot u = v'u$$

If $v \cdot u = 0$, the vectors are said to be *orthogonal* to each other. In two- and three-dimensional vector spaces, this means that they are in right angels to each other.

Determinant

This important concept is defined for quadratic matrices, $i = j$. The determinant of a matrix **A** is denoted

detA or $|A|$

The determinant is a *number* which is defined as a sum of products of the matrix elements. The factors in each of these product terms are such one element from each row and from each column is represented once and only once. The summation is taken over all possible permutations.

$$\text{detA} = \begin{bmatrix} a_{11}\dots & a_{1n} \\ \vdots & \vdots \\ a_{n1}\dots & a_{nn} \end{bmatrix} = \Sigma \pm a_{1r(1)} a_{1r(2)} \dots a_{nr(n)}$$

in which {r(1), r(2), ..., r(n)} represents all possible permutations of (1, 2, ..., n). The sign of the product term is dependent on if, r(1)...r(n), is an even (+), or an odd (-) permutation. The determinant is thus a sum of $n!$ terms.

The determinant can be regarded as a measure of the volume which is spanned by the column vectors (or row vectors) of the matrix in the vector space. For example, in a two dimensional space, two vectors can span a *surface area*, provided that the vectors are not parallel. If they should be parallel, the surface area would be zero. In the three-dimensional space, three vectors can span a *volume*, provided that they do not lie in the same plane. If they should do so, the volume would be zero. In that case, any of the three vectors can be expressed as a linear combination of the two others; the vectors are said to be *linear dependent*.

Computation of determinants

For (1 x 1) up to (3 x 3) matrices it is possible to compute the determinants by cross-wise multiplication of the matrix elements.

$A = [a]$ $\det A = a$

$$\det A = \begin{vmatrix} a_{11} & a_{12} \\ a_{21} & a_{22} \end{vmatrix} = a_{11}a_{22} - a_{12}a_{21}$$

$$\det A = \begin{vmatrix} a_{11} & a_{12} & a_{13} \\ a_{21} & a_{22} & a_{23} \\ a_{31} & a_{32} & a_{33} \end{vmatrix} = \begin{aligned} &+ a_{13}a_{21}a_{32} + a_{12}a_{23}a_{31} + a_{11}a_{22}a_{32} \\ &- a_{11}a_{23}a_{32} - a_{12}a_{21}a_{33} - a_{13}a_{22}a_{31} \end{aligned}$$

Example

$$\begin{vmatrix} 0 & 3 & 1 \\ 2 & 1 & 1 \\ -1 & 2 & 1 \end{vmatrix} = \begin{aligned} &1{\cdot}2{\cdot}2 + 3{\cdot}1{\cdot}(-1) + 0{\cdot}1{\cdot}1 \\ &- 0{\cdot}1{\cdot}1 - 3{\cdot}2{\cdot}1 - 1{\cdot}1{\cdot}(-1) = -4 \end{aligned}$$

General procedure for calculation of determinants

Determinants of matrices of higher dimensions than (3 x 3) must be computed by another procedure. The principle is general and is illustrated by (3 x 3) matrix. Consider the matrix

$$A = \begin{bmatrix} a_{11} & a_{12} & a_{13} \\ a_{21} & a_{22} & a_{23} \\ a_{31} & a_{32} & a_{33} \end{bmatrix}$$

If we take an element a_{ij} in a (n x n) matrix and ignore row i and column j the remaining elements will define a submatrix of dimension (n-1 x n-1). The determinant of this submatrix is called the *minor* M_{ij} of element a_{ij}. For the matrix element of column 1 in the matrix above, the following minors are obtained.

$$M_{11} = \begin{vmatrix} a_{22} & a_{23} \\ a_{32} & a_{33} \end{vmatrix} \qquad M_{21} = \begin{vmatrix} a_{12} & a_{13} \\ a_{32} & a_{33} \end{vmatrix} \qquad M_{31} = \begin{vmatrix} a_{12} & a_{13} \\ a_{22} & a_{23} \end{vmatrix}$$

Another concept is needed, viz. the *cofactor* A_{ij} to a matrix element a_{ij}. The cofactor is defined

$$A_{ij} = (-1)^{i+j} \cdot M_{ij}$$

Example:

$$A_{11} = (-1)^{1+1} \cdot M_{11} = M_{11}$$

$$A_{21} = (-1)^{2+1} \cdot M_{11} = - M_{11} \text{ etc.}$$

From the matrix element in a colum (or in a row), and the corresponding cofactors, the determinant for any matrix can be computed. The following relation apply:

$$\text{detA} = a_{1j} \cdot A_{1j} + a_{2j} \cdot A_{2j} + \dots + a_{nj} \cdot A_{nj} = \Sigma \, a_{ij} \cdot A_{ij} \; (i = 1 \text{ to } n)$$

This technique is called. expansion of detA in cofactors of column j. The same principle can be used row wise.

Some relations of determinants

*   If two columns (rows) change places in **A**, the sign of the determinant is reverted.

*   Any matrix has the same determinant as its transpose; detA = detA'.

*   If **A** and **B** are quadratic matrices (n x n), the product matrix **AB** is also (n x n), and detAB = detA · detB.

*   If each element a_{ij} of a column j (row i) has a common factor, α, this factor can be broken out, and the determinant will be α·detA, where **A** is the matrix in which a_{ij} in column j (row i) has been replaced by a_{ij}/α.

Example

$$\begin{vmatrix} 2 & 1 \\ 4 & 8 \end{vmatrix} = 2 \cdot \begin{vmatrix} 1 & 1 \\ 2 & 8 \end{vmatrix} = 12$$

Inverse matrix

A quadratic matrix **A** is said to be invertible if there is a matrix $\mathbf{A}^{-1}$ such that

$$\mathbf{AA}^{-1} = \mathbf{A}^{-1}\mathbf{A} = \mathbf{I} \; \text{(a unit matrix)}$$

A matrix **A** is invertible only, and only if, detA $\neq$ 0.
If detA = 0, the matrix **A** is said to be *singular* and has no inverse.

$$\text{If} \quad \mathbf{A} \quad = \begin{bmatrix} a_{11} & a_{12} & a_{13} \\ a_{11} & a_{12} & a_{13} \\ a_{11} & a_{12} & a_{13} \end{bmatrix}$$

$$A^{-1} = 1/\det A \cdot \begin{bmatrix} A_{11} & A_{21} & A_{31} \\ A_{12} & A_{22} & A_{32} \\ A_{13} & A_{23} & A_{33} \end{bmatrix}$$

This principle applies to any matrix. It is not necessary to make the computations involved by hand. It is more convenient to use a computer and there are software available for matrix computations.

For the determinant of a matrix A and its inverse A^{-1}, the following realtion applies

$\det A^{-1} = 1/\det A$

Determinant and inverse of a diagonal matrix

The determinant of a diagonal matrix is the product of the diagonal elements.

$A = $ diagonal $\{a_{ii}\}$

$\det A = a_{11} \cdot a_{22} \cdot ... \cdot a_{nn}$

Its inverse is also a diagonal matrix.

$A^{-1} = $ diagonal $\{1/a_{ii}\}$.

The elements are the inverted values of the elements of the parent diagonal matrix.

Eigenvectors, eigenvalues

Let A be a quadratic matrix (n x n), and let v be a vector different from the null vector.

If there is a number, λ, such that

$Av = \lambda \cdot v$

v is said to be an eigenvector to A, and the number λ the corresponding eigenvalue.

$Av = \lambda \cdot v$

$Av - \lambda \cdot v = o$

$(Av - \lambda \cdot I) v = o$

This has a nontrivial solution $v \neq o$ only if $\det(Av - \lambda \cdot I) = 0$.

This determinant, $\det(Av - \lambda \cdot I)$, is called the *characteristic determinant*. It is computed after substracting λ from each diagonal element in A. The equation obtained by equalizing the secular determinant to zero is called the characteristic equation. For (n x n) matrices it will be defined by a polynomial of degree n. Its roots are the eigenvalues (sometimes also called the *characteristic roots*, or the *singular values*). The eigenvalues to a symmetric, real, matrix, are all real and ≥ 0.

Each eigenvalue, λ_i, corresponds to an eigenvector, v_i. Eigenvectors corresponding to different eigenvalues are orthogonal to each other.

$$v_i'v_j = 0, \text{ if } \lambda_i \neq \lambda_j$$

Example

$$A = \begin{bmatrix} 3 & 1 \\ 1 & 3 \end{bmatrix}$$

The characteristic equation is therefore

$$\begin{vmatrix} 3 - \lambda & 1 \\ 1 & 3 - \lambda \end{vmatrix} = 0$$

$$(3 - \lambda)^2 - 1 = 0$$

which has the roots $\quad \lambda_1 = 4$

$$\lambda_2 = 2$$

Determine the eigenvector v_1 for $\lambda_1 = 4$.

$$\begin{bmatrix} (3 - 4) & 1 \\ 1 & (3 - 4) \end{bmatrix} \begin{bmatrix} v_{11} \\ v_{11} \end{bmatrix} = \begin{bmatrix} 0 \\ 0 \end{bmatrix}$$

$$\begin{bmatrix} -v_{11} & v_{12} \\ v_{11} & -v_{12} \end{bmatrix} = \begin{bmatrix} 0 \\ 0 \end{bmatrix}$$

Any vector

$$v_1 = \alpha \cdot \begin{bmatrix} v_{11} \\ v_{12} \end{bmatrix} = \alpha \cdot \begin{bmatrix} 1 \\ 1 \end{bmatrix}$$

with $\alpha \neq 0$ is an eigenvector.

The norm if v is

$$\|v_1\| = [(\alpha \cdot 1)^2 + (\alpha \cdot 1)^2]^{1/2} = \alpha \sqrt{2}$$

and the normalized eigenvector is

$$(1/\|v_1\|) \cdot v_1 \quad = \quad \begin{bmatrix} 1/\sqrt{2} \\ 1/\sqrt{2} \end{bmatrix}$$

The normalized eigenvector v_2 for $\lambda_2 = 2$ is determined analogously.

$$(1/\|v_2\|) \cdot v_2 \quad = \quad \begin{bmatrix} 1/\sqrt{2} \\ -1/\sqrt{2} \end{bmatrix}$$

The eigenvectors are orthogonal to each other which gives

$$v_1'v_2 = [\ 1/\sqrt{2} \ \ 1/\sqrt{2}] \begin{bmatrix} 1/\sqrt{2} \\ -1/\sqrt{2} \end{bmatrix} = 0$$

Orthogonal matrices

An orthogonal matrix S is a real quadratic matrix for which the transpose S' is equal to the inverse S^{-1}.

$$S' = S^{-1}$$

Orthogonal matrices are useful and may be used to accomplish variable transformations which lead to simplification of many different type of problems. Orthogonal matrices were used in the canonical transformation of response surface models. Such matrices are also used in Factor Analysis.

The following applies.

If A is a real and symmetric matrix ($n \times n$) there is always an orthogonal matrix S by which A can be transformed into a diagonal matrix in which the diagonal elements are eigenvalues to A.

$$S'AS = \text{diagonal}\{\lambda_i\}$$

A worked-out example of the above technique was given in Chapter 12 (pp. 276–280) in the context of canonical analysis of a response surface model.

Examples of other problems related to eigenvalues and eigenvectors discussed in this book are:

For the descriptor matrix X used in principal components and factor analysis, the matrix X'X is symmetric. Its eigenvalues are related to the sum of squares of the descriptors. The corresponding eigenvectors are the loading vectors of the principal component model.

The model matrix X in least squares modelling describes the variation of the variables included in the model. The matrix X'X is symmetric, and hence also the dispersion matrix, $(X'X)^{-1}$. The eigenvalues of the dispersion matrix are related to the precision of the estimated model parameters. The determinant of the dispersion matrix is the product of its eigenvalues. The "volume" of the joint confidence region of the estimated model parameters is proportional to the square root of the determinant of the dispersion matrix.

The column vectors in **S** are the normalized eigenvectors to **A**.

$$
\begin{bmatrix}
v_{11} & v_{12} \cdots\cdots\cdots & v_{1n} \\
v_{21} & v_{22} \cdots\cdots\cdots & v_{2n} \\
\cdot & \cdot & \cdot \\
\cdot & \cdot & \cdot \\
\cdot & \cdot & \cdot \\
v_{n1} & v_{n2} \cdots\cdots\cdots & v_{nn}
\end{bmatrix} = \mathbf{S}
$$

$$
\begin{array}{ccc}
\uparrow & \uparrow & \uparrow \\
\lambda_1 & \lambda_2 & \lambda_n
\end{array}
$$

Eigenvectors to λ_1 to λ_n

$$
\begin{bmatrix}
\lambda_1 & 0 \cdots\cdots\cdots & 0 \\
0 & \lambda_2 \cdots\cdots\cdots & 0 \\
\cdot & \cdot & \cdot \\
\cdot & \cdot & \cdot \\
0 & 0 \cdots\cdots\cdots & \lambda_n
\end{bmatrix} = \mathbf{S'AS}
$$

Differentiation using matrices

It is often the case that we wish to differentiate a function $f(x_1, x_2,..., x_k)$ with respect to the variables $x_1, x_2,..., x_k$. If we consider the variables in vector form, i.e.

$$
x = \begin{bmatrix}
x_1 \\
x_2 \\
\cdot \\
\cdot \\
x_k
\end{bmatrix}
$$

By the derivative $\partial f/\partial x$, we mean the column vector

$$
\partial f/\partial x = \begin{bmatrix}
\partial f/\partial x_1 \\
\partial f/\partial x_2 \\
\cdot \\
\cdot \\
\partial f/\partial x_k
\end{bmatrix}
$$

This is the column vector defined by the partial derivatives. Similarily, the derivative $\partial f/\partial x'$ is the row vector of the partial derivatives.

If b is a column vector of k constants, $b_1,..., b_k$, consider the scalar product $b'x$. It derivative with respect to x will be

$$\partial b'x/\partial x = b = \begin{bmatrix} b_1 \\ b_2 \\ . \\ . \\ b_k \end{bmatrix}$$

Similarily

$$\partial b'x/\partial x' = b' = [\, b_1 \ \ b_2 \ .... \ b_k \,]'$$

The derivatives of the scalar product $x'x$ will be

$$\partial x'x/\partial x = 2x$$

$$\partial x'x/\partial x' = 2x'$$

If A is a square matrix, the quadratic form $x'Ax$ is a scalar product. In Chapter 12 on response surface models was discussed how the stationary point on the response surface was determined as the roots of the systems of equations defined by setting all partial derivatives of the response surface model to zero. In matrix language this corresponds to determining for which values of the x variables, the vector $\partial x'Ax/\partial x$ is the null vector. This derivative is computed as

$$\partial x'Ax/\partial x = Ax + A'x = (A + A')x$$

If A is symmetric this gives

$$\partial x'Ax/\partial x = 2Ax.$$

This is the case in canonical analysis of response surface model, where the coefficient matrix is symmetric.

When the principal components method was derived, matrix differentiation was used to determine the principal component vector p which minimized the sum of squared deviation. For this the symmetric mean centred variance-covariance matrix $(X - \bar{X})'(X - \bar{X})$ was involved.

Matrix differentiation was used in Appendix 3A to show that $(X'X)^{-1}X'y = b$ gives the least squares estimation of b.

520

References

Recommended reading

A very readable, and amusing!, book on matrix calculus is

Gilbert Strang
Linear Algebra and its Applications, Third Edition
Harcourt Brace Jovanovich Publishers, Inc, San Diego 1988.

Appendix 2: Statistical tables

This appendix contains tables of the t and the F distributions. These tables are used to assess the significance of estimated parameters obtained by experiments. Two examples to illustrate how the tables can be used are given below.

Table A: t Distribution gives the critical t values. These values are used to compare estimated parameters to estimates of its standard deviation.

An example will illustrate how this can be accomplished: An experimental variable is judged to have a significant influence on the response only if it has an influence above the noise level of the random experimental error variations. The influence of the variable is measured by its response surface coefficient b_i. An estimate of its standard error can be obtained from an estimate of the standard deviation of the experimental error.

Assume that you have run experiments by a factorial design (with N_F runs) with a view to assessing the significance of the experimental variables from estimates, b_i, of the coefficients in a linear response surface model. Assume also that you have made N_0 repeated runs of one experiment to obtain an estimate of the experimental error standard deviation. From the average response, $\bar{y}$, in repeated runs, an estimate of the experimental error standard deviation, s_0, with $(N_0 - 1)$ degrees of freedom is obtained as

$$s_0 = [(y - \bar{y}) / (N_0 - 1)]^{1/2}$$

The standard error, s_e, of the estimated coefficient is then obtained by dividing the experimental error standard deviation by $\sqrt{N_F}$, see Chapter 3.

$$s_e = s_0 / \sqrt{N_F}$$

Determine the t ratio

$$t = b_i / s_e$$

For b_i to be significantly above the noise level, the value of t should be outside the range $\pm t^{Crit}$ given in Table A.

Table A has two entries: The column heads specify the value of $\alpha / 2$, where α is the significance level, i.e. the probability that the experimental t value *by pure chance* exceeds the critical t value. It is common to use a significance level of $\alpha = 5$ %. This corresponds to the column heading 0.025. The rows of the table specify the number of degrees of freedom of the estimate of the standard deviation.

A $(100 - \alpha)$ confidence interval for the "true" value, β_i, of the response surface parameter is obtained as

$$\beta_i = b_i \pm t^{Crit} \cdot s_e$$

Tables B1 – B3: F distribution give critical F values. These vaules are used to compare estimates of variances to each other.

An example is shown, on how the F distribution can be used to assess whether a response surface model can be assumed to give an adequate description of the variation of the response, or if it shows a significant lack of fit.

522

There will always be residuals, e_i, between the experimentally observed response in experiment i and the corresponding value predicted by the model. If the model is adequate, these residuals should be nothing but manifestations of the experimental error. From the sum of squared residuals, Σe_i^2, we can compute the residual mean square, RMS, as

$$RMS = \Sigma e_i^2 / (N - p)$$

where N is the number of experiments used to fit the model, and p is the number of parameters in the model. If the model is adequate RMS would be an estimate, s_1^2, of the experimental error variance with $n_1 = (N - p)$ degrees of freedom.

Another, and independent, estimate of the experimental error variance, s_2^2, with $n_2 = (N_0 - 1)$ degrees of freedom may be obtained by N_0 repeated runs of a given experiment. These two estimates of the experimental error variance have been obtained with different degrees of freedom. Even if the model is adequate, they may seem to be different. To determine whether or not they are statistically significant, we can use the F distribution. If the model is adequate, the F ratio

$$F = s_1^2 / s_2^2$$

should not exceed the critical F value for n_1 and n_2 degrees of freedom in Tables B1 - B3. These tables shows the critical F values for the significance level $\alpha = 10$, 5, and 1 %. There are two entries in the tables, corresponding to the degrees of freedom of the estimated variances.

Sources

Table A has been taken from E.S. Pearson and H.O. Hartley (Eds.) *Biometrika Tables for Statisticians*, Vol. 1, Cambridge University Press 1958.

Tables B1–B3 have been taken from M. Merrington and C.M. Thompson, Tables of percentage points of the inverted beta (*F*) distribution, *Biometrika 33* (1943) 73.

The tables have been reproduced with permission of the publishers.

Table A: t Distribution. Propability points for the significance level α and n degrees of freedom.

Degrees of	Tail area probability $\alpha/2$		
freedom	0.05	0.025	0.005
1	6.314	12.706	63.657
2	2.920	4.303	9.925
3	2.353	3.182	5.841
4	2.132	2.776	4.604
5	2.015	2.571	4.032
6	1.943	2.447	3.702
7	1.895	2.365	3.499
8	1.860	2.306	3.355
9	1.833	2.262	3.250
10	1.812	2.228	3.169
11	1.796	2.201	3.106
12	1.782	2.179	3.055
13	1.771	2.160	3.012
14	1.761	2.145	2.977
15	1.753	2.131	2.947
16	1.746	2.120	2.921
17	1.740	2.110	2.898
18	1.734	2.101	2.878
19	1.729	2.093	2.861
20	1.725	2.086	2.845
21	1.721	2.080	2.831
22	1.717	2.074	2.819
23	1.714	2.069	2.807
24	1.711	2.064	2.797
25	1.708	2.060	2.787
26	1.706	2.056	2.779
27	1.703	2.052	2.771
28	1.701	2.048	2.763
29	1.699	2.045	2.756
30	1.697	2.042	2.750
40	1.684	2.021	2.704
60	1.671	2.000	2.660
120	1.658	1.980	2.617
∞	1.645	1.960	2.576

Table B1: F Distribution. Critical $F = s_1^2/s_2^2$ ratios, upper 10 % points.
The degrees of freedom are n_1 for s_1^2 and n_2 for s_2^2

n_2	n_1 1	2	3	4	5	6	7	8	9	10
1	39.86	49.50	53.59	53.83	58.24	58.20	58.91	59.44	59.86	60.19
2	8.53	9.00	9.16	9.24	9.29	9.33	9.35	9.37	9.38	9.39
3	5.54	5.46	5.39	5.34	5.31	5.28	5.27	5.25	5.24	5.23
4	4.54	4.32	4.19	4.11	4.05	4.01	3.98	3.95	3.94	3.92
5	4.06	3.78	3.62	3.52	3.45	3.40	3.37	3.34	3.32	3.30
6	3.78	3.46	3.29	3.18	3.11	3.05	3.01	2.98	2.96	2.94
7	3.59	3.26	3.07	2.96	2.88	2.83	2.78	2.75	2.72	2.80
8	3.46	3.11	2.92	2.81	2.73	2.67	2.62	2.59	2.56	2.54
9	3.36	3.01	2.81	2.69	2.61	2.55	2.51	2.47	2.44	2.42
10	3.29	2.92	2.73	2.61	2.52	2.46	2.41	2.38	2.35	2.32
11	3.23	2.86	2.66	2.54	2.45	2.39	2.34	2.30	2.27	2.25
12	3.18	2.81	2.61	2.48	2.39	2.33	2.28	2.24	2.21	2.19
13	3.14	2.76	2.56	2.43	2.35	2.28	2.23	2.20	2.16	2.14
14	3.10	2.73	2.52	2.39	2.31	2.24	2.19	2.15	2.12	2.10
15	3.07	2.70	2.49	2.36	2.27	2.21	2.16	2.12	2.09	2.06
16	3.05	2.67	2.46	2.33	2.24	2.18	2.13	2.09	2.06	2.03
17	3.03	2.64	2.44	2.31	2.22	2.15	2.10	2.06	2.03	2.00
18	3.01	2.62	2.42	2.29	2.20	2.13	2.08	2.04	2.00	1.98
19	2.99	2.61	2.40	2.27	2.18	2.11	2.06	2.02	1.98	1.96
20	2.97	2.59	2.38	2.25	2.16	2.09	2.04	2.00	1.96	1.94
21	2.96	2.57	2.36	2.23	2.14	2.08	2.02	1.98	1.95	1.92
22	2.95	2.56	2.35	2.22	2.13	2.06	2.01	1.97	1.93	1.90
23	2.94	2.55	2.34	2.21	2.11	2.05	1.99	1.95	1.92	1.89
24	2.93	2.54	2.33	2.19	2.10	2.04	1.98	1.94	1.91	1.88
25	2.92	2.53	2.32	2.18	2.09	2.02	1.97	1.93	1.89	1.87
26	2.91	2.52	2.31	2.17	2.08	2.01	1.96	1.92	1.88	1.86
27	2.90	2.51	2.30	2.17	2.07	2.00	1.95	1.91	1.87	1.85
28	2.89	2.50	2.29	2.16	2.06	2.00	1.94	1.90	1.87	1.84
29	2.89	2.50	2.28	2.15	2.06	1.99	1.93	1.89	1.86	1.83
30	2.88	2.49	2.28	2.14	2.05	1.98	1.83	1.88	1.85	1.82
40	2.84	2.44	2.23	2.09	2.00	1.93	1.87	1.83	1.79	1.76
60	2.79	2.39	2.18	2.04	1.95	1.87	1.82	1.77	1.74	1.71
120	2.75	2.35	2.13	1.99	1.90	1.82	1.77	1.72	1.68	1.65
∞	2.71	2.30	2.08	1.94	1.85	1.77	1.72	1.67	1.63	1.60

Table B1: (continued)

n_2 \ n_1	12	15	20	24	30	40	60	120	∞
1	60.71	61.22	61.74	62.00	62.26	62.53	62.79	63.06	63.33
2	9.41	9.42	9.44	9.45	9.46	9.47	9.47	9.48	9.49
3	5.22	5.20	5.18	5.18	5.27	5.16	5.15.	5.14	5.13
4	3.90	3.87	3.84	3.83	3.82	3.80	3.79	3.78	3.76
5	3.27	3.24	3.21	3.19	3.17	3.16	3.14	3.12	3.10
6	2.90	2.87	2.84	2.82	2.80	2.78	2.76	2.74	2.72
7	2.67	2.63	2.59	2.58	2.56	2.54	2.51	2.49	2.47
8	2.50	2.46	2.42	2.40	2.38	2.36	2.34	2.32	2.29
9	2.38	2.34	2.30	2.28	2.25	2.23	2.21	2.18	2.16
10	2.28	2.24	2.20	2.18	2.16	2.13	2.11	2.08	2.06
11	2.21	2.17	2.12	2.10	2.08	2.05	2.03	2.00	1.97
12	2.15	2.10	2.06	2.04	2.01	1.99	1.96	1.93	1.90
13	2.10	2.05	2.01	1.98	1.96	1.93	1.90	1.88	1.85
14	2.05	2.01	1.96	1.94	1.91	1.89	1.86	1.83	1.80
15	2.02	1.97	1.92	1.90	1.87	1.85	1.82	1.79	1.76
16	1.99	1.94	1.89	1.87	1.84	1.81	1.78	1.75	1.72
17	1.96	1.91	1.86	1.84	1.81	1.78	1.75	1.72	1.69
18	1.93	1.89	1.84	1.81	1.78	1.75	1.72	1.69	1.66
19	1.91	1.86	1.81	1.79	1.76	1.73	1.70	1.67	1.63
20	1.89	1.84	1.79	1.77	1.74	1.71	1.68	1.64	1.61
21	1.87	1.83	1.78	1.75	1.72	1.69	1.66	1.62	1.59
22	1.86	1.81	1.76	1.73	1.70	1.67	1.64	1.60	1.57
23	1.84	1.80	1.74	1.72	1.69	1.66	1.62	1.59	1.55
24	1.83	1.78	1.73	1.70	1.67	1.64	1.61	1.57	1.53
25	1.82	1.77	1.72	1.69	1.66	1.63	1.59	1.56	1.52
26	1.81	1.76	1.71	1.68	1.65	1.61	1.58	1.54	1.50
27	1.80	1.75	1.70	1.67	1.64	1.60	1.57	1.53	1.49
28	1.79	1.74	1.69	1.66	1.63	1.59	1.56	1.52	1.48
29	1.78	1.73	1.68	1.65	1.62	1.58	1.55	1.51	1.47
30	1.77	1.72	1.67	1.64	1.61	1.57	1.54	1.50	1.46
40	1.71	1.66	1.61	1.57	1.54	1.51	1.47	1.42	1.38
60	1.66	1.60	1.54	1.51	1.48	1.44	1.40	1.35	1.29
120	1.60	1.55	1.48	1.45	1.41	1.37	1.32	1.26	1.19
∞	1.55	1.49	1.42	1.38	1.34	1.30	1.24	1.17	1.00

Table B2: F Distribution. Critical $F = s_1^2/s_2^2$ ratios, upper 5 % points.
The degrees of freedom are n_1 for s_1^2 and n_2 for s_2^2

n_2 \ n_1	1	2	3	4	5	6	7	8	9	10
1	161.4	199.5	215.7	224.6	230.2	234.0	236.8	238.9	240.5	241.9
2	18.51	19.00	19.16	19.25	19.30	19.33	19.35	19.37	19.38	19.40
3	10.13	9.55	9.28	9.12	9.01	8.94	8.89	8.85	8.81	8.79
4	7.71	6.94	6.59	6.39	6.26	6.16	6.09	6.04	6.00	5.96
5	6.61	5.79	5.41	5.19	5.05	4.95	4.88	4.82	4.77	4.74
6	5.99	5.14	4.76	4.53	4.39	4.28	4.21	4.15	4.10	4.06
7	5.59	4.74	4.35	4.12	3.97	3.87	3.79	3.73	3.68	3.64
8	5.32	4.46	4.07	3.84	3.69	3.58	3.50	3.44	3.39	3.35
9	5.12	4.26	3.86	3.63	3.48	3.37	3.39	3.23	3.18	3.14
10	4.96	4.10	3.71	3.48	3.33	3.22	3.14	3.07	3.02	2.98
11	4.84	3.98	3.59	3.36	3.20	3.09	3.01	2.95	2.90	2.85
12	4.75	3.89	3.49	3.26	3.11	3.00	2.91	2.85	2.80	2.75
13	4.68	3.81	3.41	3.18	3.03	2.92	2.83	2.77	2.71	2.67
14	4.60	3.74	3.34	3.11	2.96	2.85	2.76	2.70	2.65	2.60
15	4.54	3.68	3.29	3.06	2.90	2.79	2.71	2.64	2.59	2.54
16	4.49	3.63	3.24	3.01	2.85	2.74	2.66	2.59	2.54	2.49
17	4.45	3.59	3.20	2.96	2.81	2.70	2.61	2.55	2.49	2.45
18	4.41	3.55	3.16	2.93	2.77	2.66	2.58	2.51	2.46	2.41
19	4.38	3.52	3.13	2.90	2.74	2.63	2.54	2.48	2.42	2.38
20	4.35	3.49	3.10	2.87	2.71	2.60	2.51	2.45	2.39	2.35
21	4.32	3.47	3.07	2.84	2.68	2.57	2.49	2.42	2.37	2.32
22	4.30	3.44	3.05	2.82	2.66	2.55	2.46	2.40	2.34	2.30
23	4.28	3.42	3.03	2.80	2.64	2.53	2.44	2.37	2.32	2.27
24	4.26	3.40	3.01	2.78	2.62	2.51	2.42	2.36	2.30	2.25
25	4.24	3.39	2.99	2.76	2.60	2.49	2.40	2.34	2.28	2.24
26	4.23	3.37	2.98	2.74	2.59	2.47	2.39	2.32	2.27	2.22
27	4.21	3.35	2.96	2.73	2.57	2.46	2.37	2.31	2.25	2.20
28	4.20	3.34	2.95	2.71	2.56	2.45	2.36	2.29	2.24	2.19
29	4.18	3.33	2.93	2.70	2.55	2.43	2.35	2.28	2.22	2.18
30	4.17	3.32	2.92	2.69	2.53	2.42	2.33	2.27	2.21	2.16
40	4.08	3.23	2.84	2.61	2.45	2.34	2.25	2.18	2.12	2.08
60	4.00	3.15	2.76	2.53	2.37	2.25	2.17	2.10	2.04	1.99
120	3.92	3.07	2.68	2.45	2.29	2.17	2.09	2.02	1.96	1.91
∞	3.84	3.00	2.60	2.37	2.21	2.10	2.01	1.94	1.88	1.83

Table B2: (continued)

n_2	n_1 12	15	20	24	30	40	60	120	∞
1	243.9	245.9	248.0	249.1	250.1	251.1	252.2	253.3	254.3
2	19.41	19.43	19.45	19.45	19.46	19.47	19.48	19.49	19.50
3	8.74	8.70	8.66	8.64	8.62	8.59	8.57	8.55	8.43
4	5.91	5.86	5.80	5.77	5.75	5.72	5.69	5.66	5.63
5	4.68	4.62	4.56	4.53	4.50	4.46	4.43	4.40	4.36
6	4.00	3.94	3.87	3.84	3.81	3.77	3.74	3.70	3.67
7	3.57	3.51	3.45	3.41	3.38	3.34	3.30	3.27	3.23
8	3.28	3.22	3.15	3.12	3.08	3.04	3.01	2.97	2.93
9	3.07	3.01	2.94	2.90	2.86	2.83	2.79	2.75	2.71
10	2.91	2.85	2.77	2.74	2.70	2.66	2.62	2.58	2.54
11	2.79	2.72	2.65	2.61	2.58	2.53	2.49	2.45	2.40
12	2.69	2.62	2.54	2.51	2.48	2.43	2.38	2.34	2.30
13	2.60	2.53	2.46	2.42	2.38	2.34	2.30	2.25	2.21
14	2.53	2.46	2.39	2.35	2.31	2.27	2.22	2.18	2.13
15	2.48	2.40	2.33	2.29	2.25	2.20	2.16	2.11	2.07
16	2.42	2.35	2.28	2.24	2.19	2.15	2.11	2.06	2.01
17	2.38	2.31	2.23	2.19	2.15	2.10	2.06	2.01	1.96
18	2.34	2.27	2.19	2.15	2.11	2.06	2.02	1.97	1.92
19	2.31	2.23	2.16	2.11	2.07	2.03	1.98	1.93	1.88
20	2.28	2.20	2.12	2.08	2.04	1.99	1.95	1.90	1.84
21	2.25	2.18	2.10	2.05	2.01	1.96	1.92	1.87	1.81
22	2.23	2.15	2.07	2.03	1.98	1.94	1.89	1.84	1.78
23	2.20	2.13	2.05	2.01	1.96	1.91	1.86	1.81	1.76
24	2.18	2.11	2.03	1.98	1.94	1.89	1.84	1.79	1.73
25	2.16	2.09	2.01	1.96	1.92	1.87	1.82	1.77	1.71
26	2.15	2.07	1.99	1.95	1.90	1.85	1.80	1.75	1.69
27	2.13	2.06	1.97	1.93	1.88	1.84	1.79	1.73	1.67
28	2.12	2.04	1.96	1.91	1.87	1.82	1.77	1.71	1.65
29	2.10	2.03	1.94	1.90	1.85	1.81	1.75	1.70	1.64
30	2.09	2.01	1.93	1.89	1.84	1.79	1.74	1.68	1.62
40	2.00	1.92	1.84	1.79	1.74	1.69	1.64	1.58	1.51
60	1.92	1.84	1.75	1.70	1.65	1.59	1.53	1.48	1.39
120	1.83	1.75	1.66	1.61	1.55	1.50	1.43	1.35	1.25
∞	1.75	1.67	1.57	1.52	1.46	1.39	1.32	1.22	1.00

Table B3: F Distribution. Critical $F = s_1^2/s_2^2$ ratios, upper 1 % points.
The degrees of freedom are n_1 for s_1^2 and n_2 for s_2^2

n_2	n_1 1	2	3	4	5	6	7	8	9	10
1	40.52*	49.99*	54.03*	56.25*	57.64*	58.59*	59.28*	59.82*	69.22*	60.56*
2	98.50	99.00	99.17	99.25	99.30	99.33	99.36	99.37	99.39	99.40
3	34.12	30.82	29.46	28.71	28.24	27.91	27.67	27.49	27.35	27.23
4	21.20	18.00	16.69	15.98	15.52	15.21	14.98	14.80	14.66	14.55
5	16.26	13.27	12.06	11.29	10.97	10.67	10.46	10.29	10.16	10.05
6	13.75	10.92	9.78	9.15	8.75	8.47	8.26	8.10	7.98	7.87
7	12.25	9.55	8.45	7.85	7.46	7.19	6.99	6.84	6.72	6.62
8	11.26	8.65	7.59	7.01	6.63	6.37	6.18	6.03	5.91	5.81
9	10.56	8.02	6.99	6.42	6.06	5.80	6.61	5.47	5.35	5.26
10	10.04	7.56	6.55	5.99	5.64	5.39	5.20	5.06	4.94	4.85
11	9.65	7.21	6.22	5.67	5.32	5.07	4.89	4.74	4.63	4.54
12	9.33	6.93	5.95	5.41	5.06	4.82	4.64	4.50	4.39	4.30
13	9.07	6.70	5.74	5.21	4.86	4.62	4.44	4.30	4.19	4.10
14	8.86	6.51	5.56	5.04	4.69	4.46	4.28	4.14	4.03	3.94
15	8.68	6.36	5.42	4.89	4.56	4.32	4.14	4.00	3.89	3.80
16	8.53	6.23	5.29	4.77	4.44	4.20	4.03	3.89	3.78	3.69
17	8.40	6.11	5.18	4.67	4.34	4.10	3.93	3.79	3.68	3.59
18	8.29	6.01	5.09	4.58	4.25	4.01	3.84	3.71	3.60	3.51
19	8.18	5.93	5.01	4.50	4.17	3.94	3.77	3.63	3.52	3.43
20	8.10	5.85	4.94	4.43	4.10	3.87	3.70	3.56	3.46	3.37
21	8.02	5.78	4.87	4.37	4.04	3.81	3.64	3.51	3.40	3.31
22	7.95	5.72	4.82	4.31	3.99	3.76	3.59	3.45	3.35	3.26
23	7.88	5.66	4.76	4.26	3.94	3.71	3.54	3.41	3.30	3.21
24	7.82	5.61	4.72	4.22	3.90	3.67	3.50	3.36	3.26	3.17
25	7.77	5.57	4.68	4.18	3.85	3.63	3.46	3.32	3.22	3.13
26	7.72	5.53	4.64	4.14	3.82	3.59	3.42	3.29	3.18	3.09
27	7.68	5.49	4.60	4.11	3.78	3.56	3.39	3.26	3.15	3.06
28	7.64	5.45	4.57	4.07	3.75	3.53	3.36	3.23	3.12	3.03
29	7.60	5.42	4.54	4.04	3.73	3.50	3.33	3.20	3.09	3.00
30	7.56	5.39	4.51	4.02	3.70	3.47	3.30	3.17	3.07	2.98
40	7.31	5.18	4.31	3.83	3.51	3.29	3.12	2.99	2.89	2.80
60	7.08	4.98	4.13	3.65	3.34	3.12	2.95	2.82	2.72	2.63
120	6.85	4.79	3.95	3.48	3.17	2.96	2.79	2.66	2.56	2.47
∞	6.63	4.61	3.78	3.32	3.02	2.80	2.64	2.51	2.41	2.32

* Multiply these entries by 100.

Table B3: (continued)

n_2	n_1 12	15	20	24	30	40	60	120	∞
1	61.06*	61.57*	62.09*	62.35*	62.61*	62.87*	63.13*	63.39*	63.66*
2	99.42	99.43	99.45	99.46	99.47	99.47	99.48	99.49	99.50
3	27.05	26.87	26.69	26.60	26.50	26.41	26.32	26.22	26.13
4	14.37	14.20	14.02	13.93	13.84	13.75	13.65	13.56	13.46
5	9.89	9.72	9.55	9.47	9.38	9.29	9.20	9.11	9.02
6	7.72	7.56	7.40	7.31	7.23	7.14	7.06	6.97	6.88
7	6.47	6.31	6.16	6.07	5.99	5.91	5.82	5.74	5.65
8	5.67	5.52	5.36	5.28	5.20	5.12	5.03	4.95	4.86
9	5.11	4.96	4.81	4.73	4.65	4.57	4.48	4.40	4.31
10	4.71	4.56	4.41	4.33	4.25	4.17	4.08	4.00	3.91
11	4.40	4.25	4.10	4.02	3.94	3.86	3.78	3.69	3.60
12	4.16	4.01	3.86	3.78	3.70	3.62	3.54	3.45	3.36
13	3.96	3.82	3.66	3.59	3.51	3.43	3.34	3.25	3.17
14	3.80	3.66	3.51	3.43	3.35	3.27	3.18	3.09	3.00
15	3.67	3.52	3.37	3.29	3.21	3.13	3.05	2.96	2.87
16	3.55	3.41	3.26	3.18	3.10	3.02	2.93	2.84	2.75
17	3.46	3.31	3.16	3.08	3.00	2.92	2.83	2.75	2.65
18	3.37	3.23	3.08	3.00	2.92	2.84	2.75	2.66	2.57
19	3.30	3.15	3.00	2.92	2.84	2.76	2.67	2.58	2.49
20	3.23	3.09	2.94	2.86	2.78	2.69	2.61	2.52	2.42
21	3.17	3.03	2.88	2.80	2.72	2.64	2.55	2.46	2.36
22	3.12	2.98	2.83	2.75	2.68	2.58	2.50	2.40	2.31
23	3.07	2.93	2.78	2.70	2.62	2.54	2.45	2.35	2.26
24	3.03	2.89	2.74	2.66	2.58	2.49	2.40	2.31	2.21
25	2.99	2.85	2.70	2.62	2.54	2.45	2.36	2.27	2.17
26	2.96	2.81	2.66	2.58	2.50	2.42	2.33	2.23	2.13
27	2.93	2.78	2.63	2.55	2.47	2.38	2.28	2.20	2.10
28	2.90	2.75	2.60	2.52	2.44	2.35	2.26	2.17	2.06
29	2.87	2.73	2.57	2.49	2.41	2.33	2.23	2.14	2.03
30	2.84	2.70	2.55	2.47	2.39	2.30	2.21	2.11	2.01
40	2.66	2.52	2.37	2.29	2.20	2.11	2.02	1.92	1.80
60	2.50	2.35	2.20	2.12	2.03	1.94	1.84	1.73	1.60
120	2.34	2.19	2.03	1.95	1.86	1.76	1.66	1.53	1.38
∞	2.18	2.04	1.88	1.79	1.70	1.59	1.47	1.32	1.00

* Multiply these entries by 100.

Index